# Scaffolds and Surfaces with Biomedical Applications

# Scaffolds and Surfaces with Biomedical Applications

Editors

**Andrada Serafim**
**Stefan Ioan Voicu**

MDPI • Basel • Beijing • Wuhan • Barcelona • Belgrade • Manchester • Tokyo • Cluj • Tianjin

*Editors*
Andrada Serafim
Advanced Polymer Materials
Group
University Politehnica
of Bucharest
Bucharest
Romania

Stefan Ioan Voicu
Analytical Chemistry and
Environmental Engineering
University Politehnica
of Bucharest
Bucharest
Romania

*Editorial Office*
MDPI
St. Alban-Anlage 66
4052 Basel, Switzerland

This is a reprint of articles from the Special Issue published online in the open access journal *Polymers* (ISSN 2073-4360) (available at: www.mdpi.com/journal/polymers/special_issues/scaffolds_surfaces_biomed_appli).

For citation purposes, cite each article independently as indicated on the article page online and as indicated below:

LastName, A.A.; LastName, B.B.; LastName, C.C. Article Title. *Journal Name* **Year**, *Volume Number*, Page Range.

**ISBN 978-3-0365-7833-0 (Hbk)**
**ISBN 978-3-0365-7832-3 (PDF)**

# Contents

# Preface to "Scaffolds and Surfaces with Biomedical Applications"

The engineering of scaffolds and surfaces with improved properties for biomedical applications represents an ever-expanding field of research that is continuously gaining momentum. As technology and society evolve, the golden standard of autografts has been contested due to their lack of availability. Tremendous efforts have been dedicated to developing nature-inspired materials that are able to either undertake the functions of damaged tissues or significantly contribute to their repair. To this end, multidisciplinary research aiming at designing new or improved materials has been conducted with the purpose of finding suitable candidates that can replicate the characteristics of natural tissues with regard to their functions, mechanical behaviors, and micro-architectural features, among others. The present reprint, entitled "Scaffolds and Surfaces with Biomedical Applications", published 13 papers (10 research and 3 review papers) that describe the synthesis of new materials for biomedical applications and their thorough characterization using conventional and state-of-the-art techniques.

**Andrada Serafim and Stefan Ioan Voicu**
*Editors*

*Editorial*

# Scaffolds and Surfaces with Biomedical Applications

Andrada Serafim and Stefan Ioan Voicu *

Advanced Polymers Materials Group, University Politehnica of Bucharest, Gheorghe Polizu Str. 1-7, 011061 Bucharest, Romania; andrada.serafim0810@upb.ro
* Correspondence: stefan.voicu@upb.ro

**Citation:** Serafim, A.; Voicu, S.I. Scaffolds and Surfaces with Biomedical Applications. *Polymers* **2023**, *15*, 2126. https://doi.org/10.3390/polym15092126

Received: 26 April 2023
Accepted: 27 April 2023
Published: 29 April 2023

The engineering of scaffolds and surfaces with enhanced properties for biomedical applications represents an ever-expanding field of research that is continuously gaining momentum. As technology and society evolve, the golden standard of autografts has been contested due to their lack of availability, and tremendous efforts have been dedicated to developing nature-inspired materials that are able to either undertake the functions of damaged tissues or significantly contribute to their repair. To this end, multidisciplinary research that aims to design novel or upgraded materials has been conducted with the purpose of locating suitable candidates that replicate the characteristics of natural tissues with regard to their function, mechanical behavior, microarchitectural features, etc. The present Special Issue entitled 'Scaffolds and Surfaces with Biomedical Applications' published 13 papers (10 research and 3 review papers) that describe the synthesis of new materials with biomedical applications and their thorough characterization using conventional and emerging techniques.

Istratov et al. [1] synthesized new biodegradable implants, starting with the polymerization of L-lactide, catalyzed by tin (II) 2-ethylhexanoate in the presence of 2,2-bis(hydroxymethyl)propionic acid, and an ester of polyethylene glycol monomethyl ester and 2,2-bis(hydroxymethyl)propionic acid; these were accompanied by the introduction of a pool of hydrophilic groups, which minimize the contact angle that leads to the formation of branched pegylated copolylactides with a narrow molecular weight distribution.

Soria et al. evaluated the cytotoxic capacity of a new instillation technology via a biodegradable ureteral stent/scaffold coated with a silk fibroin matrix for application in the controlled release of mitomycin C as an anti-cancer drug [2]. They concluded that the coating of a biodegradable ureteral stent with a silk fibroin matrix impregnated in layers of mitomycin C allows the release of the cytostatic into artificial urine; this finding is crucial to the treatment of patients with a urothelial tumor of the upper urinary tract.

A 3D scaffold structure, comprising thermoplastic polyurethane and maghemite ($\gamma$-Fe$_2$O$_3$) nanoparticles mixed with ultra-hard and tough bio-resin, was reported by Fallahiarezoudar et al. [3]. Lastly, the presence of $\gamma$-Fe$_2$O$_3$ in the structure enhanced the proliferation rate of HSF1148 due to the ability of numerous magnetic nanoparticles to integrate with the cellular matrix.

Filipov et al. [4] presented the development of antimicrobial surfaces that combat implant-associated infections and simultaneously promote the host cell response; this was performed in order to enhance current therapies for orthopedic injuries. The authors reported the modification of 3D-printed polycaprolactone scaffolds with a femtosecond laser and their applicative potential in the production of patterns that resemble microchannels or microprotrusions. The parallel microchannels enabled successful guidance and enhanced the osteogenic potential of MG63 cells. In combination with the improved cytocompatibility, the same microtopography exhibited potent antibacterial effects against *S. aureus*. By developing a biodegradable scaffold that has the potential to simultaneously promote bone tissue regeneration and prevent bacterial biofilm formation, we come a step closer to overcoming the current problems encountered in bone tissue engineering.

Olaret et al. [5] proposed a simple method that can be utilized to obtain nanostructured hydrogels with enhanced mechanical characteristics and relevant antibacterial behavior for application in articular cartilage regeneration and repair; this method was based on low quantities of silver-decorated carbon-nanotubes that were used as reinforcing agents in the semi-interpenetrating polymer network. The main challenge encountered when utilizing hydrogels in applications related to tissue regeneration is represented by their inadequate mechanical properties; although hydrogels are usually soft, most of them are not able to withstand high values of stress. This research aimed to design nanocomposites that exhibit both elasticity and toughness by simultaneously employing two distinct approaches: (1) embedding the linear polyacrylamide in the 3D network of the corresponding monomer and cross-linker with the aim of improving the dispersion of carbon nanotubes in the precursor and the scaffolds' elasticity, and (2) using low ratios of nanoparticles as fillers, with the aim of providing toughness to the obtained nanostructured system.

Morales-Guadarrama et al. [6] reported the ability of diffusion tensor imaging to evaluate the evolution of spinal cord injuries in nonhuman primates via a fraction of anisotropy analysis and the diffusion tensor tractography calculus. The results revealed that the interrupted nerve fibers can be differentiated from intact regions and that the method can be applied as a qualitative indicator of spinal cord injury in order to represent nerve fibers and to observe the spinal cord evolution after an injury.

Hu et al. [7] evaluated the adult human smooth muscle cell release of angiogenic/growth factor-enriched exosomes in the coronary artery when cultured on *Bombyx mori* 3D silk fibroin nonwoven scaffolds in vitro.

In addition, Neto et al. [8] reported the development of multifunctional biphasic calcium phosphate scaffolds coated with two biopolymers—poly($\varepsilon$-caprolactone) or poly(ester urea) —loaded with the antibiotic drug Rifampicin, which possesses antibacterial properties, for application in bone regeneration.

Martinez-Garcia et al. [9] presented the synthesis of hydrogels based on the decellularized proteins and polysaccharides of the extracellular matrix that can replicate in vivo functions. Their results revealed that the elasticity of extracellular matrix hydrogels, but also their viscoelastic relaxation and gelling behavior, are organ dependent. Some these physical features are correlated with their biochemical composition and ultrastructure.

Lu et al. [10] presented a facile fabrication method for application in the protection of respiratory masks via electrospinning and the utilization of a nontoxic polyvinyl butyral polymeric matrix with the antibacterial component Thymol, a natural phenol monoterpene. Based on the results of Japanese Industrial Standards and American Association of Textile Chemists and Colorists methods, the maximum antibacterial values of the mask against Gram-positive and Gram-negative bacteria were 5.6 and 6.4, respectively.

This Special Issue also contains three review papers. The first, contributed by Arifin et al. [11], one aims to review the applicative potential of the photo-polymerization 3D printing technique in the fabrication of tissue engineering scaffolds. The review also highlights the comprehensive comparative study of photo-polymerization 3D printing and other scaffold fabrication techniques. Various parameter settings that influence mechanical properties, biocompatibility and porosity behavior are also discussed in detail. The second review paper, contributed by Radu et al. [12], is related to functionalized polysulfone membranes with enhanced hemocompatibility and applicative potential in hemodialysis. The final review paper, reported by Mustafa et al. [13] provides an overview of the utilization of a computational method in designing a unit cell of a bone tissue engineering scaffold.

**Funding:** This research received no external funding.

**Institutional Review Board Statement:** Not applicable.

**Data Availability Statement:** Not applicable.

**Acknowledgments:** We acknowledge all the authors and reviewers who have contributed to completing this Special Issue. In addition, we would like to thank the technical support team for their assistance in preparing this Special Issue.

**Conflicts of Interest:** The authors declare no conflict of interest.

## References

1. Istratov, V.; Gomzyak, V.; Vasnev, V.; Baranov, O.V.; Mezhuev, Y.; Gritskova, I. Branched Amphiphilic Polylactides as a Polymer Matrix Component for Biodegradable Implants. *Polymers* **2023**, *15*, 1315. [CrossRef] [PubMed]
2. Soria, F.; Martínez-Pla, L.; Aznar-Cervantes, S.D.; de la Cruz, J.E.; Fernández, T.; Pérez-Fentes, D.; Llanes, L.; Sánchez-Margallo, F.M. Cytotoxicity Assessment of a New Design for a Biodegradable Ureteral Mitomycin Drug-Eluting Stent in Urothelial Carcinoma Cell Culture. *Polymers* **2022**, *14*, 4081. [CrossRef] [PubMed]
3. Fallahiarezoudar, E.; Ngadiman, N.H.A.; Yusof, N.M.; Idris, A.; Ishak, M.S.A. Development of 3D Thermoplastic Polyurethane (TPU)/Maghemite ($\gamma$-Fe$_2$O$_3$) Using Ultra-Hard and Tough (UHT) Bio-Resin for Soft Tissue Engineering. *Polymers* **2022**, *14*, 2561. [CrossRef] [PubMed]
4. Filipov, E.; Angelova, L.; Vig, S.; Fernandes, M.H.; Moreau, G.; Lasgorceix, M.; Buchvarov, I.; Daskalova, A. Investigating Potential Effects of Ultra-Short Laser-Textured Porous Poly-ε-Caprolactone Scaffolds on Bacterial Adhesion and Bone Cell Metabolism. *Polymers* **2022**, *14*, 2382. [CrossRef] [PubMed]
5. Olăreț, E.; Voicu, Ș.I.; Oprea, R.; Miculescu, F.; Butac, L.; Stancu, I.-C.; Serafim, A. Nanostructured Polyacrylamide Hydrogels with Improved Mechanical Properties and Antimicrobial Behavior. *Polymers* **2022**, *14*, 2320. [CrossRef] [PubMed]
6. Morales-Guadarrama, A.; Salgado-Ceballos, H.; Grijalva, I.; Morales-Corona, J.; Hernández-Godínez, B.; Ibáñez-Contreras, A.; Ríos, C.; Diaz-Ruiz, A.; Cruz, G.J.; Olayo, M.G.; et al. Evolution of Spinal Cord Transection of Rhesus Monkey Implanted with Polymer Synthesized by Plasma Evaluated by Diffusion Tensor Imaging. *Polymers* **2022**, *14*, 962. [CrossRef] [PubMed]
7. Hu, P.; Chiarini, A.; Wu, J.; Wei, Z.; Armato, U.; Dal Prà, I. Adult Human Vascular Smooth Muscle Cells on 3D Silk Fibroin Nonwovens Release Exosomes Enriched in Angiogenic and Growth-Promoting Factors. *Polymers* **2022**, *14*, 697. [CrossRef] [PubMed]
8. Neto, A.S.; Pereira, P.; Fonseca, A.C.; Dias, C.; Almeida, M.C.; Barros, I.; Miranda, C.O.; de Almeida, L.P.; Morais, P.V.; Coelho, J.F.J.; et al. Highly Porous Composite Scaffolds Endowed with Antibacterial Activity for Multifunctional Grafts in Bone Repair. *Polymers* **2021**, *13*, 4378. [CrossRef] [PubMed]
9. Martinez-Garcia, F.D.; de Hilster, R.H.J.; Sharma, P.K.; Borghuis, T.; Hylkema, M.N.; Burgess, J.K.; Harmsen, M.C. Architecture and Composition Dictate Viscoelastic Properties of Organ-Derived Extracellular Matrix Hydrogels. *Polymers* **2021**, *13*, 3113. [CrossRef] [PubMed]
10. Lu, W.-C.; Chen, C.-Y.; Cho, C.-J.; Venkatesan, M.; Chiang, W.-H.; Yu, Y.-Y.; Lee, C.-H.; Lee, R.-H.; Rwei, S.-P.; Kuo, C.-C. Antibacterial Activity and Protection Efficiency of Polyvinyl Butyral Nanofibrous Membrane Containing Thymol Prepared through Vertical Electrospinning. *Polymers* **2021**, *13*, 1122. [CrossRef] [PubMed]
11. Arifin, N.; Sudin, I.; Ngadiman, N.H.A.; Ishak, M.S.A. A Comprehensive Review of Biopolymer Fabrication in Additive Manufacturing Processing for 3D-Tissue-Engineering Scaffolds. *Polymers* **2022**, *14*, 2119. [CrossRef]
12. Radu, E.R.; Voicu, S.I. Functionalized Hemodialysis Polysulfone Membranes with Improved Hemocompatibility. *Polymers* **2022**, *14*, 1130. [CrossRef]
13. Mustafa, N.S.; Akhmal, N.H.; Izman, S.; Ab Talib, M.H.; Shaiful, A.I.M.; Omar, M.N.B.; Yahaya, N.Z.; Illias, S. Application of Computational Method in Designing a Unit Cell of Bone Tissue Engineering Scaffold: A Review. *Polymers* **2021**, *13*, 1584. [CrossRef] [PubMed]

*Article*

# Branched Amphiphilic Polylactides as a Polymer Matrix Component for Biodegradable Implants

Vladislav Istratov [1,2,*], Vitaliy Gomzyak [3], Valerii Vasnev [1], Oleg V. Baranov [1], Yaroslav Mezhuev [1,4,*] and Inessa Gritskova [3]

[1] A.N. Nesmeyanov Institute of Organoelement Compounds of Russian Academy of Sciences, Vavilov Street, 28, 119991 Moscow, Russia
[2] Bauman Moscow State Technical University, Baumanskaya 2-ya Str., 5/1, 105005 Moscow, Russia
[3] Department of Chemistry and Technology of Macromolecular Compounds, MIREA—Russian Technological University (RTU MIREA), Vernadskogo Avenue 78, 119454 Moscow, Russia
[4] Department of Biomaterials, Mendeleev University of Chemical Technology of Russia, Miusskaya Sq., 9, 125047 Moscow, Russia
* Correspondence: slav@ineos.ac.ru (V.I.); valsorja@mail.ru (Y.M.)

Citation: Istratov, V.; Gomzyak, V.; Vasnev, V.; Baranov, O.V.; Mezhuev, Y.; Gritskova, I. Branched Amphiphilic Polylactides as a Polymer Matrix Component for Biodegradable Implants. *Polymers* 2023, *15*, 1315. https://doi.org/10.3390/polym15051315

Academic Editors: Andrada Serafim and Stefan Ioan Voicu

Received: 21 December 2022
Revised: 2 March 2023
Accepted: 2 March 2023
Published: 6 March 2023

**Abstract:** The combination of biocompatibility, biodegradability, and high mechanical strength has provided a steady growth in interest in the synthesis and application of lactic acid-based polyesters for the creation of implants. On the other hand, the hydrophobicity of polylactide limits the possibilities of its use in biomedical fields. The ring-opening polymerization of L-lactide, catalyzed by tin (II) 2-ethylhexanoate in the presence of 2,2-bis(hydroxymethyl)propionic acid, and an ester of polyethylene glycol monomethyl ester and 2,2-bis(hydroxymethyl)propionic acid accompanied by the introduction of a pool of hydrophilic groups, that reduce the contact angle, were considered. The structures of the synthesized amphiphilic branched pegylated copolylactides were characterized by $^1$H NMR spectroscopy and gel permeation chromatography. The resulting amphiphilic copolylactides, with a narrow MWD (1.14–1.22) and molecular weight of 5000–13,000, were used to prepare interpolymer mixtures with PLLA. Already, with the introduction of 10 wt% branched pegylated copolylactides, PLLA-based films had reduced brittleness, hydrophilicity, with a water contact angle of 71.9–88.5°, and increased water absorption. An additional decrease in the water contact angle, of 66.1°, was achieved by filling the mixed polylactide films with 20 wt% hydroxyapatite, which also led to a moderate decrease in strength and ultimate tensile elongation. At the same time, the PLLA modification did not have a significant effect on the melting point and the glass transition temperature; however, the filling with hydroxyapatite increased the thermal stability.

**Keywords:** amphiphilic block copolymers; polylactide; thermal properties; mechanical properties; hydrophilicity; hydroxyapatite

## 1. Introduction

One of the most urgent problems of bioengineering is the development of new biodegradable polymeric devices. Despite intensive research in recent decades, the number of polymers available for this application is extremely limited. One of their most prominent representatives is poly(L-lactide) (PLLA). It is an aliphatic semi-crystalline polyester, that can be produced from renewable sources such as corn or sugar cane [1]. Due to its biocompatibility, biodegradability up to lactic acid, and sufficiently high mechanical characteristics, as well as a biodegradation rate comparable in time to the healing of damaged human tissues, products made from this polymer are of great interest for use in regenerative medicine [2–5]. Unfortunately, due to the fragility and hydrophobicity of PLLA [6], its use in medicine is limited [7]. One possible way to vary the properties of PLLA devices, including the rate of their biodegradation, is to manufacture these devices from PLLA plasticized with other polymers [8–12]. A promising component of PLLA matrices for biodegradable devices are amphiphilic copolyesters, in most cases represented by block copolymers of polylactide or polycaprolactone with polyethylene oxide [13–16]. They have been used to create capable biomedical devices such as stents, films, thermosensitive polymer hydrogels, electrospun fibers, etc. [17–22].

Despite the wide variety of polymers used to modify PLLA, the vast majority of them have a linear structure. At the same time, in recent decades [23–27], interest has been growing in the synthesis and study of the properties of biocompatible, biodegradable, branched and hyperbranched polymers, the properties of which differ significantly from linear, star-shaped, and crosslinked analogs. As a rule, their main feature is a smaller size of molecules compared to linear analogs, a higher density of the structure of macromolecules, in combination with a spatially unloaded core, as well as a large number of free functional groups located on the surface [28–30]. All these types of macromolecular substances differ significantly in properties from their linear counterparts, and their main feature lies in the possibility of consistent regulation of the structure and, accordingly, properties. From this point of view, branched and hyperbranched polymers are of great interest; by modifying their free reactive functional groups, one can change the properties of polymers over a very wide range [31–34], and obtain biodegradable amphiphilic copolymers for targeted drug delivery [35–38]. An additional modification of functionalized polymers makes it possible to obtain copolymers with controlled colloidal chemical properties [39,40]. The use of amphiphilic branched polymers as a component of the polymer matrix in hydroxyapatite-containing systems also leads to a significant improvement in the properties of composite materials. In these systems, branched amphiphilic polylactides act not only as PLLA plasticizers, which change the mechanical properties of the polymer matrix, but also as compatibilizers, which increase adhesion between the polymer matrix and the mineral filler. The latest advances in the field of regulation of mechanical properties, and the use of branched polylactides in biomedical fields, were summarized in review [41].

Ring-opening polymerization of lactones (ROP), catalyzed by tin carboxylates in the presence of dihydroxycarboxylic acids as co-initiators and branching agents of the AB2 type, were discussed in the literature in connection with the synthesis of branched copolyesters [42–45]. Kinetic data indicate that ROP proceeds via the sequential insertion of lactones into the –Sn–OR bond, after the formation of tin alcoholates from the starting tin carboxylates [42]. Bis(hydroxymethyl)butyric acid is traditionally used as an AB2 agent [44,45].

In the present paper, in order to introduce long-chain hydrophilic fragments and reduce the wetting angle of the surface of the resulting branched copolylactides, an ester obtained by esterification of 2,2-bis(hydroxymethyl)-propyonic acid and poly(ethylene glycol) monomethyl ethers, was used as a macroinitiator. The synthesized pegylated branched copolylactides were compatible, when mixed with linear PLLA, providing its hydrophilization, which also contributed to the good compatibility of the prepared interpolymer mixtures with a hydrophilic filler, i.e., hydroxyapatite. The filling of polylactide interpolymer blends with hydroxyapatite contributed to an additional decrease in the water contact angle; however, this effect was achieved at the cost of a decrease in strength and ultimate tensile elongation (Figure 1). At the same time, the mechanical characteristics of the obtained materials were acceptable for use in biomedical fields, in particular, for bone tissue regenerative surgery.

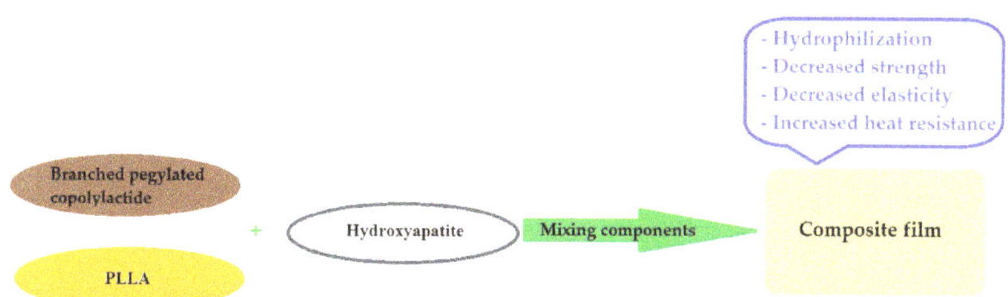



**Figure 1.** The preparation and development of properties of polylactide composite materials filled with hydroxyapatite.

## 2. Materials and Methods

### 2.1. Materials

Polylactic acid (PLLA, Mw ~60 kDa, Aldrich Chemical Co., St. Louis, MO, USA), stannous (II) 2-ethylhexanoate (Sn(Oct)$_2$, 95%, Aldrich Chemical Co., St. Louis, MO, USA), 1,1'-carbonyldiimidazole (CDI, 98%, Aldrich Chemical Co., St. Louis, MO, USA), and hydroxyapatite (HA, particle size ~1 micron, "R&P" "POLYSTOM", Moscow, Russia) were used as received. L-lactide (98%, ABCR, Karlsruhe, Germany) was recrystallized from ethyl acetate before use; 2,2-bis(hydroxymethyl)propyonic acid (BHP, 98%, Aldrich Chemical Co., St. Louis, MO, USA), poly(ethylene glycol) monomethyl ethers, with Mn ~1900 and 5000 Da (MPEG1900 and MPEG 5000, respectively, Merck, Darmstadt, Germany) were dried in a vacuum (10 Pa, 30 °C) for 24 h prior to use. Methylene chloride, THF, and DMSO ("Himmed", Moscow, Russia) were purified by standard methods [46].

### 2.2. Polymerization

Linear copolylactide was prepared by copolymerizing L-lactide and MPEG 1900; synthesis was carried out at 140 °C for 24 h in bulk, using Sn(Oct)$_2$ as a catalyst (molar ratio L-lactide:catalyst = 2000:1) according to Scheme 1 [47].

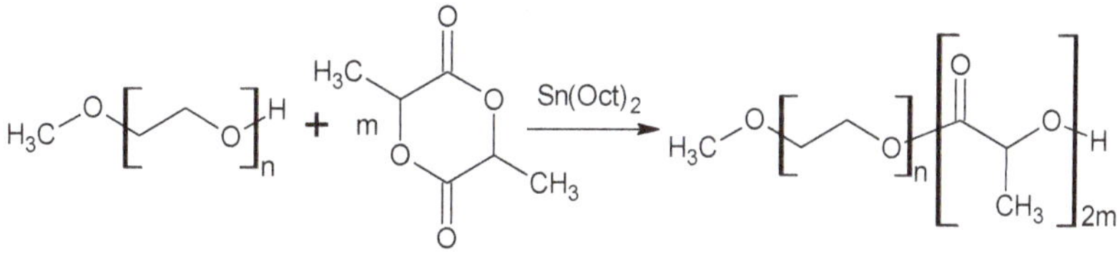

**Scheme 1.** Synthesis of linear copolylactide.

Branched copolylactides were prepared by ROP of L-lactide in the presence of BHP, as well as the esterification product of BHP and MPEG [48,49], as per Scheme 2.

Thus, to obtain linear copolylactide 1, the Sn(Oct)$_2$ catalyst (0.008 g, 0.01 mmol) was placed in a two-necked glass flask in the form of a 10% solution in THF, after which the solvent was distilled off under reduced pressure. Then, 2.883 g, 20.0 mmol of L-lactide and 0.150 g, 0.40 mmol of MPEG 1900 were added to the flask. The flask was equipped with a mechanical overhead stirrer and heated to 140 °C using a heating mantle, for 24 h. The resulting crude product was dialyzed in CHCl$_3$ (Roth "ZelluTrans" membrane, MWCO = 1000 Da, Karlsruhe, Germany) for 48 h. Then, the solvent was evaporated and the polymer was dried under vacuum for 48 h [47].

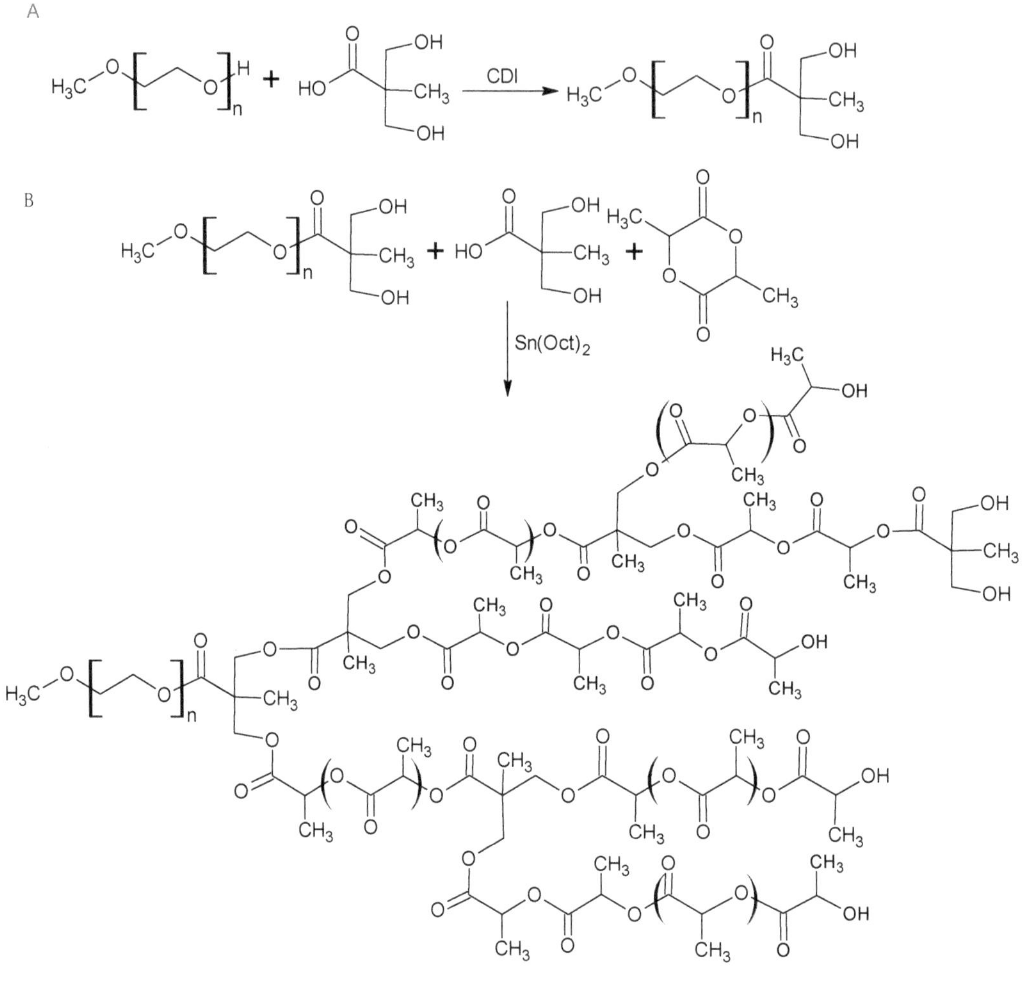

**Scheme 2.** Synthesis of pegylated branched copolylactides: esterification of BHP upon interaction with MPEG (**A**); polymerization of L-lactide in the presence of BHP and BHP ester with MPEG (**B**).

For the synthesis of branched copolylactides in the first stage (Scheme 2A), MPEG (10 mmol) and CDI (16.21 g, 10 mmol) were reacted in 50 mL of THF, for 30 min at room temperature, after which this solution was slowly added to BHP (14.75 g, 11 mmol) in 50 mL THF, at 30 °C, and reacted for 2 h. The resulting crude product was purified by column chromatography on silica gel (eluent—THF) with a yield of 68%. At the next stage, an ester of BHP and MPEG was introduced into copolymerization with L-lactide and BHP (Scheme 2B, Table 1). The reaction was carried out at 140 °C, as described in article [47]. The resulting crude product was dialyzed in CHCl₃ (Roth "ZelluTrans" membrane, MWCO = 1000 Da, Karlsruhe, Germany) for 48 h. Then the solvent was evaporated and the polymer was dried under vacuum for 48 h [47].

**Table 1.** Molecular parameters for PEG-co-PLLA copolylactides (BHP was used in an equimolar amount with respect to PEG units).

| Copoly-Lactide | PEG Mn | Yield | PEG/PLLA Units | | $M_n$ * | $M_n$ ** | $M_w/M_n$ ** |
| --- | --- | --- | --- | --- | --- | --- | --- |
| | | | Calculated | Measured * | | | |
| 1 | 1900 | 92 | 30:70 | 31.2:68.8 | 8640 | 8400 | 1.17 |
| 2 | 1900 | 87 | 30:70 | 30.3:69.7 | 9540 | 8700 | 1.22 |
| 3 | 1900 | 84 | 50:50 | 51.5:48.5 | 5030 | 4600 | 1.14 |
| 4 | 5000 | 85 | 50:50 | 51.8:48.2 | 13,220 | 12,400 | 1.20 |

*—Obtained from $^1$H NMR spectra of the polymers; **—Obtained by GPC.

## 2.3. Characterization

The $^1$H NMR spectra were recorded on a Brucker spectrometer, with operating frequency $^1$H—400.22 MHz at 25 °C, using CDCl$_3$ as solvent and TMS as an internal standard.

FTIR spectra were recorded on a Nicolet 380 spectrometer (Thermo Fisher Scientific Inc., Waltham, MA, USA).

Size exclusion chromatography (SEC) of the copolymers was carried out on a "Waters 150" chromatograph (eluent—THF (1 mL/min), column—PL-GEL 5u MIXC (300 mm × 7.5 mm) ("Waters", Milford, MA, USA).

The differential scanning calorimetry (DSC) and thermogravimetric analysis (TGA) were performed using a NETZSCH STA 449F3 Jupiter thermal analyzer (Selb, Bavaria, Germany). The $T_g$ was determined as the peak or midpoint of the heat capacity increase at the second heating cycle, and Tm as the endothermal peak in the DSC curve of the second heating cycle. From the TGA curves were obtained the temperature required for a 5% weight loss ($T_{5\%}$), which is representative of the onset of decomposition, the temperature of maximum decomposition rate ($T_{deg}$), and the amount of residual mass at 600 °C.

Copolymer films were prepared by the solvent casting method, from polymer composition samples dissolved in methylene chloride. Polymer samples were prepared at 10% ($w/v$) concentration. HA was added to the polymer solution at 20% ($w/v$) concentration, with respect to the contents of the polymer composition. Polymer compositions were blended by a magnetic stirrer at room temperature and cast on the glass flat surface by a manual film applicator, then allowed to evaporate in a desiccator, to obtain a homogeneous film.

The tensile test was performed using an Instron-3366 universal mechanical testing machine. Samples were cut into rectangular pieces, about 0.15 mm thick, 10 mm wide, and 30 mm long. The sample gauge length was 10 mm and the testing speed was 1 mm/min. The results were taken as an average of five tests.

The SEM were obtained on a Hitachi TM300 electron microscope (Hitachi, Tokyo, Japan).

Water contact angle (WCA) measurements, between the polymer films and MilliQ water drops, were measured by the sessile drop method, on a "Tracker" automatic drop shape analyzer (Teclis-Scientific, Civrieux-d'Azergues, France), for drops of 100 s age, at a temperature of 25 °C and about 100% relative humidity. All measurements were repeated at least four times for each polymer sample.

The water absorption studies were performed in MilliQ water (pH = 7.00), at 25 °C. The amount of fluid absorbed through time was monitored gravimetrically. Dry 20 mm × 20 mm film samples (approximately 40 mg) were immersed in 50 mL of water and at regular time intervals were removed, dried with filter paper, weighed, and placed in the same bath. The water absorption (q) was calculated by the following equation:

$$q = (m_t - m_0)/m_0 \tag{1}$$

where $m_t$ is the weight of the absorbed polymer film at the time of measuring, and $m_0$ is the initial weight of the polymer film [50].

## 3. Results

### 3.1. The Structure of Linear and Branched Pegylated Copolylactides

In this work, a number of MPEG-containing amphiphilic copolylactides were obtained; PLLA blocks were either linear (for copolylactide 1) or branched containing BHP monomer units (for copolylactides 2–4). The polymerization of L-lactide was carried out at 140 °C, as recommended in [51]. By systematic variation of composition and molecular weight of the polylactide block, a series of oligomeric amphiphilic copolymers, with varied ratios of PEG and PLLA molecular units, were prepared. The synthesized branched pegylated copolylactides are waxy substances, soluble in $CH_2Cl_2$, DMSO, and THF.

The structure of the synthesized linear and branched pegylated copolylactides was confirmed by $^1$H NMR spectroscopy; the assignment is shown in Figure 2 and is consistent with the literature data [28,44,52].

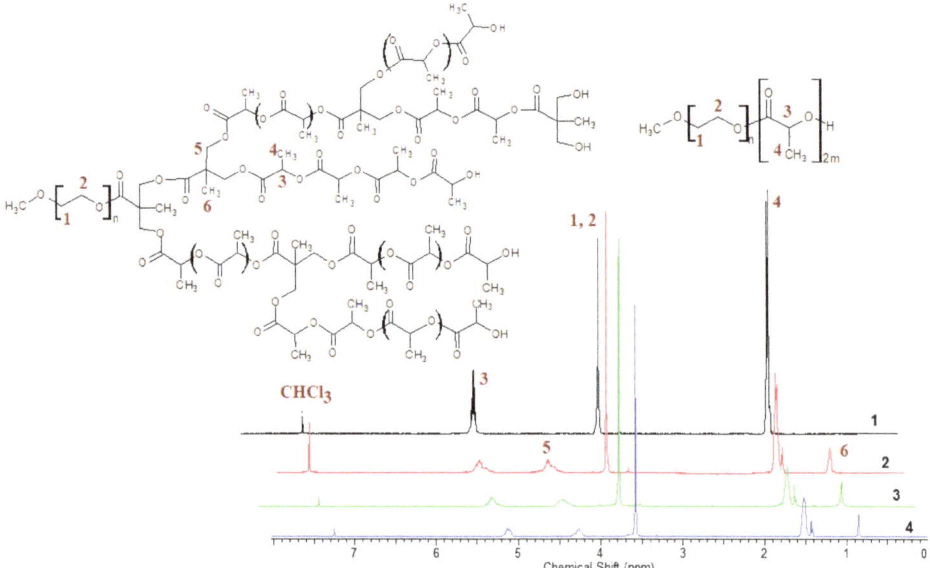

**Figure 2.** $^1$H NMR spectra of synthesized polymers. The indices at the spectra correspond to the designations of the polymers in Table 1.

The incorporation of the MPEG was confirmed by the $^1$H NMR spectroscopy for both the linear and branched copolylactides. The ratio of incorporated PEG and PLLA units was determined from the respective $^1$H NMR signal intensity: thus, PEG units show characteristic signals at 3.56–3.70 ppm, that can be used to quantify their fraction, by comparison with the signals of the (>CH–) group of the lactide monomer units located at 5.1 ppm [47].

The synthesized copolylactides, the molecular parameters of which are summarized in Table 1, were characterized by narrow unimodal MWDs, with dispersity indices in the range of 1.17–1.22. According to the GPC data, the values of the average molecular weights of the copolylactides were in the range 8400–12,400 (Table 1), and this is probably somewhat on the low side, since polystyrene standards were used to calibrate the columns. It is important that in polymers changed either structure of PLLA block (linear or branched), or ratios of PEG/PLLA units (30:70 or 50:50), or molecular weight of the PEG block (1900 or 5000Da). The agreement between the calculated and experimental ratios of PEG and PLLA blocks was satisfactory. A decrease in the yield of pegylated copolylactides, with an increase in the amount of BHP in the reaction mixture, was observed (Table 1).

*3.2. The Properties of PLLA Films Plasticized with Branched Pegylated Copolylactides Filled with Hydroxyapatite*

Prior to the examination of PLLA plasticized with copolymers, the DSC analysis of pure polymers was performed. The obtained data are listed in Table 2.

**Table 2.** Characteristics of polymer materials used.

| Polymer | PEG | | PLLA | |
|---|---|---|---|---|
| | $T_g$ | $T_m$ | $T_g$ | $T_m$ |
| PLLA | - | - | 60.5 | 163.8 |
| Copolylactide 1 | 45.3 | 48.7 | 46.8 | 153.6 |
| Copolylactide 2 | 45.1 | 47.6 | 46.2 | 154.0 |
| Copolylactide 3 | 45.4 | 39.7 | 38.4 | 148.5 |
| Copolylactide 4 | 52.1 | 57.9 | 54.7 | 158.0 |

All the studied copolylactides had a wide bimodal peak at 38.7–54.7 °C, the maxima of which corresponded to the Tg of the PEG and PLLA blocks. Along with this, pronounced monomodal peaks, corresponding to the Tm of PEG and PLA, were observed. While the Tm peaks of PEG and PLLA did not change significantly for all samples, the temperature of Tg peaks for PEG and PLLA blocks increased with their molecular weight.

Five PLLA films, plasticized with 10 wt% of different copolylactides, were obtained by the solvent casting technique. Furthermore, five composite films, containing 20% by weight of hydroxyapatite, were obtained using polymer matrices with 10 wt% of different copolylactides.

The IR spectra of the PLLA 1, PLLA 2, and PLLA 3 samples, shown in Figure 3, contained all the signals characteristic of copolylactides [53], and were close to the IR spectra of the PLLA 4 and PLLA 5 samples. At the same time, filling with hydroxyapatite lead to the appearance of characteristic absorption bands in the 500–700 cm$^{-1}$ region.

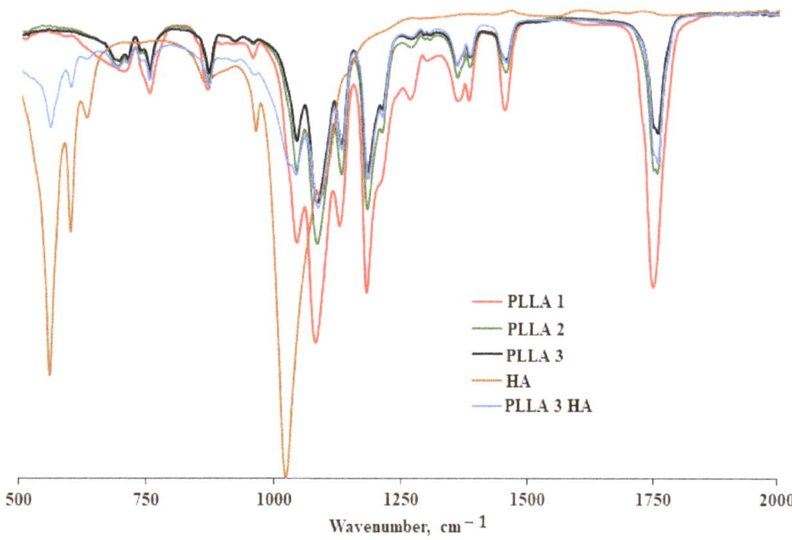


**Figure 3.** IR spectra of unfilled and hydroxyapatite-filled copolylactide films.

The thermal transitions of the prepared PLLA blends were assessed by DSC experiments. Typical DSC thermograms of PLLA 1, PLLA 1 HA, PLLA 5, and PLLA 5 HA composites (second heating run), are given in Figure 4. The designations of the obtained samples of polyester composites, as well as their compositions and thermal properties, are given in Table 3.

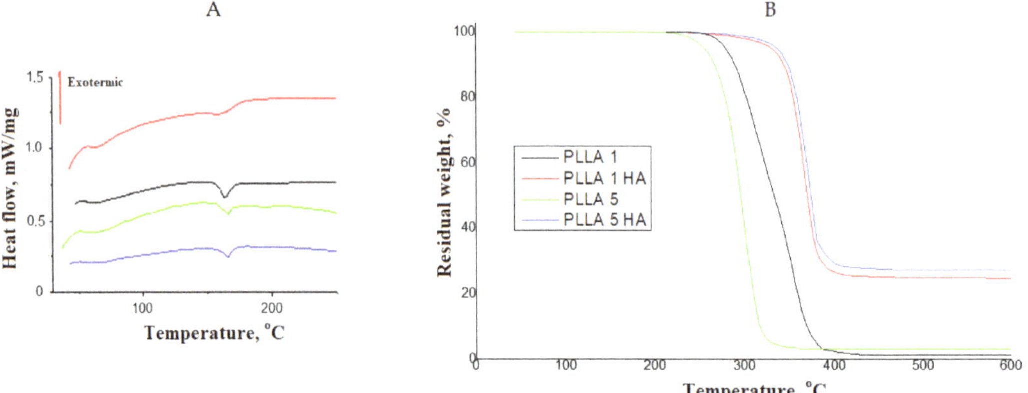



**Figure 4.** Representative thermal analysis data for composite samples: DSC curves (**A**) and TGA curves (**B**).

**Table 3.** Thermal properties of PLLA films.

| Sample | Copolylactide | Composition (%) | | | $T_g$ (°C) | $T_m$ (°C) | $T_{5\%}$ (°C) | $T_{deg}$ (°C) | Residual Mass (%) |
|---|---|---|---|---|---|---|---|---|---|
| | | PLLA | Copo Lylac Tide | HA | | | | | |
| PLLA 1 | - | 100 | 0 | 0 | 61.2 | 164.2 | 279.3 | 361.4 | 2.1 |
| PLLA 2 | Copolylactide 1 | 90 | 10 | 0 | 59.8 | 166.4 | 280.7 | 295.3 | 1.9 |
| PLLA 3 | Copolylactide 2 | 90 | 10 | 0 | 56.2 | 165.1 | 254.1 | 299.1 | 2.2 |
| PLLA 4 | Copolylactide 3 | 90 | 10 | 0 | 60.3 | 166.7 | 257.2 | 307.5 | 2.7 |
| PLLA 5 | Copolylactide 4 | 90 | 10 | 0 | 60.7 | 165.3 | 255.8 | 314.7 | 3.1 |
| PLLA 1 HA | - | 80 | 0 | 20 | 64.3 | 158.8 | 330.5 | 368.1 | 24.8 |
| PLLA 2 HA | Copolylactide 1 | 72 | 8 | 20 | 64.4 | 163.0 | 350.8 | 360.2 | 25.2 |
| PLLA 3 HA | Copolylactide 2 | 72 | 8 | 20 | 63.4 | 162.3 | 332.4 | 365.6 | 27.7 |
| PLLA 4 HA | Copolylactide 3 | 72 | 8 | 20 | 59.1 | 163.7 | 332.7 | 364.3 | 26.2 |
| PLLA 5 HA | Copolylactide 4 | 72 | 8 | 20 | 63.1 | 165.8 | 336.6 | 369.7 | 27.4 |

The glass transition temperature ($T_g$) and the melting point ($T_m$) of the PEG and PLLA phases, were evaluated for the polyester composites shown in Table 3. The samples of composite films had a wide glass transition temperature range, that sits between the glass transition temperatures of pure PLLA and pegylated copolylactides. The monomodal nature of DSC in the temperature range corresponding to the relaxation transition (glass transition), indicates good compatibility of pegylated copolylactides and pure PLLA for all studied compositions [54].

The addition of copolylactides to PLA only slightly changed the $T_g$ and $T_m$ values of neat PLLA; a correlation was observed between the molecular weights of the copolylactide blocks of PLLA and the Tg of the PLLA films plasticized by them. Addition of HA slightly increased the glass transition values of the PLA composites, perhaps due to a reduction of the chain segment mobility of the polymeric amorphous phase around the HA particles [55].

To study the thermal stability of the obtained films, thermogravimetric analysis (TGA) was performed. The experimental data are summarized in Table 3.

One can observe that HA containing samples underwent thermal decomposition in a single step, ranging from approximately 260 to 400 °C, which correlates well with values reported elsewhere [56–58]. Interestingly, the films plasticized by linear copolylactide 1 showed the highest $T_{5\%}$ value (Table 3), but faster decomposition, which was reflected in a lower $T_{deg}$ value. In the same case, the addition of copolylactides 2–4 neither caused significant changes in $T_{5\%}$ (for films without HA), nor did they remarkably reduce the thermal stability (for films containing HA). The temperature of the maximum decomposition rate for films plasticized with branched copolylactides was generally close to the temperature of the maximum decomposition rate of neat PLLA films, only the use of copolylactide 4 with the largest polylactide block led to a slight excess of $T_{deg}$ over this value of neat polylactide.

The addition of 20% HA to the PLLA matrix improved the thermal stability of the polymer, as can be seen in Figure 4. A rise in the $T_{5\%}$ and in the $T_{deg}$ was observed for all composite samples (Table 3), suggesting an insulating effect of HA incorporation. We did not observe an increase in the degradation rate of the polymer matrix with the presence of the HA, as was stated in [55].

One can also observe that the residual masses for samples without HA, were relatively low, and progressively increased with the molecular weight of copolylactides up to 3.1%. For compositions with 20% HA, since the filler is thermally stable at high temperatures and can also act as a heat barrier, it could enhance the formation of char after thermal decomposition. Thus, the residual weights of samples containing hydroxyapatite were higher than those of samples without hydroxyapatite.

The tensile curves of PLLA films plasticized with copolylactides 1–4, as well as films of composite materials containing HA, are shown in Figure 5, and specific data are also given in Table 4. Pure PLLA 1 film showed mechanical properties of a hard but brittle material, with values of tensile modulus ($E_t$) of 1290.8 MPa, tensile strength at break ($\sigma_b$) of 25.4 MPa, and elongation at break ($\varepsilon_b$) of 4.3% (Figure 5, Table 4).

**Table 4.** Properties of composite polyester materials: tensile modulus ($E_t$), elongation at break ($\varepsilon_b$), tensile strength (MPa) ($\sigma_b$), and water contact angle (WCA) on different PLLA substrates.

| Sample | $E_t$ (MPa) | $\varepsilon_b$ (%) | $\sigma_b$ (MPa) | WCA (°) |
|---|---|---|---|---|
| PLLA 1 | 1290.8 | 4.3 | 25.4 | 93.3 |
| PLLA 2 | 1076.1 | 5.1 | 27.4 | 76.8 |
| PLLA 3 | 951.1 | 6.0 | 18.9 | 71.9 |
| PLLA 4 | 672.3 | 5.4 | 13.9 | 88.5 |
| PLLA 5 | 897.0 | 8.9 | 19.1 | 86.6 |
| PLLA 1 HA | 1295.0 | 2.4 | 17.9 | 75.9 |
| PLLA 2 HA | 1116.2 | 2.7 | 17.3 | 71.5 |
| PLLA 3 HA | 798.5 | 3.3 | 13.0 | 75.8 |
| PLLA 4 HA | 691.2 | 4.0 | 10.2 | 66.1 |
| PLLA 5 HA | 1042.4 | 5.9 | 18.1 | 66.4 |

The tensile modulus for blends of PLLA with copolylactides was lower than for pure PLLA. At the same time, plasticization of PLLA with linear copolylactide 1 led to mixtures with the tensile modulus only slightly lower than that of unplasticized PLLA. The presence of copolylactides with a branched PLLA block in mixtures, as a rule, led to a decrease in the tensile modulus of the polymer mixture. This effect decreased with increasing molecular weight of the copolylactides.

The filling of polyester films with hydroxyapatite had almost no effect on the elastic modulus of composite materials; however, the tensile strength and elongation at break decreased markedly. The low sensitivity of the elastic modulus of polyester materials to filling with hydroxyapatite indicated high compatibility of the matrix and filler. At the same time, it can be assumed that the interface between the polyester matrix and the filler (hydroxyapatite) becomes the center of the formation of defects and cracks under high mechanical loads. The decrease in the tensile strength of polyester materials when filled with hydroxyapatite, apparently, limits the achievable ultimate deformation corresponding to the rupture of composites.

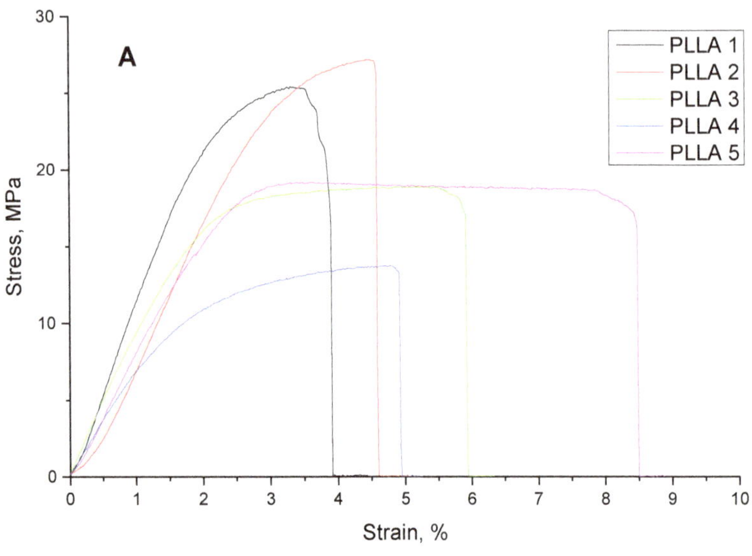

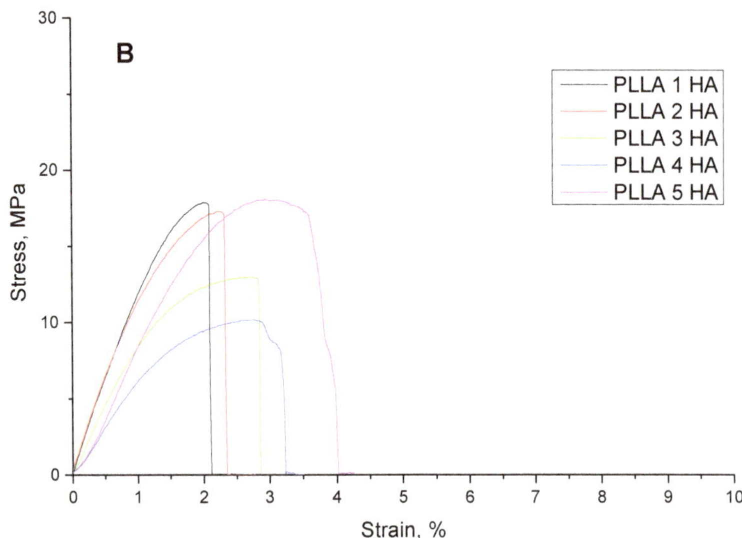

**Figure 5.** Tensile test curves for PLLA films without (**A**) and with (**B**) addition of 20 wt% of hydroxyapatite.

The addition of copolylactides tended to plasticize the PLA film, resulting in a significant increase in flexibility and a decrease in tensile strength, as noted in various previous studies. Nevertheless, we found that the addition of linear copolylactide 1 caused only a slight increase in the value of elongation at break, which contradicts previously published data [59]. The yield point is observed only for copolylactides with high molecular weight branched PLLA blocks. Since the stage of plastic deformation is due to the parallel alignment of PLLA polymer chains, the presence of bulky high-molecular weight branched PLLA blocks increases the mobility of the polymer chains.

Typically, the addition of hydroxyapatite to PLLA in a 20:80 mix ratio, resulted in a decrease in ductility in all samples. However, the elongation at break for films plasticized with copolyesters was always higher than for unplasticized PLLA. Here, copolylactide 4 is of particular interest: the values of elongation at break for films plasticized with it were two times higher than for unplasticized films, regardless of the presence of HA in the film.

Tensile strength values for films plasticized with copolylactides were slightly lower than for pure PLLA films; at the same time, the PLLA 5HA sample plasticized with copolylactide 4 showed a tensile strength value characteristic of the unplasticized sample. This may be due to the good miscibility of the PLLA macromolecules and the plasticizer, as well as to the fact that HA particles are able to interact with the polar groups in branched PLLA polymer blocks [60,61]. As different polar groups, such as -OH, -CO, -CH$_2$-O-CH$_2$-, and -COOH, appear in the copolylactide polymer chains, they can react or interact with the -OH groups in commercial HA, as demonstrated by Zhang et al. [62].

The SEM data indicate the compatibility of the copolylactide matrices with hydroxyapatite. The SEMs of the PLLA5 and PLLA 5 HA samples, shown in Figure 6, were typical for all obtained composite materials, and indicate a uniform distribution of the filler in the matrix volume. On the other hand, the pronounced surface roughness of the mixture of branched copolylactide 4 and PLLA (sample PLLA5) may be due to microphase separation of hydrophilic PEG fragments and hydrophobic lactide residues.

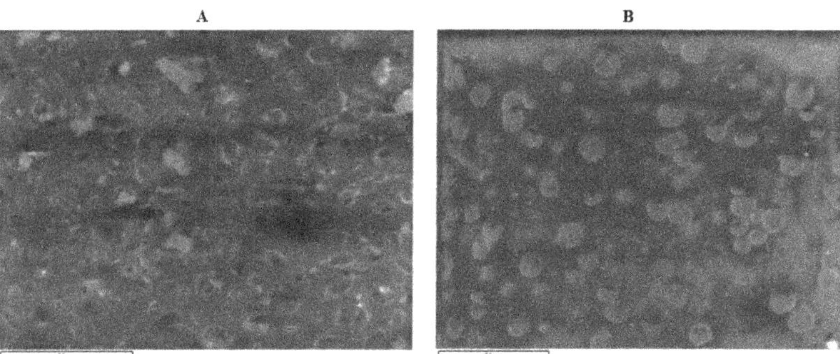

**Figure 6.** SEM of a polymer matrix (sample PLLA5)—(**A**) and a composite obtained by filling of the matrix with hydroxyapatite (sample PLLA 5 HA)—(**B**).

The hydrophilicity of the surface of the material is considered as a paramount requirement for its use in living organisms. Various studies have shown that wettability plays a key role in determining the ability of a polymeric material to serve as a scaffold for bone osteointegration [63,64]. However, the hydrophobicity of pure PLLA puts a limit on its compatibility with tissues and body fluids, and requires reduction. In this work, the PLLA hydrophilization was achieved by introducing hydrophilic polyethylene oxide fragments and a hydrophilic filler, i.e., hydroxyapatite. As can be seen in Table 4, the introduction of polyethylene oxide blocks and filling with hydroxyapatite led to a decrease in the water contact angle. Apparently, the value of water contact angle for the obtained composite materials is an integral characteristic of intermolecular interactions between hydrophilic and hydrophobic blocks in the composition of macromolecules, as well as hydrophilic polyethylene oxide blocks with hydroxyl groups on the surface of hydroxyapatite, used for filling. It can be assumed that if the main contribution comes from the interactions of the hydrophilic polyethylene oxide blocks with the hydroxyl groups on the surface of hydroxyapatite, then the surface layer is enriched with the hydrophobic polylactide fragments, which leads to an increase in the water contact angle, as was observed in the case of filling with PLLA3. Nevertheless, in the vast majority of cases, the contribution of hydrophobic interactions of polylactide fragments and displacement of hydrophilic polyethylene oxide fragments to the surface, is predominant, which leads to the desired effect of reducing the water contact angle.

The WCA observation can be correlated with the data on water absorption by polymer films shown in Figure 7. It is well known that the wetting and water absorption can play an important role in the release of a drug from the polymer matrix into an aqueous solution [65]. It is known that upon degradation of PLLA, the hydrophilicity of the polymer increases and the weight of the polymer decreases. Since these processes can occur simultaneously with water absorption, a relatively short time of 1 week was used to evaluate water absorption, in which polymer degradation processes do not have a large effect on the weight of the samples [66,67].

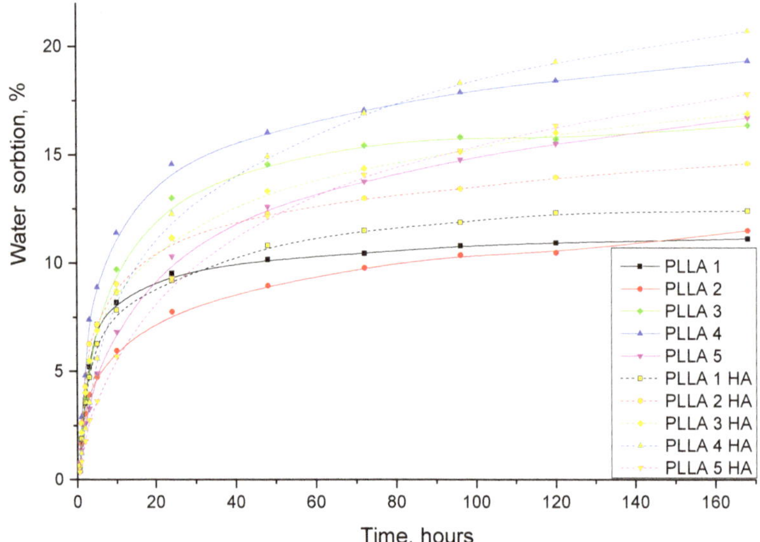


**Figure 7.** Kinetic curves for water absorption by different PLLA film samples.

The amount of water absorbed by plasticized PLLA films is generally higher than by non-plasticized ones. The water absorption of all films steadily increased with time as ever deeper layers of the material became available for water. The water absorption of the films increased along with the rise in the mole fraction of hydrophilic pegylated fragments in the compositions of the branched copolylactides, as expected. The PLLA films not plasticized with copolylactides, as expected, absorbed water at the lowest rate. Among plasticized films, the sample containing linear copolylactide 1 had the lowest water absorption rate. Films plasticized with branched copolylactides showed the highest water absorption rates. Samples containing copolylactide 3, with the lowest molecular weight, showed the fastest water absorption. This may be due to different accessibility of the interior of the polymer matrix for water. Whereas linear copolylactides form fairly dense polymeric structures, branched copolylactides resulted in looser, permeable polymeric structures. In addition, low molecular weight copolylactide 3 may have the lowest adhesion to the polymer matrix and be removed from the polymer with water. All samples containing hydroxyapatite absorbed water faster than samples without hydroxyapatite of the same polymer composition, perhaps due to the hydrophilic nature of hydroxyapatite. In this case, the same tendencies are observed as for films not containing HA: films plasticized with linear copolylactides absorb less water than films plasticized with pegylated branched copolylactides. Low molecular weight copolylactides absorb more water than films containing high molecular weight copolylactides.

It is known that, unlike linear polylactides, the hydrolytic degradation of which proceeds in bulk, the hydrolytic degradation of branched polylactides has a surface nature [68]. Although the determination of the dynamics and kinetics of the biodegradation of the obtained composite materials will be the subject of subsequent studies, it can be assumed that the rate of weight loss due to hydrolysis will hardly differ from that characteristic of linear PLLA, since the minimum content of the latter was 72 wt% at the maximum content of branched pegylated copolylactides (10 wt%) (Table 3).

## 4. Conclusions

It was established that the polymerization of L-lactide, catalyzed by tin (II) 2-ethylhexanoate in the presence of both BHP and BHP ester with MPEG, leads to the formation of branched pegylated copolylactides with a narrow molecular weight distribution. It was shown that the introduction of PEG fragments into linear and branched copolylactides leads to a decrease in the glass transition temperatures and melting points, as compared to PLLA. The resulting pegylated copolylactides were compatible with PLLA when blended, and formed films with higher tensile strength, lower strength, and lower tensile modulus. The introduction of 10 wt% of the synthesized pegylated copolylactides into PLLA did not have a significant effect on the melting points and the glass transition temperatures of the resulting interpolymer mixtures, but lead to a decrease in the water contact angle and increased water absorption. Additional hydrophilization of PLLA interpolymer blends and pegylated copolylactides can be achieved by filling with hydroxyapatite. In this case, the strength and ultimate elongation in tension decreased. Since HA-containing scaffolds are very important for medicine, the presented study will provide more information about their behavior and may lead to their wider use.

**Author Contributions:** Conceptualization, V.I., V.V.; methodology, V.V.; validation, V.G. and I.G.; formal analysis, V.I.; investigation, V.I., V.V., V.G., I.G., Y.M., O.V.B.; resources, V.I.; data curation, V.V.; writing—original draft preparation, V.I., Y.M.; writing—review and editing, V.I., V.V., I.G., Y.M.; visualization, V.I., V.G.; supervision, V.V.; project administration, V.I. All authors have read and agreed to the published version of the manuscript.

**Funding:** This work was supported by the Ministry of Science and Higher Education of the Russian Federation (Contract No. 075-03-2023-642), and was performed employing the equipment of the Center for molecular composition studies of INEOS RAS.

**Institutional Review Board Statement:** Not applicable.

**Data Availability Statement:** The data presented in this study are available on request from the corresponding author (the obtained initial data are fully presented in the article).

**Acknowledgments:** The study of polymer composites was carried out within the framework of the program of state support for the centers of the National Technology Initiative (NTI), on the basis of educational institutions of higher education and scientific organizations (Center NTI "Digital Materials Science: New Materials and Substances" on the basis of the Bauman Moscow State Technical University).

**Conflicts of Interest:** The authors declare no conflict of interest.

## References

1. Jayasekara, T.; Wickrama Surendra, Y.; Rathnayake, M. Polylactic Acid Pellets Production from Corn and Sugarcane Molasses: Process Simulation for Scaled-Up Processing and Comparative Life Cycle Analysis. *J. Polym. Environ.* **2022**, *30*, 4590–4604. [CrossRef]
2. Capuana, E.; Lopresti, F.; Ceraulo, M.; La Carrubba, V. Poly-L-Lactic Acid (PLLA)-Based Biomaterials for Regenerative Medicine: A Review on Processing and Applications. *Polymers* **2022**, *14*, 1153. [CrossRef]
3. Dai, Y.; Lu, T.; Shao, M.; Lyu, F. Recent advances in PLLA-based biomaterial scaffolds for neural tissue engineering: Fabrication, modification, and applications. *Front. Bioeng. Biotechnol.* **2022**, *10*, 1011783. [CrossRef] [PubMed]
4. Pawar Rajendra, P.; Tekale Sunil, U.; Shisodia Suresh, U.; Totre Jalinder, T.; Domb, A.J. Biomedical Applications of Poly(Lactic Acid). *Recent Pat. Regen. Med.* **2014**, *4*, 40–51. [CrossRef]
5. Shah, T.V.; Vasava, D.V. A glimpse of biodegradable polymers and their biomedical applications. *e-Polymers* **2019**, *19*, 385–410. [CrossRef]
6. Bastekova, K.; Guselnikova, O.; Postnikov, P.; Elashnikov, R.; Kunes, M.; Kolska, Z.; Švorčik, V.; Lyutakov, O. Spatially selective modification of PLLA surface: From hydrophobic to hydrophilic or to repellent. *Appl. Surf. Sci.* **2017**, *397*, 226–234. [CrossRef]

7.  Huang, Y.; Wang, Y.; Wen, J. A Study on modification of polylactic acid and its biomedical application. *E3S Web Conf.* **2021**, *308*, 02008. [CrossRef]

8.  Tian, J.; Cao, Z.; Qian, S.; Xia, Y.; Zhang, J.; Kong, Y.; Sheng, K.; Zhang, Y.; Wan, Y.; Takahashi, J. Improving tensile strength and impact toughness of plasticized poly(lactic acid) biocomposites by incorporating nanofibrillated cellulose. *Nanotechnol. Rev.* **2022**, *11*, 2469–2482. [CrossRef]

9.  Murariu, M.; Paint, Y.; Murariu, O.; Laoutid, F.; Dubois, P. Tailoring and Long-Term Preservation of the Properties of PLA Composites with "Green" Plasticizers. *Polymers* **2022**, *14*, 4836. [CrossRef]

10. Omar, A.A.; Mohd Hanafi, M.H.; Razak, N.H.; Ibrahim, A.; Ab Razak, N.A. A Best-Evidence Review of Bio-based Plasticizer and the Effects on the Mechanical Properties of PLA. *Chem. Engineer. Transact.* **2021**, *89*, 241–246.

11. Xuan, W.; Odelius, K.; Hakkarainen, M. Tailoring Oligomeric Plasticizers for Polylactide through Structural Control. *ACS Omega* **2022**, *7*, 14305–14316. [CrossRef] [PubMed]

12. López-Rodríguez, N.; Sarasua, J.R. Plasticization of Poly-L-lactide with L-lactide, D-lactide, and D,L-lactide monomers. *Polym. Engineer. Sci.* **2013**, *53*, 2073–2080. [CrossRef]

13. Lee, H.J.; Jeong, B. ROS-Sensitive Degradable PEG-PCL-PEG Micellar Thermogel. *Small* **2020**, *16*, 1903045. [CrossRef] [PubMed]

14. Dethe, M.R.; Prabakaran, A.; Ahmed, H.; Agrawal, M.; Roy, U.; Alexander, A. PCL-PEG copolymer based injectable thermosensitive hydrogels. *J. Control Release* **2022**, *343*, 217–236. [CrossRef] [PubMed]

15. Hwang, J.-J.; Huang, S.-M.; Lin, W.-Y.; Liu, H.-J.; Chuang, C.-C.; Chiu, W.-H. The Synthesis of Biodegradable Poly(L-Lactic Acid)-Polyethylene Glycols Copolymer/Montmorillonite Nanocomposites and Analysis of the Crystallization Properties. *Minerals* **2022**, *12*, 14. [CrossRef]

16. Kutikov, A.B.; Song, J. Biodegradable PEG-Based Amphiphilic Block Copolymers for Tissue Engineering Applications. *ACS Biomater. Sci. Eng.* **2015**, *1*, 463–480. [CrossRef] [PubMed]

17. Ilyas, R.A.; Zuhri, M.Y.M.; Norrrahim, M.N.F.; Misenan, M.S.M.; Jenol, M.A.; Samsudin, S.A.; Nurazzi, N.M.; Asyraf, M.R.M.; Supian, A.B.M.; Bangar, S.P.; et al. Natural Fiber-Reinforced Polycaprolactone Green and Hybrid Biocomposites for Various Advanced Applications. *Polymers* **2022**, *14*, 182. [CrossRef] [PubMed]

18. Han, L.-Y.; Wu, Y.-L.; Zhu, C.-Y.; Wu, C.-S.; Yang, C.-R. Improved Pharmacokinetics of Icariin (ICA) within Formulation of PEG-PLLA/PDLA-PNIPAM Polymeric Micelles. *Pharmaceutics* **2019**, *11*, 51. [CrossRef]

19. Cokelaere, S.M.; Groen, W.M.G.A.C.; Plomp, S.G.M.; de Grauw, J.C.; van Midwoud, P.M.; Weinans, H.H.; van de Lest, C.H.A.; Tryfonidou, M.A.; van Weeren, P.R.; Korthagen, N.M. Sustained Intra-Articular Release and Biocompatibility of Tacrolimus (FK506) Loaded Monospheres Composed of [PDLA-PEG1000]-b-[PLLA] Multi-Block Copolymers in Healthy Horse Joints. *Pharmaceutics* **2021**, *13*, 1438. [CrossRef]

20. Shimanouchi, T.; Iwamura, M.; Deguchi, S.; Kimura, Y. Fibril Growth Behavior of Amyloid β on Polymer-Based Planar Membranes: Implications for the Entanglement and Hydration of Polymers. *Appl. Sci.* **2021**, *11*, 4408. [CrossRef]

21. Pan, L.; Yang, J.; Xu, L. Preparation and Characterization of Simvastatin-Loaded PCL/PEG Nanofiber Membranes for Drug Sustained Release. *Molecules* **2022**, *27*, 7158. [CrossRef]

22. Wu, D.T.; Munguia-Lopez, J.G.; Cho, Y.W.; Ma, X.; Song, V.; Zhu, Z.; Tran, S.D. Polymeric Scaffolds for Dental, Oral, and Craniofacial Regenerative Medicine. *Molecules* **2021**, *26*, 7043. [CrossRef] [PubMed]

23. Shen, B.; Lu, S.; Sun, C.; Song, Z.; Zhang, F.; Kang, J.; Cao, Y.; Xiang, M. Effects of Amino Hyperbranched Polymer-Modified Carbon Nanotubes on the Crystallization Behavior of Poly (L-Lactic Acid) (PLLA). *Polymers* **2022**, *14*, 2188. [CrossRef] [PubMed]

24. Xu, P.; Huang, X.; Pan, X.; Li, N.; Zhu, J.; Zhu, X. Hyperbranched Polycaprolactone through RAFT Polymerization of 2-Methylene-1,3-dioxepane. *Polymers* **2019**, *11*, 318. [CrossRef]

25. Peng, Q.; Cheng, J.; Lu, S.; Li, Y. Electrospun hyperbranched polylactic acid–modified cellulose nanocrystals/polylactic acid for shape memory membranes with high mechanical properties. *Polym. Adv. Technol.* **2020**, *31*, 15–24. [CrossRef]

26. Christodoulou, E.; Notopoulou, M.; Nakiou, E.; Kostoglou, M.; Barmpalexis, P.; Bikiaris, D.N. Branched Poly(ε-caprolactone)-Based Copolyesters of Different Architectures and Their Use in the Preparation of Anticancer Drug-Loaded Nanoparticles. *Int. J. Mol. Sci.* **2022**, *23*, 15393. [CrossRef] [PubMed]

27. Fang, F.; Niu, D.; Xu, P.; Liu, T.; Yang, W.; Wang, Z.; Li, X.; Ma, P. A Quantitative Study on Branching Density Dependent Behavior of Polylactide Melt Strength. *Macromol. Rapid Commun.* **2023**, *44*, e2200858. [CrossRef]

28. Nim, B.; Rahayu, S.S.; Thananukul, K.; Eang, C.; Opaprakasit, M.; Petchsuk, A.; Kaewsaneha, C.; Polpanich, D.; Opaprakasit, P. Sizing down and functionalizing polylactide (PLA) resin for synthesis of PLA-based polyurethanes for use in biomedical applications. *Sci. Rep.* **2023**, *13*, 2284. [CrossRef]

29. Zamboulis, A.; Nakiou, E.A.; Christodoulou, E.; Bikiaris, D.N.; Kontonasaki, E.; Liverani, L.; Boccaccini, A.R. Polyglycerol Hyperbranched Polyesters: Synthesis, Properties and Pharmaceutical and Biomedical Applications. *Int. J. Mol. Sci.* **2019**, *20*, 6210. [CrossRef]

30. Shen, B.; Xu, Y.; Zhang, Y.; Xie, Z.; Zhang, F.; Kang, J.; Cao, Y.; Xiang, M. Effects of Different End Functional Groups Hyperbranched Polymers-Modified Carbon Nanotubes on the Crystallization and Mechanical Properties of Poly(l-lactic acid) (PLLA). *ACS Omega* **2022**, *7*, 42939–42948. [CrossRef]

31. Zhang, X.; Dai, Y.; Dai, G. Advances in amphiphilic hyperbranched copolymers with an aliphatic hyperbranched 2,2-bis(methylol)propionic acid-based polyester core. *Polym. Chem.* **2020**, *11*, 964–973. [CrossRef]

32. Prabaharan, M.; Grailer, J.J.; Pilla, S.; Steeber, D.; Gong, S. Folate- conjugated amphiphilic hyperbranched block copolymers based on Boltorn H40, poly(l-lactide) and poly(ethylene glycol) for tumor-targeted drug delivery. *Biomaterials* **2009**, *30*, 3009–3019. [CrossRef]
33. Perše, L.S.; Huskić, M. Rheological characterization of multiarm star copolymers. *Eur. Polym. J.* **2016**, *76*, 188–195. [CrossRef]
34. Hawker, C.J.; Lee, R.; Fréchet, J.M.J. One-Step Synthesis of Hyperbranched Dendritic Polyesters. *J. Am. Chem. Soc.* **1991**, *113*, 4583–4588. [CrossRef]
35. Korake, S.; Shaikh, A.; Salve, R.; Gajbhiye, K.R.; Gajbhiye, V.; Pawar, A. Biodegradable dendritic BoltornTM nanoconstructs: A promising avenue for cancer theranostics. *Int. J. Pharm.* **2021**, *594*, 120177. [CrossRef]
36. Istratov, V.V.; Krupina, T.V.; Gomzyak, V.I.; Vasnev, V.A. Development and characterization of bioresorbable polyglycerol esters and drug-loaded microparticles. *High Perf. Polym.* **2017**, *29*, 708–715. [CrossRef]
37. Di Martino, A.; Trusova, M.E.; Postnikov, P.S.; Sedlarik, V. Branched poly (lactic acid) microparticles for enhancing the 5-aminolevulinic acid phototoxicity. *J. Photochem. Photobiol. B.* **2018**, *181*, 80–88. [CrossRef]
38. Hadar, J.; Skidmore, S.; Garner, J.; Park, H.; Park, K.; Wang, Y.; Qin, B.; Jiang, X. Characterization of branched poly(lactide-co-glycolide) polymers used in injectable, long-acting formulations. *J. Control Release* **2019**, *304*, 75–89. [CrossRef] [PubMed]
39. Žagar, E.; Žigon, M. Characterization of a Commercial Hyperbranched Aliphatic Polyester Based on 2,2-Bis(methylol) propionic Acid. *Macromolecules* **2002**, *35*, 9913–9925. [CrossRef]
40. Kricheldorf, H.R.; Kreiser-Saunders, I.; Boettcher, C. Polylactones: Sn(II)octoate-initiated polymerization of L-lactide: A mechanistic study. *Polymer* **1995**, *36*, 1253–1259. [CrossRef]
41. Zhao, X.; Li, J.; Liu, J.; Zhou, W.; Peng, S. Recent progress of preparation of branched poly(lactic acid) and its application in the modification of polylactic acid materials. *Int. J. Biol. Macromol.* **2021**, *193 Pt A*, 874–892. [CrossRef]
42. Kowalski, A.; Duda, A.; Penczek, S. Kinetics and Mechanism of Cyclic Esters Polymerization Initiated with Tin(II) Octoate. 3. Polymerization of L,L-Dilactide. *Macromolecules* **2000**, *33*, 7359–7370. [CrossRef]
43. Bednarek, M. Branched aliphatic polyesters by ring-opening (co)polymerization. *Prog. Polym. Sci.* **2016**, *58*, 27–58. [CrossRef]
44. Smet, M.; Gottschalk, C.; Skaria, S.; Frey, H. Aliphatic Hyperbranched Copolyesters by Combination of ROP and AB2-Polycondensation. *Macromol. Chem. Phys.* **2005**, *206*, 2421–2428. [CrossRef]
45. Gottschalk, C.; Frey, H. Hyperbranched Polylactide Copolymers. *Macromolecules* **2006**, *39*, 1719–1723. [CrossRef]
46. Armarego, W.L.F.; Li, C.; Chai, L. (Eds.) *Purification of Laboratory Chemicals*, 5th ed.; Butterworth Heinemann: Amsterdam, The Netherlands, 2003. [CrossRef]
47. Istratov, V.V.; Polezhaev, A.V. Branched copolylactides: The effect of the synthesis method on their properties. *J. Phys. Conf. Ser.* **2021**, *1990*, 012046. [CrossRef]
48. Siengalewicz, P.; Mulzer, J.; Rinner, U. Synthesis of Esters and Lactones. In *Comprehensive Organic Synthesis*, 2nd ed.; Knochel, P., Ed.; Elsevier: Amsterdam, The Netherlands, 2014; pp. 355–410; ISBN 9780080977430. [CrossRef]
49. Heller, S.T.; Sarpong, R. On the reactivity of imidazole carbamates and ureas and their use as esterification and amidation reagents. *Tetrahedron* **2011**, *67*, 8851–8859. [CrossRef]
50. Bell, C.L.; Peppas, N.A. Measurement of the swelling force in ionic polymer networks. III. Swelling force of interpolymer complexes. *J. Control. Release* **1995**, *37*, 277–280. [CrossRef]
51. Zhou, Z.H.; Liu, X.P.; Liu, L.H. Synthesis of Ultra-high Weight Average Molecular Mass of Poly-L-lactide. *Int. J. Polym. Mater.* **2008**, *57*, 532–542. [CrossRef]
52. Huh, K.M.; Bae, Y.H. Synthesis and characterization of poly(ethylene glycol)/poly(l-lactic acid) alternating multiblock copolymers. *Polymer* **1999**, *40*, 6147–6155. [CrossRef]
53. Chieng, B.W.; Ibrahim, N.A.; Yunus, W.M.Z.W.; Hussein, M.Z. Poly(lactic acid)/Poly(ethylene glycol) Polymer Nanocomposites: Effects of Graphene Nanoplatelets. *Polymers* **2014**, *6*, 93–104. [CrossRef]
54. Schliecker, G.; Schmidt, C.; Fuchs, S.; Wombacher, R.; Kissel, T. Hydrolytic degradation of poly(lactide-co-glycolide) films: Effect of oligomers on degradation rate and crystallinity. *Int. J. Pharmaceut.* **2003**, *266*, 39–49. [CrossRef] [PubMed]
55. Albano, C.; González, G.; Palacios, J.; Karam, A.; Castillo, R.V.; Covis, M. Characterization of poly l-lactide/hydroxyapatite composite:chemical, thermal and thermomechanical properties. *Rev. Fac. Ing. Univ. Cent. Venez.* **2013**, *28*, 97–108.
56. Ferri, J.; Jordá, J.; Montanes, N.; Fenollar, O.; Balart, R. Manufacturing and characterization of poly(lactic acid) composites with hydroxyapatite. *J. Thermoplast. Compos. Mater.* **2017**, *31*, 865–881. [CrossRef]
57. Kesenci, K.; Fambri, L.; Migliaresi, C.; Piskin, E. Preparation and properties of poly(L-lactide)/hydroxyapatite composites. *J. Biomater. Sci. Polym. Ed.* **2000**, *11*, 617–632. [CrossRef]
58. Chieng, B.W.; Ibrahim, N.A.; Then, Y.Y.; Loo, Y.Y. Epoxidized Vegetable Oils Plasticized Poly(lactic acid) Biocomposites: Mechanical, Thermal and Morphology Properties. *Molecules* **2014**, *19*, 16024–16038. [CrossRef]
59. Ji, L.; Gong, M.; Qiao, W.; Zhang, W.; Liu, Q.; Dunham, R.E.; Gu, J. A gelatin/PLA-b-PEG film of excellent gas barrier and mechanical properties. *J. Polym. Res.* **2018**, *25*, 210. [CrossRef]
60. Bai, H.; Huang, C.; Xiu, H.; Gao, Y.; Zhang, Q.; Fu, Q. Toughening of poly(l-lactide) with poly($\varepsilon$-caprolactone): Combined effects of matrix crystallization and impact modifier particle size. *Polymer* **2013**, *54*, 5257–5266. [CrossRef]
61. Lin, Y.; Zhang, K.-Y.; Dong, Z.-M.; Dong, L.-S.; Li, Y.-S. Study of Hydrogen-Bonded Blend of Polylactide with Biodegradable Hyperbranched Poly(ester amide). *Macromolecules* **2007**, *40*, 6257–6267. [CrossRef]

62.  Zhang, H.; Lu, X.; Leng, Y.; Fang, L.; Qu, S.; Feng, B.; Fang, L.; Qu, S.; Feng, B.; Weng, J.; et al. Molecular dynamics simulations on the interaction between polymers and hydroxyapatite with and without coupling agents. *Acta Biomater.* **2009**, *5*, 1169–1181. [CrossRef] [PubMed]
63.  Ganeles, J.; Zöllner, A.; Jackowski, J.; Bruggenkate, C.T.; Beagle, J.; Guerra, F. Immediate and Early Loading of Straumann Implants With a Chemically Modified Surface (SLActive) in the Posterior Mandible and Maxilla: 1-Year Results from a Prospective Multicenter Study. *Clin. Oral Implant. Res.* **2008**, *19*, 1119–1128. [CrossRef] [PubMed]
64.  Rani, V.D.; Vinoth-Kumar, L.; Anitha, V.; Manzoor, K.; Deepthy, M.; Shantikumar, V.N. Osteointegration of Titanium Implant Is Sensitive to Specific Nanostructure Morphology. *Acta Biomater.* **2012**, *8*, 1976–1989. [CrossRef]
65.  Hu, X.; Liu, S.; Zhou, G.; Huang, Y.; Xie, Z.; Jing, X. Electrospinning of polymeric nanofibers for drug delivery applications. *J. Control. Release* **2014**, *185*, 12–21. [CrossRef] [PubMed]
66.  Trofimchuk, E.S.; Moskvina, M.A.; Nikonorova, N.I.; Efimov, A.V.; Garina, E.S.; Grokhovskaya, T.E.; Ivanova, O.A.; Bakirov, A.V.; Sedush, N.G.; Chvalun, S.N. Hydrolytic degradation of polylactide films deformed by the environmental crazing mechanism. *Eur. Polym. J.* **2020**, *139*, 110000. [CrossRef]
67.  Behera, K.; Sivanjineyulu, V.; Chang, Y.-H.; Chiu, F.-C. Thermal properties, phase morphology and stability of biodegradable PLA/PBSL/HAp composites. *Polym. Degrad. Stabil.* **2018**, *154*, 248–260. [CrossRef]
68.  Scoponi, G.; Francini, N.; Paradiso, V.; Donno, R.; Gennari, A.; d'Arcy, R.; Capacchione, C.; Athanassiou, A.; Tirelli, N. Versatile Preparation of Branched Polylactides by Low-Temperature, Organocatalytic Ring-Opening Polymerization in *N*-Methylpyrrolidone and Their Surface Degradation Behavior. *Macromolecules* **2021**, *54*, 9482–9495. [CrossRef]

Article

# Cytotoxicity Assessment of a New Design for a Biodegradable Ureteral Mitomycin Drug-Eluting Stent in Urothelial Carcinoma Cell Culture

Federico Soria [1,]*, Luna Martínez-Pla [1], Salvador D. Aznar-Cervantes [2], Julia E. de la Cruz [1], Tomás Fernández [3], Daniel Pérez-Fentes [4], Luis Llanes [5] and Francisco Miguel Sánchez-Margallo [1]

[1] Jesus Uson Minimally Invasive Surgery Centre Foundation, Endoscopy-Endourology Department, 10071 Cáceres, Spain
[2] Biotechnology, Genomics and Plant Breeding Department, Instituto Murciano de Investigación y Desarrollo Agrario y Ambiental (IMIDA), La Alberca, 30150 Murcia, Spain
[3] Urology Department, Morales Meseguer University Hospital, 30008 Murcia, Spain
[4] Urology Department, Santiago de Compostela University Hospital, 15706 Santiago de Compostela, Spain
[5] Urology Department, Getafe University Hospital, 28905 Madrid, Spain
* Correspondence: fsoria@ccmijesususon.com

**Abstract:** Urothelial tumour of the upper urinary tract is a rare neoplasm, but unfortunately, it has a high recurrence rate. The reduction of these tumour recurrences could be achieved by the intracavitary instillation of adjuvant chemotherapy after nephron-sparing treatment in selected patients, but current instillation methods are ineffective. Therefore, the aim of this in vitro study is to evaluate the cytotoxic capacity of a new instillation technology through a biodegradable ureteral stent/scaffold coated with a silk fibroin matrix for the controlled release of mitomycin C as an anti-cancer drug. Through a comparative study, we assessed, in urothelial carcinoma cells in a human cancer T24 cell culture for 3 and 6 h, the cytotoxic capacity of mitomycin C by viability assay using the CCK-8 test (Cell counting Kit-8). Cell viability studies in the urothelial carcinoma cell line confirm that mitomycin C embedded in the polymeric matrix does not alter its cytotoxic properties and causes a significant decrease in cell viability at 6 h versus in the control groups. These findings have a clear biomedical application and could be of great use to decrease the recurrence rate in patients with upper tract urothelial carcinomas by increasing the dwell time of anti-cancer drugs.

**Keywords:** ureteral stent; biodegradable stent; cell viability assessment; chemotherapy; UTUC

**Citation:** Soria, F.; Martínez-Pla, L.; Aznar-Cervantes, S.D.; de la Cruz, J.E.; Fernández, T.; Pérez-Fentes, D.; Llanes, L.; Sánchez-Margallo, F.M. Cytotoxicity Assessment of a New Design for a Biodegradable Ureteral Mitomycin Drug-Eluting Stent in Urothelial Carcinoma Cell Culture. *Polymers* **2022**, *14*, 4081. https://doi.org/10.3390/polym14194081

Academic Editors: Stefan Ioan Voicu and Andrada Serafim

Received: 25 July 2022
Accepted: 24 September 2022
Published: 29 September 2022

**Publisher's Note:** MDPI stays neutral with regard to jurisdictional claims in published maps and institutional affiliations.

## 1. Introduction

Urothelial carcinomas are the sixth most common tumour in developed countries [1]. These carcinomas are mainly located at the bladder level (90–95%), and only 5–10% correspond to upper tract urothelial carcinoma (pyelocaliceal cavities and ureter) (UTUC). The highest incidence of UTUC is in the age group between 70 and 90 years, and UTUC is diagnosed in men twice as often as in women [2,3]. In contrast to bladder cancer, two-thirds of patients at the time of their diagnosis of UTUC have invasive musculo-infiltrative disease [4].

The gold standard for UTUC treatment is radical nephroureterectomy (RNU), which is mainly for tumours classified as high-risk. However, one-third of UTUC cases may benefit from nephron-sparing surgery: those that are classified as low-risk according to the current European Association of Urology (EAU) guidelines [5]. In addition, selected patients with high-risk UTUC, such as patients with a solitary kidney or advanced chronic kidney disease, patients who refuse RNU, or patients with bilateral UTUC, are also candidates for endoscopic management, as they are not candidates for nephrectomy [6]. Unfortunately, unlike the well-established protocol for bladder urothelial tumours, the instillation of chemotherapy as an adjuvant treatment for UTUC in the upper urinary tract following endoscopic treatment of UTUC by ureteroscopic laser ablation is uncommon [7]. The aim of chemotherapy instillation is the reduction of the recurrence rate of UTUC, as evidenced by some authors in the scientific literature [8]. However, due to the particularities of the upper urinary tract, intracavitary instillation is very challenging. This is mainly due to the small volume of the upper urinary tract as well as to the inadequate drug dwell time of topical agents [7].

Current research seeks to develop new upper urinary tract chemotherapy delivery systems that can overcome the limitations of traditional techniques, mainly using drug-eluting stents and thermosensitive polymers [7,9]. The most recent advance has been UGN-101, a novel formulation of mitomycin C (MMC) containing a reverse thermal gel designed to increase urinary dwell time and thereby the efficacy of treatment. It showed a 56% complete response rate after 12 months in its first clinical trial, but with significant related side effects, including 44% of patients experiencing ureteral stricture [10]. However, the development of chemotherapy-eluting ureteral stents has not yet made enough progress due to difficulties in drug-delivery control through the stent coating. In this regard, one of the possible solutions that could help to overcome some of the biggest shortcomings in stent technology may be silk fibroin (SF) from *Bombyx mori* as a carrier and delivery system for drug-eluting stents due to its biocompatibility and easy processing [11]. SF's features make this protein a great candidate for biomaterial-controlled release technology [11].

For these reasons, the aim of our in vitro study is to assess cell viability in a urothelial carcinoma cell culture using a new design for biodegradable ureteral stents coated with SF for the controlled release of mitomycin C (BraidStent-SF-MMC) for topical instillation in the adjuvant treatment of UTUC. We need to confirm whether the MMC released by the stent maintains its cytotoxic capacity after embedding in the SF polymeric matrix.

## 2. Materials and Methods

### 2.1. BraidStent-SF-MMC

#### 2.1.1. Materials for Biodegradable Ureteral Stent Preparation

To perform the in vitro study, 24 fragments of a 10 mm long BraidStent® (JUMISC, Cáceres, Spain) biodegradable ureteral stent were made by combining biodegradable polymers and copolymers—Glycomer™ 631 and polyglycolic acid (PGA), at a ratio of 54% and 46%, respectively [12–14]. Two 0.17 mm thick threads of Glycomer™ 631 and two 0.14 mm thick threads of PGA, all 10 mm long, were used to manufacture the stent. The threads were then braided together to constitute the central core of the stent. These 24 stent fragments all underwent coating with SF, and 12 of them subsequently underwent the addition of a chemotherapeutic agent widely used for the intracavitary instillation in the UTUC, namely MMC [8].

#### 2.1.2. Materials for Stent SF and MMC Coating

Cocoons of *Bombyx mori* were obtained from worms reared in the sericulture facilities of the Imida Instituto Murciano de Investigación y Desarrollo Agrario y Ambiental (IMIDA), Biotechnology, Genomics and Plant Breeding Department (La Alberca, Murcia, Spain). Cocoons were chopped up and boiled in 0.02 M $Na_2CO_3$ for 30 min to eliminate the sericin. Then, the raw SF was rinsed with distilled water and dried at room temperature for 3 days. Subsequently, SF was dissolved in 9.3 M LiBr (Acros Organics) for 3 h at 60 °C, yielding a 20% weight per volume (*w/v*) dissolution that was dialysed against distilled water for 3 days (Snakeskin Dialysis Tubing 3.5 KDa MWCO, Thermo Scientific, Rockford, IL, USA), with eight total water changes (at 4 °C) [15]. The resultant 7–8% *w/v* SF solution was recovered and used for the preparation of the coated stents, adjusting the concentration to 7% *w/v* before use [16,17].

The coating of the BraidStent-SF-MMC stent was carried out with powdered 70 mg pure MMC (Mitomycin CRS, Sigma-Aldrich, Darmstadt, Germany). To this end, the concentration employed for both the fibroin and methanol solutions was 10 mg/mL, and 10 dip-coating cycles were performed by dissolving the drug (10 mg/mL MMC) for 30 min under orbital agitation by stirring at 120 rpm. These two solutions were used to alternately coat the stent fragments by dipping them in the first SF solution for 5 s and then in the methanol solution for 5 s, allowing them to dry for 1 min before repeating the procedure. The stents were completely stable in room air. However, the MMC coating needs special care (i.e., it must be packaged in lightproof packaging) [17].

In previous studies by our research group, stent coating characterisation was performed to determine whether there was adequate SF coating on the BraidStent-SF-MMC surface by adding a fluorescent aqueous marker (sulforhodamine B). The results of the fluorescence microscopy study confirmed the homogeneity of the SF coating on the stent surface [17].

### 2.1.3. Determination and Assessment of the MMC Release from BraidStent-SF-MMC

To determine MMC release, we tested 10 fragments of BraidStent-SF-MMC that were 10 mm in length and coated as described above in artificial urine (AU) (Human Synthetic Urine, BioIVT, West Sussex, UK). The stent fragments were incubated in an orbital shaker incubator under mimicked biological conditions (36.5 °C with 5% $CO_2$ at 90 rpm). The concentration of MMC released in the AU at 3 and 6 h was determined by HPLC-DAD. Urine from each follow-up was replaced and analysed. The permeability and release kinetics of the MMC depends on the SF coating and is related to the percentage of beta-sheet structure, which is controlled by dipping the SF solution in methanol [17].

The HPLC-DAD method is an isocratic method with an acetonitrile mobile phase. Ultrapure water that was 80:20 volume per volume ($v/v$) at a flow rate of 1 mL/min was used, and separation was performed on a LUNA C18 using 250 mm, 4.6 mm, 5 µm columns at 30 °C. The MMC was detected with a diode array detector (DAD) at 365 nm (1260 Infinity II Prime LC System, Agilent Technologies, Santa Clara, CA, USA) [17].

### 2.2. T24 Cell Culture Line

Urothelial carcinoma cells from human bladder cancer T24 cells were used as a model cell line (EP-CL-0227, BioNova científica®, Madrid, Spain) [18,19]. This is a tumour cell line from human transitional bladder cell carcinoma and is frequently used to assess the cellular cytotoxicity of chemotherapeutics in the urinary tract in vitro [18–20].

Human T24 cells were seeded in plates containing McCoy's 5a (Thermo Fisher Scientific®®®, Madrid, Spain) and supplemented with 10% foetal bovine serum (FBS), 1% penicillin/streptomycin, and 1% glutamine and were incubated at 37 °C at 95% relative humidity and 5% $CO_2$. Cells were maintained at 37 °C in a humidified 5% $CO_2$ atmosphere for 48 h.

### 2.3. In vitro Cytotoxicity of the BraidStent-SF-MMC

At 80% confluence, adherent cells were detached with Trypsin EDTA solution (Lonza Bioscience, Pontevedra-Spain) and seeded in four 24-well plates at a final concentration of 20,000 cells per well. Once adherent, four groups of studies were performed. In group 1 (G1), cell viability was assessed after the instillation of pure mitomycin C was added at a concentration of 0.66 mg/mL, the concentration recommended for the adjuvant treatment of UTUC in medical practice [8]. Group 2 (G2), the BraidStent-SF-MMC stent group, is the subject of the current study. The negative controls used were the BraidStent-SF without MMC coating (group 3, G3) and T24 cells in an MMC-free medium (group 4, G4). The sample size was six samples per group. All of these groups were assessed at baseline (T0), at 3 h (T3), and at 6 h (T6) (Table 1; Figure 1).

**Table 1.** Experimental study groups.

| | Groups |
|---|---|
| G1 (3 h-G13; 6 h-G16) | T24 cell culture + Recommended MMC oce in UTUC (0.66 mg/mL) [8]. No stent. Positive control. |
| G2 (3 h-G23; 6 h-G26) | T24 cell culture + BraidStent-SF-MMC. |
| G3 (3 h-G33; 6 h-G36) | T24 cell culture + BraidStent-SF. Negative control. |
| G4 (3 h-G43; 6 h-G46) | T24 cell culture. No stent. Untreated cells were used as negative control. |

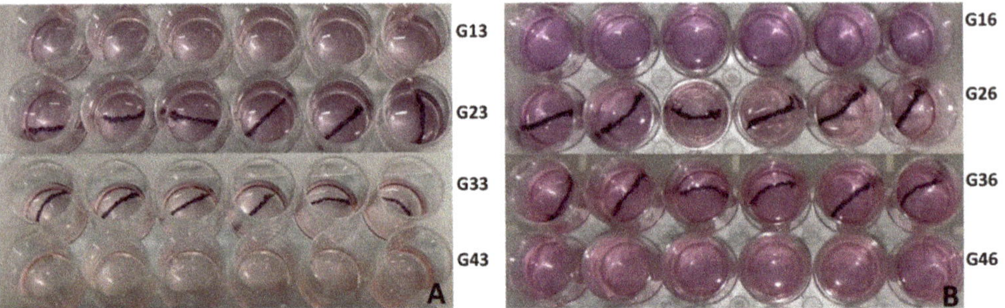


**Figure 1.** The figure shows the plates that constitute the experimental trial. Each row corresponds to an experimental group. (**A**) refers to T3 (3 h), and (**B**) refers to follow-up at T6 (6 h).

Mitomycin C cytotoxicity was assessed by a viability assay using CCK-8 (Cell Counting Kit-8, Boster Biological Technology, Pleasanton, CA, USA). The protocol was carried out according to the manufacturer's recommendations, and the absorbance at 450 nm was recorded using a Synergy™ Mx microplate reader (BioTek Instruments, Winooski-Vermont-USA) [21]. Briefly, after the first 24 h of incubation, the cell viability assay was performed with CCK-8 at T0 in two wells from each experimental group. Next, 10 μL of CCK-8 solution was added to each well of the plates together with Gibco™ DMEM complete FBS (10% FBS, 1% penicillin/streptomycin, 1% glutamine) (ThermoFisher Scientific, Madrid, Spain). Subsequently, they were kept in the incubator at 37 °C for 45 min. Finally, absorbance was measured using a Synergy™ Mx microplate reader (BioTek Instruments, Winooski-Vermont-USA) at 450 nm. After ending the CCK-8 assay at T0, the study groups were maintained in culture on their respective plates for 3 h and 6 h at 37 °C in the HeraCell™ 150i $CO_2$ incubator (Thermo Scientific™, Waltham, MA, USA), with six replicates per experimental group [21]. The absorbance of formazan was measured at 450 nm using the plate reader mentioned above. The percentage (%) of cell viability was calculated as follows [22]:

$$\text{Cell viability } (\%) = \frac{\text{Absorbance of sample}}{\text{Absorbance of control}} \times 100\%$$

### 2.4. Statistical Analysis

Statistical analysis was performed with the SPSS 25.0 program for Windows (IBM, USA). The variable of study was cell viability expressed as the % of cell viability. The normality of the data was analysed using the Shapiro–Wilk test. Student's *t*-test was used to compare the concentration of MMC released by the BraidStent-SF-MMC at 3 and 6 h. A comparison between groups at 3 and 6 h was carried out using the Kruskal–Wallis test, and, in the case of statistical significance, a corresponding post hoc analysis was carried out using the Bonferroni test. The trend of % of cell viability over time at 3 and 6 h was analysed via the Wilcoxon signed-rank test. The confidence interval set at 95% (95% CI), and significance was determined with $p < 0.05$.

## 3. Results

The results of the artificial urine study for the determination of the concentration of MMC released by the BraidStent-SF-MMC at 3 and 6 h are shown in Figure 2. No significance was found between groups (Student's test). Regarding the percentage of release with respect to the MMC encapsulated in the stent, 81.7% was released at 3 h, and 100% was released at 6 h. The urine study demonstrated the stability of the SF coating in a physiological environment.

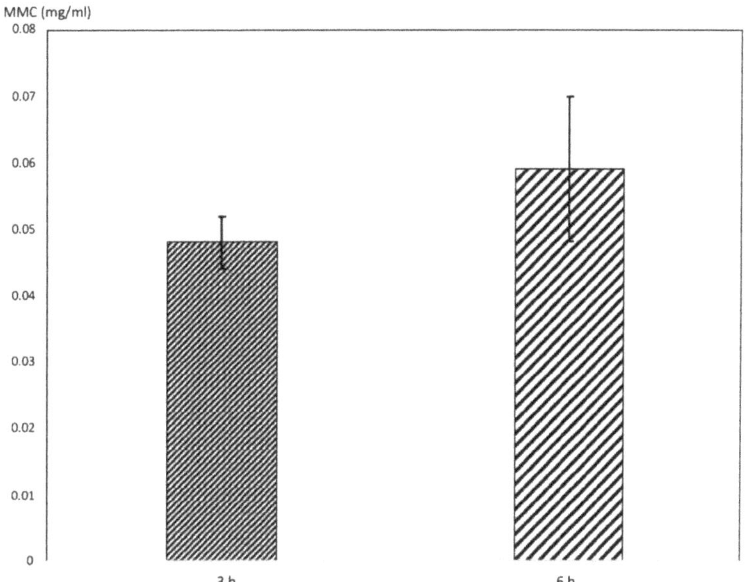


**Figure 2.** Concentration of MMC (mg/mL) released by BraidStent-SF-MMC at 3 h and 6 h. No significance found between the two groups (Student's test).

Regarding the results of the cell viability % studies according to the determination of absorbance at T3 and T6, the data do not follow a normal distribution.

As for the comparison of the cell viability % between groups over time, two different trends were observed. On the one hand, a trend of G3 and G4 corresponding to the non-MMC groups with BraidStent-SF and T24 cell culture alone (negative control), and on the other hand, the trend of the cultures in which MMC was present in G1 and G2, with a lower cell viability % with respect to the two control groups (Figure 3).

From the point of view of the tendency over time within each group, we found statistical significance between 3 and 6 h regarding the groups with MMC (G1 and G2), with a decrease in cell viability related to exposure time. In the non-MMC groups, only G3 showed statistical significance over time, increasing its cell viability % (Figure 3). In detail, the cell viability of the T24 cells in the presence of G1 and G2 at 3 h were $62.21 \pm 2.04\%$ and $65.40 \pm 5.26\%$, respectively. Additionally, at 6 h, it was $52.29 \pm 2.42\%$ and $47.66 \pm 3.78\%$, respectively.

Regarding the inter-group comparison of the percentage of cell viability at T24, G1 showed statistical significance at T3 compared to G3 ($p = 0.013$) and G4 ($p = 0.001$), and G2 only showed statistical significance with G4 ($p = 0.012$). At T6, significance was found for the two groups with MMC, G1, and G2 compared to the two negative controls: the G3 (bare BraidStent-SF) culture and G4 (T24 cell culture alone) ($p < 0.005$) (Figure 3).

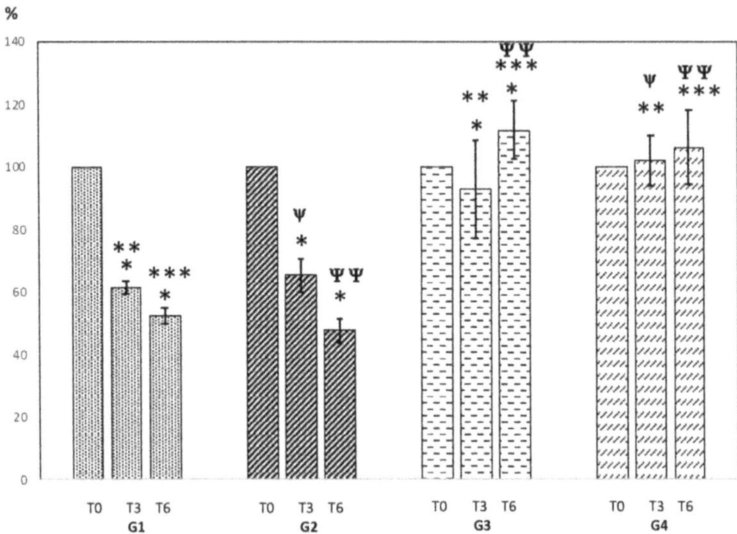

**Figure 3.** Cell viability % assessment in T24 cell culture (%). T0 (start of study); T3 (3 h); and T6 (6 h). Values indicate mean ± SD. Significance of the values among tested groups within each group and follow-up was determined via a Wilcoxon signed-rank test. The inter-groups at 3 and 6 h were carried out using the Kruskal–Wallis test and Bonferroni post hoc analysis. Statistical significance is depicted as follows: intra-groups (* $p < 0.05$); inter-groups (** $p < 0.05$) (T3: G1 vs. G2, G3, and G4); inter-groups ($^{\Psi}$ $p < 0.05$) (T3: G2 vs. G3 and G4); inter-groups (*** $p < 0.05$) (T6: G1 vs. G2, G3, and G4), and inter-groups ($^{\Psi\Psi}$ $p < 0.05$) (T6: G2 vs. G3 and G4).

## 4. Discussion

The search for new technologies for the instillation of adjuvant chemotherapy in urothelial carcinoma of the upper urinary tract is one of the current challenges in urological oncology due to the shortcomings of current systems [3,7]. Therefore, the development of a ureteral stent that delivers chemotherapy on a scheduled basis may be of great benefit in patients with low-grade UTUC or in those patients who are not suitable candidates for radical nephroureterectomy despite their tumours being high-grade [17].

The main advantages of the development of a biodegradable ureteral stent such as BraidStent-SF-MMC are mainly that it allows the release of anti-cancer drugs such as MMC into the upper urinary tract; this release system allows a longer dwell time of the chemotherapy drug with the urothelium, improving its cytotoxic effect [19]. A second surgical intervention is not required to remove the stent, as it is biodegradable [12–14]. Additionally, it is not related to the possible increases in intrarenal pressure that may cause urosepsis, which occurs in cases of intracavitary instillation via ureteral catheters. The latter currently requires prior assessment of renal volume so as not to exceed the intrarenal pressure that could cause pyelovenous or pyelolymphatic reflux, which is a limiting factor in achieving the recommended chemotherapy dose [3,7,9].

The preliminary results confirm that the BraidStent coated with the SF matrix and embedded MMC in its 10 dips (dip-coating technique), called BraidStent-SF-MMC, allows the controlled release of MMC into the urinary environment at 3 and 6 h (Figure 2). In view of these results, future studies will be carried out to determine the release rate at earlier phases, such as 1 h. The aim of the stent is to release the effective dose over a prolonged period of time to increase the dwell time of the MMC with the urothelium.

The objectives of the current study are very straightforward, as we need to assess whether the mitomycin C embedded in the SF matrix layers of the BraidStent-SF-MMC maintain their cytotoxic capacity, as well as to compare its inhibitory effect on cell viability versus the dose administered in patients today. The results of the cell viability study show that the MMC groups (G1, G2) show statistical significance compared to the non-MMC groups (G3, G4) at 6 h (Figure 3).

From the first follow-up, the MMC in T24 cell culture, regardless of whether the release is through direct instillation or whether it is impregnated in a SF matrix (G2), shows cytotoxicity (Figure 3). The results indicate that G3 and G4, the negative control groups, show no significance between the two groups at the two follow-ups. In addition, there was no perceptible cytotoxicity exhibited by the non-MMC BraidStent-SF (G3), ruling out any harmful effects related to the SF matrix, and there was statistical significance compared to G4. A very important finding in this study is that there is no statistical difference between G1 and G2, demonstrating that the concentration of MMC released by the BraidStent-SF-MMC is adequate and comparable to that currently used in patients administered MMC as adjuvant therapy to UTUC (Figure 3). It is important to highlight that we chose a suitable test: CCK8. The main characteristics of this test are that it is an easy and sensitive colorimetric assay for the determination of in vitro cell viability in cytotoxicity assays. It is also true that it is possible to use other types of cell viability assays, such as flow cytometry, which allows for a more specific phenotypic quantitative cell viability analysis. Perhaps the use of more than one cell viability assay in our study would have allowed us to obtain more precise details regarding cell viability.

Since 1993, when silk fibroin was recognised by the US Food and Drug Administration (FDA) as a biomaterial, its use has become extremely widespread due to its interesting characteristics. It has excellent biocompatibility, high strength, mechanical toughness, robust flexibility, high processability, tuneable degradation, ease of processing, and ease acquisition [11,23]. Due to its excellent features, SF has been employed in a wide variety of medical applications, such as in drug-eluting stents [23]. It is the characteristics of its processability that allow the control of the crystalline state, Beta-sheet content, and use in biomaterial-controlled-release applications that regulate the kinetics of the incorporated drugs. The stability of SF coatings can be modified by immersion in methanol, which allows the Beta sheet content to be modified. This is related to the rate of degradation and to the release of the drugs embedded in the SF coating. In our case, we used 10 dip-coating cycles to obtain controlled and early release, with the idea of providing a treatment similar to that currently used in patients but that increases the exposure time of MMC.

On the other hand, the MMC groups showed a significant association with exposure time. The longer the exposure time, the greater the cytotoxic effect. This capacity was more important in the BraidStent-SF-MMC group, but no statistical significance was observed compared to G1, with a cell viability of 47.66% compared to 52.29% for G1 (Figure 3). These findings are very encouraging for the use of this stent in patients, as they confirm that the cytotoxic effects increase when the exposure time of the anti-cancer drugs is increased. Currently, the few clinical groups using the intracavitary instillation of MMC for adjuvant therapy for UTUC and to treat carcinoma in situ only provide this therapy for one hour [7,8]. This limited upper tract dwell time in patients could be the cause of recurrence and progression after UTUC kidney-sparing surgery and is a strength of the stent evaluated in this in vitro study [3,7,9].

Our results are consistent with previous studies using the same T24 cell culture and in which different research groups evaluated biodegradable drug-eluting ureteral stents. For example, Barros et al. evaluated four different anti-cancer drugs (epirubicin, paclitaxel, doxorubicin, and gemcitabine) impregnated via a supercritical fluid $CO_2$ technique [19]. Wang et al. also developed an in vitro and animal model study of a new design for gradiently degraded electrospun polyester scaffolds with an epirubicin coating [18]. The three in vitro studies are similar despite the different cytostatics evaluated, mainly because there is currently no consensus as to which is the most suitable—although MMC is the most widely used as the clinical level [8,9]. Regarding the incubation time of stents in cell cultures, different ranges are shown: from 3 to 6 h in our study to 24 h in Wang et al. [18] and from 4 to 72 h in Barros et al. [19]. However, in the assessment of these new stent designs, the viability of cancer cells decreases similarly: 62% to 46% [18], 65% [19], and 65% to 47% in our study. Similar to Barros et al., our study found a greater cytostatic effect in the cytostatic-eluting stent group (G2) than in the groups where the anti-cancer drug was instilled in the wells, possibly due to the longer release effect by the SF matrix [19].

## 5. Conclusions

The coating of a biodegradable ureteral stent with a silk fibroin matrix impregnated in layers of mitomycin C allows the release of the cytostatic in artificial urine. Cell viability studies in a human urothelial carcinoma cell line confirm that mitomycin C embedded in the polymeric matrix does not alter its cytotoxic properties and causes a significant decrease in cell viability at 6 h. These findings could be of great use to decrease the recurrence rate in patients with UTUC.

**Author Contributions:** Conceptualization, F.S. and S.D.A.-C.; methodology, F.S., S.D.A.-C., L.M.-P., and J.E.d.l.C.; funding acquisition, F.S.; investigation, F.S., S.D.A.-C., L.M.-P., J.E.d.l.C., T.F., D.P.-F., L.L. and F.M.S.-M.; data curation, F.S.; writing—original draft preparation, F.S. and S.D.A.-C.; project administration, F.S.; writing—review and editing, F.S. All authors have read and agreed to the published version of the manuscript.

**Funding:** This research was funded by Instituto de Salud Carlos III (ISCIII, Spain) through the projects PI16/01707 and PI20/01188. Additionally, project IB18107 was funded by Consejería de Economía, Ciencia y Agenda Digital-Junta de Extremadura (Spain) and co-funded by the European Union. This study has been partially supported (80%) by the European Commission ERDF/FEDER Operational Programme 'Murcia' CCI Nº 2007ES161PO001 (Project No. 14–20/20).

**Institutional Review Board Statement:** Not applicable.

**Informed Consent Statement:** Not applicable.

**Data Availability Statement:** Not applicable.

**Acknowledgments:** We thank Esther López-Nieto and Verónica Álvarez for their important assistance in the cell culture methodology.

**Conflicts of Interest:** The authors declare no conflict of interest. The funders had no role in the design of the study; in the collection, analyses, or interpretation of data; in the writing of the manuscript; or in the decision to publish the results.

## References

1. Siegel, R.L.; Miller, K.D.; Fuchs, H.E.; Jemal, A. Cancer Statistics, 2021. *CA Cancer J. Clin.* **2021**, *71*, 7–33. [CrossRef] [PubMed]
2. Rouprêt, M.; Yates, D.R.; Comperat, E.; Cussenot, O. Upper urinary tract urothelial cell carcinomas and other urological malignancies involved in the hereditary nonpolyposis colorectal cancer (lynch syndrome) tumor spectrum. *Eur. Urol.* **2008**, *54*, 1226–1236. [CrossRef] [PubMed]
3. Leow, J.J.; Liu, Z.; Tan, T.W.; Lee, Y.M.; Yeo, E.K.; Chong, Y.L. Optimal Management of Upper Tract Urothelial Carcinoma: Current Perspectives. *Onco. Targets* **2020**, *13*, 1–15. [CrossRef] [PubMed]
4. Margulis, V.; Shariat, S.F.; Matin, S.F.; Kamat, A.M.; Zigeuner, R.; Kikuchi, E.; Lotan, Y.; Weizer, A.; Raman, J.D.; Wood, C.G. Upper Tract Urothelial Carcinoma CollaborationThe Upper Tract Urothelial Carcinoma Collaboration. Outcomes of radical nephroureterectomy: A series from the Upper Tract Urothelial Carcinoma Collaboration. *Cancer* **2009**, *115*, 1224–1233. [CrossRef]
5. Rouprêt, M.; Babjuk, M.; Burger, M.; Capoun, O.; Cohen, D.; Compérat, E.M.; Cowan, N.C.; Dominguez-Escrig, J.L.; Gontero, P.; Hugh Mostafid, A.; et al. European association of urology guidelines on upper urinary tract urothelial carcinoma: 2020 Update. *Eur. Urol.* **2021**, *79*, 62–79. [CrossRef]
6. Petros, F.G.; Li, R.; Matin, S.F. Endoscopic Approaches to Upper Tract Urothelial Carcinoma. *Urol. Clin. North. Am.* **2018**, *45*, 267–286. [CrossRef]
7. Knoedler, J.J.; Raman, J.D. Intracavitary therapies for upper tract urothelial carcinoma. *Expert Rev. Clin. Pharmacol.* **2018**, *11*, 487–493. [CrossRef]
8. Gallioli, A.; Boissier, R.; Territo, A.; Vila Reyes, H.; Sanguedolce, F.; Gaya, J.M.; Regis, F.; Subiela, J.D.; Palou, J.; Breda, A. Adjuvant Single-Dose Upper Urinary Tract Instillation of Mitomycin C After Therapeutic Ureteroscopy for Upper Tract Urothelial Carcinoma: A Single-Centre Prospective Non-Randomized Trial. *J. Endourol.* **2020**, *34*, 573–580. [CrossRef]
9. Foerster, B.; D'Andrea, D.; Abufaraj, M.; Broenimann, S.; Karakiewicz, P.I.; Rouprêt, M.; Gontero, P.; Lerner, S.P.; Shariat, S.F.; Soria, F. Endocavitary treatment for upper tract urothelial carcinoma: A meta-analysis of the current literature. *Urol. Oncol.* **2019**, *37*, 430–436. [CrossRef]

10. Matin, S.F.; Pierorazio, P.M.; Kleinmann, N.; Gore, J.L.; Shabsigh, A.; Hu, B.; Chamie, K.; Godoy, G.; Hubosky, S.G.; Rivera, M.; et al. Durability of response to primary chemoablation of low-grade upper tract urothelial carcinoma using UGN-101, a mitomycin-containing reverse thermal gel: OLYMPUS Trial Final Report. *J. Urol* **2022**, *207*, 779–788. [CrossRef]
11. Qi, Y.; Wang, H.; Wei, K.; Yang, Y.; Zheng, R.Y.; Kim, I.S.; Zhang, K.Q. A review of structure construction of silk fibroin biomaterials from single structures to multi-level structures. *Int. J. Mol. Sci.* **2017**, *18*, 237. [CrossRef]
12. Soria, F.; de La Cruz, J.E.; Fernandez, T.; Budia, A.; Serrano, Á.; Sanchez-Margallo, F.M. Heparin coating in biodegradable ureteral stents does not decrease bacterial colonization-assessment in ureteral stricture endourological treatment in animal model. *Transl. Androl. Urol.* **2021**, *10*, 1700–1710. [CrossRef]
13. Soria, F.; de La Cruz, J.E.; Caballero-Romeu, J.P.; Pamplona, M.; Pérez-Fentes, D.; Resel-Folskerma, L.; Sanchez-Margallo, F.M. Comparative assessment of biodegradable-antireflux heparine coated ureteral stent: Animal model study. *BMC Urol.* **2021**, *21*, 32. [CrossRef]
14. Soria, F.; de La Cruz, J.E.; Budia, A.; Cepeda, M.; Álvarez, S.; Serrano, Á.; Sanchez-Margallo, F.M. Iatrogenic ureteral injury treatment with biodegradable antireflux heparin-coated ureteral stent-animal model comparative study. *J. Endourol.* **2021**, *35*, 1244–1249. [CrossRef]
15. Aznar-Cervantes, S.D.; Pagan, A.; Monteagudo Santesteban, B.; Cenis, J.L. Effect of different cocoon stifling methods on the properties of silk fibroin biomaterials. *Sci. Rep.* **2019**, *9*, 6703. [CrossRef]
16. Rockwood, D.N.; Preda, R.C.; Yücel, T.; Wang, X.; Lovett, M.L.; Kaplan, D.L. Materials fabrication from Bombyx mori silk fibroin. *Nat. Protoc.* **2011**, *22*, 1612–1631. [CrossRef]
17. Soria, F.; Aznar-Cervantes, S.D.; de la Cruz, J.E.; Budia, A.; Aranda, J.; Caballero, J.P.; Serrano, Á.; Sánchez Margallo, F.M. Assessment of a coated mitomycin-releasing biodegradable ureteral stent as an adjuvant therapy in upper urothelial carcinoma: A Comparative in vitro study. *Polymers* **2022**, *14*, 3059. [CrossRef]
18. Wang, J.; Wang, G.; Shan, H.; Wang, X.; Wang, C.; Zhuang, X.; Ding, J.; Chen, X. Gradiently degraded electrospun polyester scaffolds with cytostatic for urothelial carcinoma therapy. *Biomater. Sci.* **2019**, *7*, 963–974. [CrossRef]
19. Barros, A.A.; Browne, S.; Oliveira, C.; Lima, E.; Duarte, A.R.; Healy, K.E.; Reis, R.L. Drug-eluting biodegradable ureteral stent: New approach for urothelial tumors of upper urinary tract cancer. *Int. J. Pharm.* **2016**, *513*, 227–237. [CrossRef]
20. Pinto-Leite, R.; Botelho, P.; Ribeiro, E.; Oliveira, P.A.; Santos, L. Effect of sirolimus on urinary bladder cancer T24 cell line. *J. Exp. Clin. Cancer Res.* **2009**, *28*, 3. [CrossRef]
21. Marinaro, F.; Silva, J.M.; Barros, A.A.; Aroso, I.M.; Gómez-Blanco, J.C.; Jardin, I.; Lopez, J.J.; Pulido, M.; de Pedro, M.Á.; Reis, R.L.; et al. A fibrin coating method of polypropylene meshes enables the adhesion of menstrual blood-derived mesenchymal stromal cells: A new delivery strategy for stem cell-based therapies. *Int. J. Mol. Sci.* **2021**, *22*, 13385. [CrossRef]
22. Iqbal, H.; Razzaq, A.; Uzair, B.; Ul Ain, N.; Sajjad, S.; Althobaiti, N.A.; Albalawi, A.E.; Menaa, B.; Haroon, M.; Khan, M.; et al. Breast cancer inhibition by biosynthesized titanium dioxide nanoparticles is comparable to free doxorubicin but appeared safer in Balb/c Mice. *Materials* **2021**, *14*, 3155. [CrossRef]
23. Wang, X.; Zhang, X.; Castellot, J.; Herman, I.; Iafrati, M.; Kaplan, D.L. Controlled release from multilayer silk biomaterial coatings to modulate vascular cell responses. *Biomaterials* **2008**, *29*, 894–903. [CrossRef]

*Article*

# Development of 3D Thermoplastic Polyurethane (TPU)/Maghemite ($\gamma$-Fe$_2$O$_3$) Using Ultra-Hard and Tough (UHT) Bio-Resin for Soft Tissue Engineering

Ehsan Fallahiarezoudar [1], Nor Hasrul Akhmal Ngadiman [2,*], Noordin Mohd Yusof [2], Ani Idris [3] and Mohamad Shaiful Ashrul Ishak [4]

1   Department of Industrial Engineering, Faculty of Engineering, East of Guilan, University of Guilan, Roudsar 44918, Guilan, Iran; fallahi.ehsan@guilan.ac.ir
2   School of Mechanical Engineering, Faculty of Engineering, Universiti Teknologi Malaysia, Johor Bahru 81310, Johor, Malaysia; noordin@utm.my
3   School of Chemical Engineering, Faculty of Engineering, c/o Institute of Bioproduct Development, Universiti Teknologi Malaysia, Johor Bahru 81310, Johor, Malaysia; aniidris@utm.my
4   Faculty of Mechanical Engineering Technology, Universiti Malaysia Perlis, Kampus Pauh Putra, Arau 02600, Perlis, Malaysia; mshaiful@unimap.edu.my
*   Correspondence: norhasrul@utm.my

**Citation:** Fallahiarezoudar, E.; Ngadiman, N.H.A.; Yusof, N.M.; Idris, A.; Ishak, M.S.A. Development of 3D Thermoplastic Polyurethane (TPU)/Maghemite ($\gamma$-Fe$_2$O$_3$) Using Ultra-Hard and Tough (UHT) Bio-Resin for Soft Tissue Engineering. *Polymers* **2022**, *14*, 2561. https://doi.org/10.3390/polym14132561

Academic Editors: Andrada Serafim and Stefan Ioan Voicu

Received: 17 May 2022
Accepted: 21 June 2022
Published: 23 June 2022

**Publisher's Note:** MDPI stays neutral with regard to jurisdictional claims in published maps and institutional affiliations.



**Abstract:** The use of soft tissue engineering scaffolds is an advanced approach to repairing damaged soft tissue. To ensure the success of this technique, proper mechanical and biocompatibility properties must be taken into consideration. In this study, a three-dimensional (3D) scaffold was developed using digital light processing (DLP) and ultra-hard and tough (UHT) bio-resin. The 3D scaffold structure consisted of thermoplastic polyurethane (TPU) and maghemite ($\gamma$-Fe$_2$O$_3$) nanoparticles mixed with UHT bio-resin. The solution sample for fabricating the scaffolds was varied with the concentration of the TPU (10, 12.5, and 15% wt/v) and the amount of $\gamma$-Fe$_2$O$_3$ (1, 3, and 5% $v/v$) added to 15% wt/v of TPU. Before developing the real geometry of the sample, a pre-run of the DLP 3D printing process was done to determine the optimum curing time of the structure to be perfectly cured, which resulted in 30 s of curing time. Then, this study proceeded with a tensile test to determine the mechanical properties of the developed structure in terms of elasticity. It was found that the highest Young's Modulus of the scaffold was obtained with 15% wt/v TPU/UHT with 1% $\gamma$-Fe$_2$O$_3$. Furthermore, for the biocompatibility study, the degradation rate of the scaffold containing TPU/UHT was found to be higher compared to the TPU/UHT containing $\gamma$-Fe$_2$O$_3$ particles. However, the MTT assay results revealed that the existence of $\gamma$-Fe$_2$O$_3$ in the scaffold improved the proliferation rate of the cells.

**Keywords:** soft tissue engineering; bone scaffold; DLP 3D printing; mechanical strength; biocompatibility

## 1. Introduction

In the human body, soft tissue refers to the tissue that connects, supports, and surrounds other structures. It includes muscles, tendons, ligaments, fascia, nerves fibrous tissue, fat, blood vessels, and synovial membranes. Regularly, soft tissue damage is caused by disease, congenital defects, trauma, and aging, which lead to the inability of the tissue to self-heal [1]. Due to this matter, an alternative technology called tissue engineering was proposed to help the healing process by regenerating or replacing the damaged tissue [2]. The important part of this method is the development of the scaffold, which is able to restore, maintain, and improve the function of tissue [3]. A scaffold is a template for tissue formation that allows the cells to migrate, adhere to, and produce tissue [4]. According to Li et al. [5], in the development of a successful and well-functioning scaffold, the following properties should be taken into account; (i) the biocompatibility and biodegradability of the scaffold in order to match the cell or tissue growth in vitro/vivo; (ii) suitable mechanical properties to match the tissue at the implantation site; (iii) proper surface chemistry for cell attachment, proliferation, and differentiation activity. In addition, the architecture of the scaffold in terms of porosity is also necessary to consider as it is important for cell growth, nutrient transportation, and metabolic waste [6,7]. In applying the tissue engineering method, the scaffold must satisfy these properties.

The digital light processing (DLP) 3D printing technique is one of the best-defined techniques to produce scaffolds for soft tissue engineering. DLP is based on the basic principle of stereolithography (SLA), which offers better resolution and is more versatile than other conventional and additive manufacturing methods [8]. Practically, the DLP 3D printing process uses ultraviolet (UV) light to project the entire X and Y cross-sectional layers of the structure to be produced at one time onto a photopolymer resin, which will change the area exposed to UV light from a liquid to a solid. The solidified layer is formed on the collector, which is the Z axis, as shown in Figure 1. By using this process, the production time is reduced, which leads to higher productivity and reproducibility of the scaffold [9–11]. It is also able to create complex structures with highly accurate internal architecture as it has a high feature resolution [6]. Thus, this process can be used to develop the 3D structure of scaffolds with any kind of shape while maintaining good mechanical strength, and it can create a good environment for enhancing the biocompatibility performance [5]. However, it is worth noting that this method lacks resin as the resin must be capable of a photopolymerization reaction [12]. Therefore, UHT resin is used with the DLP 3D printer system. This resin is one of the standard bio-resins that can solidify quickly in the presence of a specific light source [11].

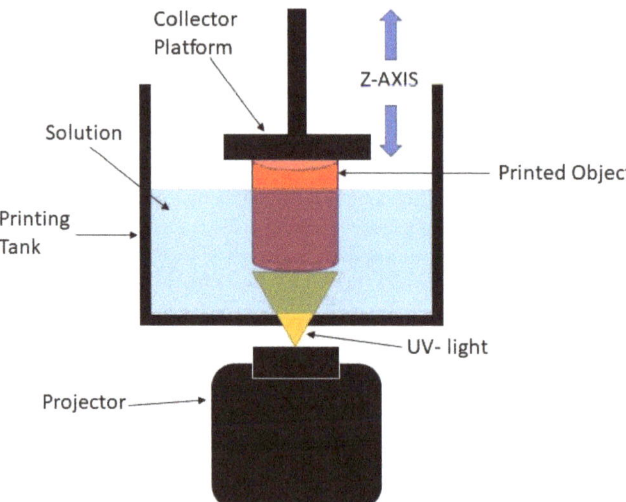



**Figure 1.** Schematic diagram of DLP 3D printing process.

In recent years, researchers have been exploring the use of nanofiber-based scaffolding systems that can act as a scaffold for tissue engineering applications. This is because the structures produced by nanofiber scaffolds mimic the structure of natural human tissue and, thus, enhance the cell growth rate [13]. Considering this, efforts in finding the best materials and techniques for developing tissue engineering scaffolds that fulfill the requirements are still ongoing. In this study, biocompatible thermoplastic polyurethane (TPU) was used as it has been widely used in medical applications. This chosen material has important characteristics required for building bone scaffold, which represents good biocompatibility and flexibility compared to other types of synthesized polymers. Thermoplastic polyurethane (TPU) is a class of PU with excellent elastic and tear-resistant properties and moderate tensile strength. In our previous studies [14–16], TPU has been used as a novel mixture for fabricating engineered tissue for an aortic heart valve using the electrospinning process.

As a continuation, this study focused on the adjustment of parameter settings for printing TPU using a DLP 3D printer before applying it to the various applications of soft tissue engineering scaffolds. Ultra-hard and tough (UHT) bio-resin acted as a curing agent to print the TPU. However, the use of the polymer itself with the resin did not meet the mechanical properties required of the scaffold [17].

Magnetic nanoparticles in the form of maghemite ($\gamma$-Fe$_2$O$_3$) have been used in biomedical applications [18–22], such as cell sheet construction, cell expansion, magnetic cell seeding, cancer hyperthermia treatment, and drug delivery. Maghemite ($\gamma$-Fe$_2$O$_3$) nanoparticles have been proven to enhance the mechanical properties of tissue engineering scaffolds. Maghemite ($\gamma$-Fe$_2$O$_3$) nanoparticles were chosen because of their low toxicity and ability to act as a reinforcing agent to the scaffold due to a larger surface area provided by the very fine nanoparticles [23–28]. In our previous work [29], the addition of maghemite ($\gamma$-Fe$_2$O$_3$) nanoparticles exhibited good biocompatibility. The presence of magnetic nanoparticles within scaffolds has also increased their rigidity favorably [30,31]. The material characterization of maghemite ($\gamma$-Fe$_2$O$_3$) nanoparticles has also been discussed in our previous works [10,11,14–17,32–35]. However, an excessive amount of maghemite used led to a decrease in the mechanical properties as the stiffness of the scaffold increased. $\gamma$-Fe$_2$O$_3$ nanoparticles also offer better cell adhesion properties and enhance the cell growth rate [11,36,37].

In addition, in a previous study by Fallahiarezoudar et al. [16], for the development of cardiac tissue engineering scaffold focusing on the aortic portion and using the electrospinning process, they established materials consisting of thermoplastic polyurethane (TPU) and maghemite ($\gamma$-Fe$_2$O$_3$) nanoparticles, and both of the materials showed good mechanical properties. However, a limitation occurred because of the process that they used, which was electrospinning. This process can only develop two-dimensional structures, and therefore, there is a limit to the strength of the scaffolds produced. Cardiac tissue engineering scaffolds need to be produced with 3D printing in order to obtain the appropriate mechanical and biocompatibility properties as well as to be implemented clinically.

In this study, a 3D structure of scaffold that consisted of TPU containing $\gamma$-Fe$_2$O$_3$ nanoparticles mixed with ultra-hard and tough (UHT) bio-resin was developed using the DLP 3D printing process. The aim of this study was to evaluate the performance of the 3D structure in terms of the mechanical and biocompatibility properties.

## 2. Materials and Methods

Almost all the chemicals used in this research were of analytical purity and no further purification was applied. The base material used for developing the 3D structure in this study was thermoplastic polyurethane (TPU). Flexible elastomer TPU granules with Shore A 60 hardness, an Mw of 90 kDa, and a glass transition temperature of $-50\ ^\circ$C were purchased from Wenzhou City Sanho Co., Ltd. Maghemite ($\gamma$-Fe$_2$O$_3$) was used as a filler to enhance the properties of the 3D structure, while ultra-hard and tough (UHT) bio-resin acted as a resin that bound the TPU to the maghemite. UHT also reacted with the ultraviolet (UV) light to form the 3D structure. The chemicals used in this study were reagent grade and included the following: iron (II) chloride (FeCl2) (98% purity, Sigma Aldrich, St. Louis, MO, USA), iron (III) chloride (FeCl3) (45% purity, Riedel-de Haen), sulfuric acid (H$_2$SO$_4$) (QRëC), nitric acid (HNO3) (65% purity, QRëC), ammonia solution (NH3) (25% purity, Merck), hydrochloric acid (HCl) (37% purity, QRëC), thermoplastic polyurethane (TPU), dichloromethane (DCM),ultra-hard and tough (UHT) resin, HSF-1184 cell line (human skin fibroblast cell line, ATCC, Manassas, VA, USA), phosphate-buffered saline (PBS, Gibco, Grand Island, NY, USA), Dulbecco's modified Eagle's medium (DMEM, Gibco, Grand Island, NY, USA), fetal bovine serum (FBS, Gibco, Grand Island, NY, USA), penicillin (Gibco, Grand Island, NY, USA), streptomycin (Gibco, Grand Island, NY, USA), and trypsinase (Sigma, USA).

### 2.1. Preparation of Thermoplastic Polyurethane Solution and Maghemite Synthesis

In order to make a 3D scaffold, TPU in a solid state must be first transformed into a liquid state. So, the TPU was first dissolved with dimethylformamide (DMF). In detail, for 10% wt/v of the TPU solution, 10 g of TPU granules was dissolved with 100 mL of DMF, which was constantly stirred using a magnetic stirrer at room temperature for at least 6 h. The procedures were repeated for 12.5% wt/v and 15% wt/v of the TPU solution by increasing the amount of TPU to 12.5 g and 15 g, respectively.

In addition, the maghemite nanoparticles were synthesized based on the method described by Idris et al. [38]. Briefly, the co-precipitation method (Massart, 1981) was used to synthesize the maghemite. Ferrous and ferric chloride were added in stoichiometric amounts to an ammonium hydroxide solution by alkaline co-precipitation. Magnetite $(Fe_3O_4)$, which is a black precipitate, was obtained, and then nitric acid was used to acidify the precipitate. The solution of the ferric nitrate was used to oxidize it at $100\,^{\circ}\text{C}$ to transform the solution into $\gamma\text{-}Fe_2O_3$. Then, citrate anions were used to coat the maghemite solution to prevent agglomeration between the particles. Later, the precipitate was washed with acetone and finally dispersed in water, resulting in the final stable state $\gamma\text{-}Fe_2O_3$ with a pH of 7.

### 2.2. Preparation of TPU/UHT Bio-Resin Mixed with Maghemite ($\gamma\text{-}Fe_2O_3$)

Maghemite ($\gamma\text{-}Fe_2O_3$) contains nanoparticles that act as filler, which could enhance the mechanical and biodegradability properties of the developed structure. So, in this study, three different concentrations of $\gamma\text{-}Fe_2O_3$, 1, 3, and 5% v/v, were added to 15% wt/v TPU solution. In order to prepare the 1% $\gamma\text{-}Fe_2O_3$ in the 15% wt/v TPU solution, 1 mL of $\gamma\text{-}Fe_2O_3$ was added to 99 mL of 15% wt/v TPU solution prepared early. This step was repeated for 3% and 5% $\gamma\text{-}Fe_2O_3$, adding 3 mL and 5 mL to the TPU solution, respectively.

### 2.3. Digital Light Processing (DLP) 3D Printing

In order to fabricate the bone scaffold, a DLP 3D printing approach was used in this study. A 3D dumbbell-shaped scaffold was designed in computer-aided design (CAD) software. The file was converted into STL format and transferred to 3D printer software named Creation Workshop. Next, the support structure was designed to ensure the connection between the product and the burn layer. Then, the sample was printed layer by layer and the TPU/UHT/$\gamma\text{-}Fe_2O_3$ was solidified as the UV light was projected onto the mixture.

### 2.4. Curing Time

It is important to determine the curing time first because the specimen must be cured perfectly. In this case, the mixed solution of 12.5% wt/v of TPU and UHT resin went through the DLP 3D printing process. Several curing times were tested in order to determine the optimum curing time for the process. The 3D structure condition was visually observed and the optimum curing time was selected when the cured body had a sharp end and precise structure.

### 2.5. Mechanical Testing

One of the major requirements of biomaterials for bone tissue engineering is the mechanical properties of the scaffold that must match the physical demand of the healthy surrounding bone tissues. Testing of the mechanical properties was conducted according to BS ISO 37:2011. In this study, the dumbbell-shaped specimen was used for the mechanical testing. The expected parameter was Young's Modulus. The tensile test was run using an LRX Tensile Machine (Lloyd, LRX, Singapore) with a load cell of 0.5 kN at a 5 mm/min crosshead. Young's modulus (which is a measurement of the material's elasticity) was calculated using Equation (1);

$$E_e = \frac{\sigma}{\varepsilon} \tag{1}$$

where "$E_e$" is Young's modulus, "$\sigma$" is the tensile stress, and "$\varepsilon$" is the tensile strain.

### 2.6. Biodegradation Testing

A degradation test was run to evaluate the weight loss of the structure over time. Samples of the scaffold with and without $\gamma$-Fe$_2$O$_3$ were weighed at their initial weights and then immersed in a gradual beaker filled with 40 mL of simulation body fluid (SBF) at 37 °C, in accordance with the optimum body temperature. The samples were weighed weekly for a month. Next, the degraded samples were rinsed and dried in an oven at 70 °C for 5 min. Then, the dried scaffolds were weighed and recorded. Equation (2) was used to calculate the degradation rate in terms of weight loss over a month.

$$Weight\ loss\ =\ \frac{(Initial\ weight\ -\ Weight\ after)}{Initial\ weight} \tag{2}$$

### 2.7. MTT Assay

In this study, an MTT assay was conducted to evaluate the cell proliferation rate of the human skin fibroblast cell line (HSF1184) on the TPU/UHT and TPU/UHT/$\gamma$-Fe$_2$O$_3$ 3D structures. Both the TPU/UHT and TPU/UHT/$\gamma$-Fe$_2$O$_3$ were sterilized under UV light for one hour on both sides of the structures. Then, the HSF1148 was grown in DMEM that contained 7 g/L of sodium bicarbonate (NaHCO$_3$), 1% penicillin–streptomycin, and 10% fetal bovine serum (FBS). The cells were cultured at 37 °C in a humidified atmosphere containing 5% CO$_2$, dissociated with 0.25% trypsin in PBS (pH 7.4), and centrifuged at 1000 r/min for 5 min at room temperature. A suitable number of normal human skin fibroblast (HSF1184) cells ($5 \times 10^4$ cells cm$^2$ seventh passage) were collected and dispersed into 20 mL of PBS, and then 200 mL of the dispersion was used for seeding the scaffold. Cell counting by the MTT assay procedure was performed after 72 h from seeding in a hemacytometer.

## 3. Results and Discussion

All the tests were performed three times, and three samples were employed for each test every week. All data are denoted as the mean ± standard deviation (SD). Significant differences were determined by performing Student's *t* test, and a *p*-value $\geq$ 0.05 was indicated to be statistically significant.

### 3.1. Curing Time

Before printing the 3D scaffold structures, the exact curing times had to be determined in order to ensure every single layer of the 3D scaffold was burned by the UV laser and cured perfectly. Different materials and concentrations will have different optimum curing times. Investigations into the curing times needed to be performed for each formulation used in the study.

The observations and evaluations were conducted optically. In this study, the curing time was varied with three sets of experiments. For the first set, the curing time varied within 12 different periods. From the results, it was observed that the structure started to develop at 5 s and above. From 1 to 4 s, it was observed that the cured body did not develop well, as the structure was not clearly seen, as shown in Figure 2.

The experiment progressed by increasing the range of the curing time. The curing time ranged from 5 to 60 s with increments of 5 s. It was observed that the structures started to develop at 30 s and were over-cured at 40 s and above, at which point the structure was observed to melt and destroy other structures. Therefore, the structure could not be inspected at above 40 s.

After knowing the specific range, the third set of experiments was conducted with the curing time starting at 30 s, with increments of 2 s, and ending at 40 s, as shown in Figure 3. From the figure, it can be observed that the best structures were cured at 30 s and 40 s. However, from a closed observation and evaluation, the cured structures at 30 s had a sharper end structure and precise structure of the inner body compared to the 40 s curing time. Thus, the optimum curing time for the TPU mixed with UHT bio-resin was 30 s. Figure 4 shows the side view of developed structure during investigation of curing times.

Figure 2. Curing times from 1 s to 12 s.

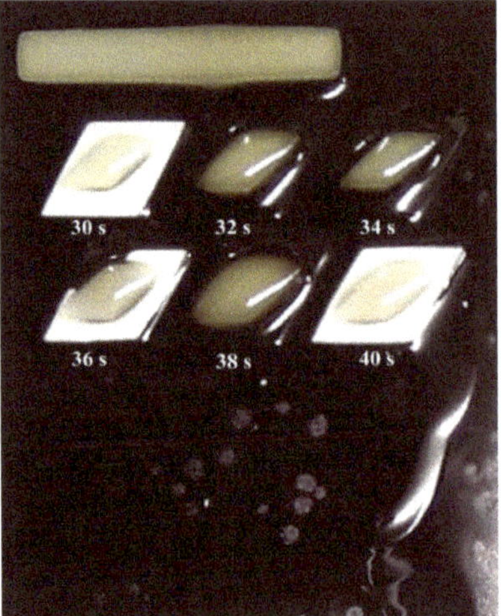

Figure 3. Curing times from 30 s to 40 s.

Figure 4. Side view of developed structure during investigation of curing times.

### 3.2. Mechanical Properties

Dumbbell-shaped scaffolds were printed using a DLP 3D printer for testing the mechanical characteristics of the specimen. Figure 5 presents the 3D scaffold printed in a dumbbell shape. The color of the dumbbell-shaped structures became more yellow and darker as the nanoparticle loading increased.

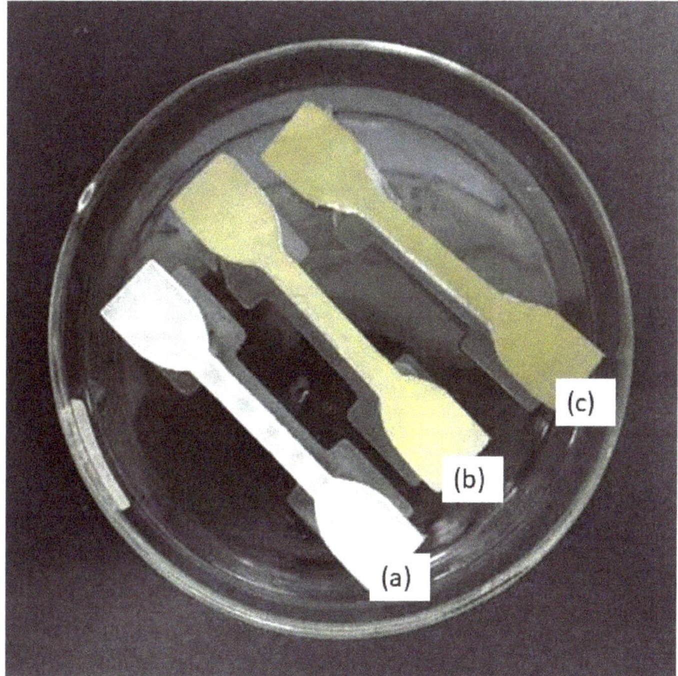


**Figure 5.** Dumbbell-shaped 3D-printed scaffold (**a**) 15% TPU + UHT; (**b**) 15% TPU+ UHT + 1% $\gamma$-Fe$_2$O$_3$; (**c**) 15% TPU+ UHT + 5% $\gamma$-Fe$_2$O$_3$.

The dimensions of the printed scaffold for mechanical testing were as follows: gauge length of 30 mm, overall length of 60 mm, width of 10 mm of the narrow parallel parts, width of 20 mm at the end, and thickness of 3 mm. Table 1 and Figure 6 indicate the results of the mechanical properties in terms of Young's Modulus for the developed 3D structure. For the first part, the mechanical test was run for TPU/UHT at three different concentrations of TPU, which were 10, 12.5, and 15% wt/v.

**Table 1.** Young's Modulus of scaffolds at different concentrations with and without $\gamma$-Fe$_2$O$_3$.

| Sample Label | Sample | Young's Modulus (MPa) | | | |
|---|---|---|---|---|---|
| | | 1 | 2 | 3 | Average |
| A | 10% TPU + UHT | 35.71 | 35.73 | 34.08 | 35.72 ± 1.001 |
| B | 12.5% TPU + UHT | 59.70 | 57.87 | 60.22 | 59.04 ± 1.871 |
| C | 15% TPU + UHT | 87.91 | 90.69 | 86.29 | 88.30 ± 1.032 |
| D | 15% TPU+ UHT + 1% $\gamma$-Fe$_2$O$_3$ | 100.65 | 115.13 | 121.08 | 112.28 ± 2.011 |
| E | 15% TPU+ UHT + 3% $\gamma$-Fe$_2$O$_3$ | 69.17 | 65.10 | 59.05 | 64.44 ± 2.210 |
| F | 15% TPU+ UHT + 5% $\gamma$-Fe$_2$O$_3$ | 54.39 | 52.29 | 55.94 | 54.39 ± 1.171 |

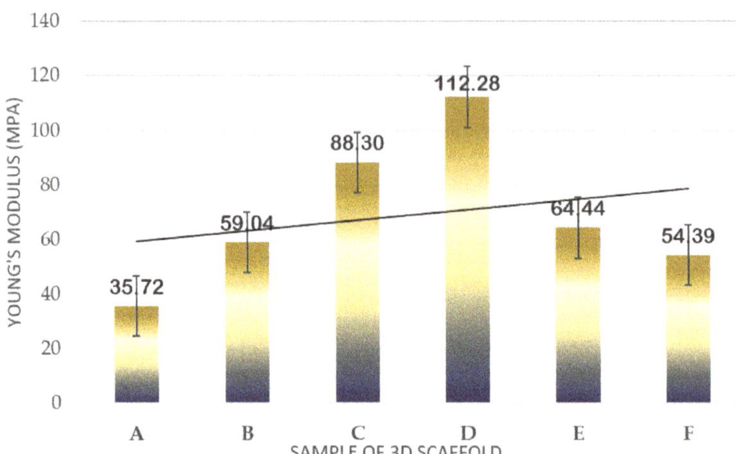



**Figure 6.** Plot of mechanical strength results in terms of Young's Modulus of samples.

In Figure 6, it can be observed that as the concentration of TPU increased, the Young's Modulus also increased. Among the three concentrations, the highest Young's Modulus resulted in 15% wt/v TPU/UHT with 88.30 $\pm$ 1.032 MPa due to the high strength of the covalent bond of TPU. In addition, the 15% wt/v of TPU was selected to test with three different concentrations (1, 3, and 5% v/v) of $\gamma$-Fe$_2$O$_3$ for further study. As mentioned before, $\gamma$-Fe$_2$O$_3$ acts as a filler, which enhances the mechanical properties of the structure. However, the correct concentration must be added to ensure the nanoparticles in the $\gamma$-Fe$_2$O$_3$ function well with the other solutions. Hence, in the graph, it can be observed that the highest Young's Modulus resulted in 15% wt/v TPU/UHT with 1% $\gamma$-Fe$_2$O$_3$, which was at 112.28 $\pm$ 2.011 MPa.

Overall, the Young's Modulus for the three concentrations of $\gamma$-Fe$_2$O$_3$ decreased as the concentration increased. This was because the materials were disrupted with agglomerations of $\gamma$-Fe$_2$O$_3$ nanoparticles. In our previous study [33], excessive nanoparticle loading in the PVA tended to agglomerate and, thus, form cracks when spun at a very rapid speed, which subsequently contributed to a decrease in the Young's Modulus of the electrospun nanofibrous mats. Similar findings were reported by [39], where it was reported that any additives tended to agglomerate when present in excessive amounts during mixing with other materials. An excessive number of nanoparticles in the composition was not suitable for this study.

### 3.3. Biodegradation Test

Figure 7 shows the degradation rates of TPU/UHT and TPU/UHT/$\gamma$-Fe$_2$O$_3$ in one month. After a month, the degradation rate of TPU/UHT was 4.60%, which was higher than the TPU/UHT/$\gamma$-Fe$_2$O$_3$ structure, with a 2.12% degradation rate. For both samples, the graph shows that the weight loss increased non-linearly during the testing period. In comparison, throughout the month, the degradation rate of the TPU/UHT structure in the absence of $\gamma$-Fe$_2$O$_3$ nanoparticles was lower than the TPU/UHT structure containing $\gamma$-Fe$_2$O$_3$. This was due to the existence of magnetic nanoparticles, which made the degradation rate slower. In addition, the strong Fe-O bonds contained a polymer chain that triggered a slower degradation rate P [33].

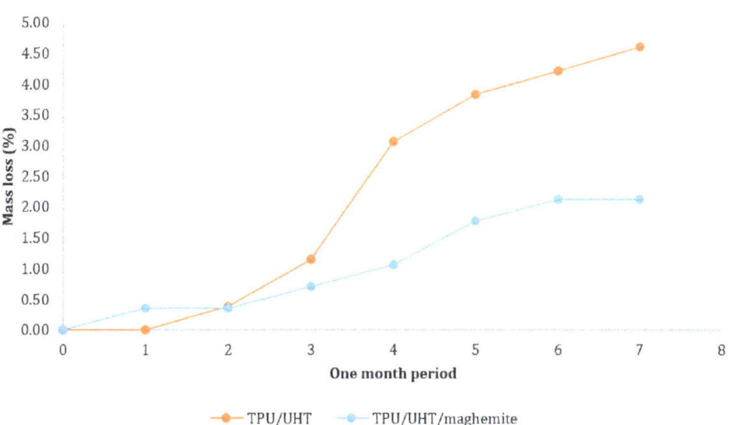

**Figure 7.** Plot of degradation rate over a period of one month.

### 3.4. MTT Assay

In this study, an MTT assay was also conducted to evaluate the cell proliferation rate of human skin fibroblast cell line (HSF1184) on the 3D structure of TPU/UHT and TPU/UHT/$\gamma$-Fe$_2$O$_3$ scaffolds. The control group had nothing except the test cells (no scaffold). Figure 8 shows the results of the relative cell viability with different structure compositions. According to the plot, the proliferation rate of the HSF1184 by TPU/$\gamma$-Fe$_2$O$_3$ was higher than the TPU structure without $\gamma$-Fe$_2$O$_3$. This was because the existence of the magnetic field in the $\gamma$-Fe$_2$O$_3$, known as the osteoinductive effect, accelerated the proliferation rate. This result is in accordance with our previous research and that of Fallahiarezoudar et al. [15]. In addition, the presence of $\gamma$-Fe$_2$O$_3$ developed a great number of tiny magnetic fields, and each nanoparticle acted as a single magnetic field that integrated with the matrix. This created a micro-environment on the surface of the blend, which made a great number of tiny magnetic fields. This situation led to an increase in the cell proliferation rate. In addition, $\gamma$-Fe$_2$O$_3$, which consists of magnetic nanoparticles has a large surface area-to-volume ratio. Thus, the existence of $\gamma$-Fe$_2$O$_3$ in TPU increased the cell area attachment, which allowed more cells to anchor [40].

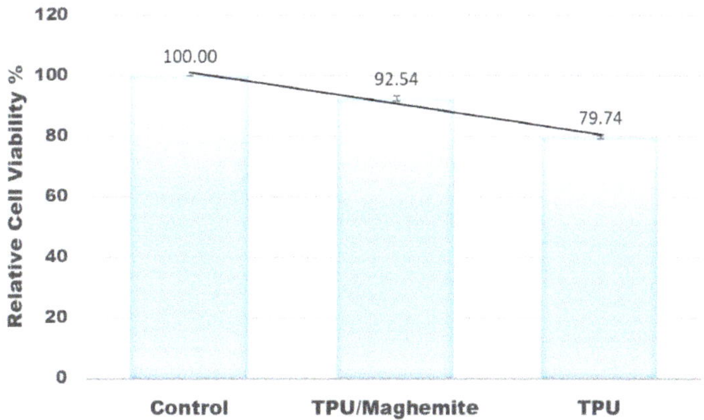

**Figure 8.** Relative values of cell viability with different composites of materials.

## 4. Conclusions

In conclusion, the objectives were successfully achieved in this study. Firstly, the curing time for the composite material of TPU/UHT containing $\gamma$-Fe$_2$O$_3$ scaffold developed using a DLP 3D printer was 30 s. A higher concentration of TPU in the composite materials provided better mechanical strength and elastic modulus. Furthermore, the addition of $\gamma$-Fe$_2$O$_3$ enhanced the mechanical properties of the 3D structure. However, an excessive amount of $\gamma$-Fe$_2$O$_3$ could decrease the mechanical properties due to the agglomeration of the $\gamma$-Fe$_2$O$_3$ nanoparticles. The degradation rate of the TPU/UHT structure containing $\gamma$-Fe$_2$O$_3$ was lower than that of the TPU/UHT structure without $\gamma$-Fe$_2$O$_3$ due to the strong Fe-O bonds in $\gamma$-Fe$_2$O$_3$ nanoparticles. Lastly, the presence of $\gamma$-Fe$_2$O$_3$ in the structure increased the proliferation rate of HSF1148 because of numerous magnetic nanoparticles that can integrate with the cellular matrix.

**Author Contributions:** Conceptualization, N.H.A.N. and E.F.; methodology, E.F. and N.H.A.N.; software, E.F., N.H.A.N. and M.S.A.I.; validation, N.H.A.N. and N.M.Y.; formal analysis, E.F. and N.H.A.N.; investigation, A.I., E.F. and N.H.A.N.; writing—original draft preparation, all authors contributed to writing; writing—review and editing, E.F. and N.H.A.N.; supervision, N.H.A.N.; project administration, N.H.A.N.; funding acquisition, N.H.A.N. All authors have read and agreed to the published version of the manuscript.

**Funding:** This research was funded by the Universiti Teknologi Malaysia (UTM) under the UTM R&D Fund (4J506) and UTMShine (09G94).

**Institutional Review Board Statement:** Not applicable.

**Informed Consent Statement:** Not applicable.

**Data Availability Statement:** The data presented in this study are available upon request from the corresponding author. The data are not publicly available due to the project being currently uncompleted.

**Acknowledgments:** The authors would like to thank the Ministry of Higher Education (MOHE) and the Universiti Teknologi Malaysia (UTM) for their financial support of this work.

**Conflicts of Interest:** The authors declare no conflict of interest.

## References

1. Iqbal, N.; Khan, A.S.; Asif, A.; Yar, M.; Haycock, J.W.; Rehman, I.U. Recent concepts in biodegradable polymers for tissue engineering paradigms: A critical review. *Int. Mater. Rev.* **2019**, *64*, 91–126. [CrossRef]
2. Janouskova, O. Synthetic polymer scaffolds for soft tissue engineering. *Physiol. Res.* **2018**, *67*, S335–S348. [CrossRef] [PubMed]
3. Chocholata, P.; Kulda, V.; Babuska, V. Fabrication of scaffolds for bone-tissue regeneration. *Materials* **2019**, *12*, 568. [CrossRef] [PubMed]
4. Dikici, B.A.; Sherborne, C.; Reilly, G.C.; Claeyssens, F. Emulsion templated scaffolds manufactured from photocurable polycaprolactone. *Polymer* **2019**, *175*, 243–254. [CrossRef]
5. Li, X.; Cui, R.; Sun, L.; Aifantis, K.E.; Fan, Y.; Feng, Q.; Cui, F.; Watari, F. 3D-printed biopolymers for tissue engineering application. *Int. J. Polym. Sci.* **2014**, *2014*, 1–13. [CrossRef]
6. Vickers, N.J. Animal communication: When im calling you, will you answer too? *Curr. Biol.* **2017**, *27*, R713–R715. [CrossRef]
7. Chen, G.; Sun, Y.; Lu, F.; Jiang, A.; Subedi, D.; Kong, P.; Wang, X.; Yu, T.; Chi, H.; Song, C.; et al. A three-dimensional (3D) printed biomimetic hierarchical scaffold with a covalent modular release system for osteogenesis. *Mater. Sci. Eng. C* **2019**, *104*, 109842. [CrossRef]
8. Nurulhuda, A.; Izman, S.; Ngadiman, N.H.A. Fabrication PEGDA/ANFs biomaterial as 3D tissue engineering scaffold by DLP 3D printing tecshnology. *Int. J. Eng. Adv. Technol.* **2019**, *8*, 751–758. [CrossRef]
9. Jia, Y.; He, H.; Peng, X.; Meng, S.; Chen, J.; Geng, Y. Preparation of a new filament based on polyamide-6 for three-dimensional printing. *Polym. Eng. Sci.* **2017**, *57*, 1322–1328. [CrossRef]
10. Ngadiman, N.H.A.; Abidin, R.Z.; Murizan, N.I.S.; Yusof, N.M.; Idris, A.; Kadir, A.Z.A. Optimization of Materials Composition and UV-VIS Light Wavelength Towards Curing Time Performance on Development of Tissue Engineering Scaffold. *Biointerface Res. Appl. Chem.* **2020**, *11*, 8740–8750.
11. Ngadiman, N.H.A.; Zulkifli, Z.; Yusof, N.M.; Idris, A.; Kadir, A.Z.A.; Pusppanathan, J. Poly-lactic acid (PLA)/maghemite ($\gamma$-Fe$_2$O$_3$) nanoparticles mixed with ultra hard and flexible (UHF) bio-resin for 3D tissue engineering scaffold. *AIP Conf. Proc.* **2019**, *2129*, 20035.

12. Nurulhuda, A.; Sudin, I.; Ngadiman, N.H.A. Fabrication a novel 3D tissue engineering scaffold of Poly (ethylene glycol) diacrylate filled with Aramid Nanofibers via Digital Light Processing (DLP) technique. *J. Mech. Eng.* **2020**, *9*, 1–12.

13. Nemati, S.; Kim, S.; Shin, Y.M.; Shin, H. Current progress in application of polymeric nanofibers to tissue engineering. *Nano Converg.* **2019**, *6*, 1–16. [CrossRef]

14. Fallahiarezoudar, E.; Ahmadipourroudposht, M.; Idris, A.; Yusof, N.M. Optimization and development of Maghemite ($\gamma$-Fe$_2$O$_3$) filled poly-L-lactic acid (PLLA)/thermoplastic polyurethane (TPU) electrospun nanofibers using Taguchi orthogonal array for tissue engineering heart valve. *Mater. Sci. Eng. C* **2017**, *76*, 616–627. [CrossRef]

15. Fallahiarezoudar, E.; Ahmadipourroudposht, M.; Idris, A.; Yusof, N.M.; Marvibaigi, M.; Irfan, M. Characterization of maghemite ($\gamma$-Fe$_2$O$_3$)-loaded poly-l-lactic acid/thermoplastic polyurethane electrospun mats for soft tissue engineering. *J. Mater. Sci.* **2016**, *51*, 8361–8381. [CrossRef]

16. Fallahiarezoudar, E.; Ahmadipourroudposht, M.; Mohd Yusof, N.; Idris, A.; Ngadiman, N.H.A. 3D biofabrication of thermoplastic polyurethane (TPU)/poly-L-lactic acid (PLLA) electrospun nanofibers containing maghemite ($\gamma$-Fe$_2$O$_3$) for tissue engineering aortic heart valve. *Polymers* **2017**, *9*, 584. [CrossRef]

17. Ngadiman, N.; Yusof, N.; Idris, A.; Fallahiarezoudar, E.; Kurniawan, D. Novel Processing Technique to Produce Three Dimensional Polyvinyl Alcohol/Maghemite Nanofiber Scaffold Suitable for Hard Tissues. *Polymers* **2018**, *10*, 353. [CrossRef]

18. Arbab, A.S.; Bashaw, L.A.; Miller, B.R.; Jordan, E.K.; Lewis, B.K.; Kalish, H.; Frank, J.A. Characterization of Biophysical and Metabolic Properties of Cells Labeled with Superparamagnetic Iron Oxide Nanoparticles and Transfection Agent for Cellular MR Imaging1. *Radiology* **2003**, *229*, 838–846. [CrossRef]

19. Wilhelm, C.; Billotey, C.; Roger, J.; Pons, J.N.; Bacri, C.-J.; Gazeau, F. Intracellular uptake of anionic superparamagnetic nanoparticles as a function of their surface coating. *Biomaterials* **2003**, *24*, 1001–1011. [CrossRef]

20. Shimizu, K.; Ito, A.; Arinobe, M.; Murase, Y.; Iwata, Y.; Narita, Y.; Kagami, H.; Ueda, M.; Honda, H. Effective cell-seeding technique using magnetite nanoparticles and magnetic force onto decellularized blood vessels for vascular tissue engineering. *J. Biosci. Bioeng.* **2007**, *103*, 472–478. [CrossRef]

21. Attaluri, A.; Ma, R.; Qiu, Y.; Li, W.; Zhu, L. Nanoparticle distribution and temperature elevations in prostatic tumours in mice during magnetic nanoparticle hyperthermia. *Int. J. Hyperth.* **2011**, *27*, 491–502. [CrossRef] [PubMed]

22. Akiyama, H.; Ito, A.; Kawabe, Y.; Kamihira, M. Genetically engineered angiogenic cell sheets using magnetic force-based gene delivery and tissue fabrication techniques. *Biomaterials* **2010**, *31*, 1251–1259. [CrossRef] [PubMed]

23. Coffel, J.; Nuxoll, E. Magnetic nanoparticle/polymer composites for medical implant infection control. *J. Mater. Chem. B* **2015**, *3*, 7538–7545. [CrossRef] [PubMed]

24. Chiapasco, M.; Casentini, P. Horizontal bone-augmentation procedures in implant dentistry: Prosthetically guided regeneration. *Periodontology 2000* **2018**, *77*, 213–240. [CrossRef]

25. Chen, Q.; Liang, S.; Thouas, G.A. Elastomeric biomaterials for tissue engineering. *Prog. Polym. Sci.* **2013**, *38*, 584–671. [CrossRef]

26. Chen, H.; Yuan, L.; Song, W.; Wu, Z.; Li, D. Biocompatible polymer materials: Role of protein-surface interactions. *Progress in Polym. Sci.* **2008**, *33*, 1059–1087. [CrossRef]

27. Cheung, H.Y.; Lau, K.T.; Lu, T.P.; Hui, D. A critical review on polymer-based bio-engineered materials for scaffold development. *Compos. Part B Eng.* **2007**, *38*, 291–300. [CrossRef]

28. Meng, L.; Watson, B.W.; Qin, Y. Hybrid conjugated polymer/magnetic nanoparticle composite nanofibers through cooperative non-covalent interactions. *Nanoscale Adv.* **2020**, *2*, 2462–2470. [CrossRef]

29. Ngadiman, N.H.A.; Idris, A.; Irfan, M.; Kurniawan, D.; Yusof, N.M.; Nasiri, R. $\gamma$-Fe$_2$O$_3$ nanoparticles filled polyvinyl alcohol as potential biomaterial for tissue engineering scaffold. *J. Mech. Behav. Biomed. Mater.* **2015**, *49*, 90–104. [CrossRef]

30. Shao, D.; Qin, L.; Sawyer, S. Optical properties of polyvinyl alcohol (PVA) coated In$_2$O$_3$ nanoparticles. *Opt. Mater.* **2013**, *35*, 563–566. [CrossRef]

31. Idris, A.; Misran, E.; Hassan, N.; Abd Jalil, A.; Seng, C.E. Modified PVA-alginate encapsulated photocatalyst ferro photo gels for Cr(VI) reduction. *J. Hazard. Mater.* **227–** **2012**, *228*, 309–316. [CrossRef]

32. Ngadiman, N.H.A.; Yusof, N.M.; Idris, A.; Misran, E.; Kurniawan, D. Development of highly porous biodegradable $\gamma$-Fe$_2$O$_3$/polyvinyl alcohol nanofiber mats using electrospinning process for biomedical application. *Mater. Sci. Eng. C* **2017**, *70*, 520–534. [CrossRef]

33. Ngadiman, N.H.A.; Mohd Yusof, N.; Idris, A.; Kurniawan, D.; Fallahiarezoudar, E. Fabricating high mechanical strength $\gamma$-Fe$_2$O$_3$ nanoparticles filled poly (vinyl alcohol) nanofiber using electrospinning process potentially for tissue engineering scaffold. *J. Bioact. Compat. Polym.* **2017**, *32*, 411–428. [CrossRef]

34. Ngadiman, N.H.A.; Mohd Yusof, N.; Idris, A.; Kurniawan, D. Mechanical properties and biocompatibility of co-axially electrospun polyvinyl alcohol/maghemite. *Proc. Inst. Mech. Eng. Part H J. Eng. Med.* **2016**, *230*, 739–749. [CrossRef]

35. Ngadiman, N.; Yusof, N.; Kurniawan, D.; Idris, A. Characterisation of electrospun magnetic nanoparticle?-Fe$_2$O$_3$/PVA nanofibers. *Trans. North Am. Manuf. Res. Inst. SME* **2014**, *42*, 396–400.

36. Nurdin, I.; Johan, M.R.; Yaacob, I.I.; Ang, B.C.; Andriyana, A. Synthesis, characterisation and stability of superparamagnetic maghemite nanoparticle suspension. *Mater. Res. Innov.* **2014**, *18*, S6-200–S6-203. [CrossRef]

37. Biehl, P.; der Lühe, M.; Dutz, S.; Synthesis, F.H. Schacher, characterization, and applications of magnetic nanoparticles featuring polyzwitterionic coatings. *Polymers* **2018**, *10*, 91. [CrossRef]

38. Idris, A.; Hassan, N.; Ismail, N.S.M.; Misran, E.; Yusof, N.M.; Ngomsik, A.-F.; Bee, A. Photocatalytic magnetic separable beads for chromium (VI) reduction. *Water Res.* **2010**, *44*, 1683–1688. [CrossRef]
39. Wong, B.A.; Nash, D.G.; Moss, O.R. Generation of nanoparticle agglomerates and their dispersion in lung serum simulant or water. *J. Phys. Conf. Ser.* **2009**, *151*, 12036. [CrossRef]
40. Zhang, H.; Xia, J.; Pang, X.; Zhao, M.; Wang, B.; Yang, L.; Wan, H.; Wu, J.; Fu, S. Magnetic nanoparticle-loaded electrospun polymeric nanofibers for tissue engineering. *Mater. Sci. Eng. C* **2017**, *73*, 537–543. [CrossRef]

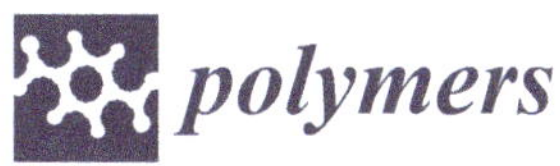

*Article*

# Investigating Potential Effects of Ultra-Short Laser-Textured Porous Poly-ε-Caprolactone Scaffolds on Bacterial Adhesion and Bone Cell Metabolism


Emil Filipov [1,*], Liliya Angelova [1], Sanjana Vig [2,3], Maria Helena Fernandes [2,3], Gerard Moreau [4], Marie Lasgorceix [4], Ivan Buchvarov [5] and Albena Daskalova [1]

[1] Institute of Electronics, Bulgarian Academy of Sciences, 72 Tzarigradsko Shousse Blvd., 1784 Sofia, Bulgaria; lily1986@abv.bg (L.A.); albdaskalova@gmail.com (A.D.)

[2] Faculdade de Medicina Dentaria, Universidade do Porto, Rua Dr. Manuel Pereira da Silva, 4200-393 Porto, Portugal; sanjana@fmd.up.pt (S.V.); mhfernandes@fmd.up.pt (M.H.F.)

[3] LAQV/REQUIMTE, University of Porto, 4160-007 Porto, Portugal

[4] Laboratoire des Matériaux Céramiques et Procédés Associés, Université Polytechnique Hauts-de-France, INSA Hauts-de-France, CERAMATHS, F-59313 Valenciennes, France; gerard.moreau@uphf.fr (G.M.); marie.lasgorceix@uphf.fr (M.L.)

[5] Faculty of Physics, St. Kliment Ohridski University of Sofia, 5 James Bourchier Blvd., 1164 Sofia, Bulgaria; ivan.buchvarov@phys.uni-sofia.bg

* Correspondence: emil.filipov95@gmail.com



Citation: Filipov, E.; Angelova, L.; Vig, S.; Fernandes, M.H.; Moreau, G.; Lasgorceix, M.; Buchvarov, I.; Daskalova, A. Investigating Potential Effects of Ultra-Short Laser-Textured Porous Poly-ε-Caprolactone Scaffolds on Bacterial Adhesion and Bone Cell Metabolism. *Polymers* **2022**, *14*, 2382. https://doi.org/10.3390/polym14122382

Academic Editors: Andrada Serafim and Stefan Ioan Voicu

Received: 12 May 2022
Accepted: 10 June 2022
Published: 12 June 2022

Publisher's Note: MDPI stays neutral with regard to jurisdictional claims in published maps and institutional affiliations.

**Abstract:** Developing antimicrobial surfaces that combat implant-associated infections while promoting host cell response is a key strategy for improving current therapies for orthopaedic injuries. In this paper, we present the application of ultra-short laser irradiation for patterning the surface of a 3D biodegradable synthetic polymer in order to affect the adhesion and proliferation of bone cells and reject bacterial cells. The surfaces of 3D-printed polycaprolactone (PCL) scaffolds were processed with a femtosecond laser ($\lambda$ = 800 nm; $\tau$ = 130 fs) for the production of patterns resembling microchannels or microprotrusions. MG63 osteoblastic cells, as well as *S. aureus* and *E. coli*, were cultured on fs-laser-treated samples. Their attachment, proliferation, and metabolic activity were monitored via colorimetric assays and scanning electron microscopy. The microchannels improved the wettability, stimulating the attachment, spreading, and proliferation of osteoblastic cells. The same topography induced cell-pattern orientation and promoted the expression of alkaline phosphatase in cells growing in an osteogenic medium. The microchannels exerted an inhibitory effect on *S. aureus* as after 48 h cells appeared shrunk and disrupted. In comparison, *E. coli* formed an abundant biofilm over both the laser-treated and control samples; however, the film was dense and adhesive on the control PCL but unattached over the microchannels.

**Keywords:** ultra-short laser processing; biomaterials; 3D printing; cell adhesion; antibacterial surfaces


## 1. Introduction

As one of the most common types of injury, bone fractures require novel approaches and constant improvements in order to build on the current tissue-repairing standards. Such standards include the use of metallic implants (titanium and its alloys, stainless steel, or cobalt chrome) that exhibit outstanding mechanical properties. However, there are still some disadvantages that limit the performance of these materials: (i) the stress-shielding effect that could lead to bone resorption; (ii) the release of ions that could be toxic to the host cells; (iii) the generally smooth surface of the implants representing a favorable site for bacterial adhesion; and (iv) the lack of biodegradability concerning non-load-bearing implants [1,2]. The need for novel materials to overcome these limitations has brought attention to the degradable biomaterials of either natural or synthetic origin. The chemistry of such materials (e.g., polymers) can be altered, which allows for the fine-tuning of their physical and mechanical properties [3]. An important aspect of the comparison between metals and polymers is that the latter can be enzymatically degraded to their building components and excreted or metabolized by the body in the process of tissue regeneration without the need for secondary surgery for removing the material [4].

Polycaprolactone (PCL) is a synthetic polymer with a wide application in tissue engineering. It is a semicrystalline polyester with a melting point between 59 °C and 64 °C [5]. Due to its low glass transition temperature of −56 °C, the polymer remains highly permeable for macromolecules under physiological conditions. Because of this property, PCL has been used as a material for drug-delivery systems [3]. Similar to PLA, the high number of methylene groups makes the polymer hydrophobic, which could affect its interaction with human cells [6]. PCL exhibits a very slow degradation rate that arises from its hydrophobicity and degree of crystallization. The molecular weight further influences these two properties of the material [7]. The degradation of PCL mainly occurs via hydrolysis either in an enzymatic (by esterases and lipases) or a non-enzymatic manner [8]. The process of polymer degradation yields intermediate products that could be metabolized via the citric acid cycle and thus eliminated from the body [5]. It has been investigated that PCL's mechanical properties are not sufficient to withstand the tension that, for example, cortical bones could exert on it [9–11]. Thus, different strategies for the improvement of these properties have been employed, e.g., creating blends with ceramic materials or producing PCL scaffolds with distinct pore sizes and geometries [11].

The initial attachment of single bacterial cells to the implant's surface is followed by the aggregation of microorganisms and the secretion of extracellular polymeric substances (EPS) [12]. The gradual formation of a thick slime layer represents the bacterial biofilm, which plays the role of a shelter for the bacterial cells. Thus, they can not only evade host immune cells but also resist antibiotic treatment that can lead to a major irreversible infection [13]. Bacteria can sense the surrounding environment via several mechanisms including chemical, biological, and physical [14]. The ionic concentration and the presence of biological molecules or antimicrobial peptides can guide the bacterial cells to the appropriate site for adhesion. Furthermore, the cells can use mechanosensors in order to make physical contact with different surfaces. The interactions between cells and surfaces fall into three main categories: non-specific physiochemical, specific, and surface mechanosensing [14]. On the material's side, the factors that have a great influence on the bacterial adhesion include the chemical composition, surface energy, wettability, and surface topography [12,15]. The morphological appearance and the roughness of a material at the micro or nano level could physically hinder the attachment of a cell. Patterns with a distinct morphology (e.g., pillar-like) can lead to a reduced contact area and points for contact, thus preventing the cell from properly expanding its mechanosensors and establishing a physical connection with the material through specific molecules called adhesins [15]. Hence, various strategies for the modification of implants' surfaces have emerged including the development of layers that either contain antimicrobial agents or aim to change the roughness of the implant.

A very prominent solution for developing a physical barrier for the inhibition of bacterial adhesion onto materials is femtosecond (fs)-laser surface processing [16]. Unlike the linear absorption of laser energy in metals, in the case of PCL, which is considered a dielectric, the laser energy is absorbed via non-linear processes, which involve multiphoton ionization [17]. Since the duration of an ultra-short pulse is less than the time (<ps) required for the pulse energy to dissipate along the structural lattice after a multiphoton ionization, the irradiation does not lead to a thermal diffusion that could, in turn, result in unwanted photomechanical damage to the material [18]. Thus, fs-laser micromachining allows for a gentle surface patterning that alters the roughness and the morphological appearance of the irradiated zone without substantially affecting its structural integrity [16]. Several research groups report the use of an fs-laser for surface texturing of materials [19–21]. Chen et al. investigated the effects of surface topography by creating parallel microchannels (width of 1–2 µm) lined with nanopillars (200 nm) ($\lambda = 1040$ nm; $\tau = 375$ fs; $\nu = 20$ kHz; $E = 80$ nJ)

on borosilicate glass [20]. The study reported that the nanoroughness greatly reduced the viability of both *E. coli* and *S. aureus* by exerting mechanical stress on their membranes. Jalil et al. studied how the increase in laser fluence (0.1 J/cm$^2$ to 3 J/cm$^2$) with a shorter pulse width (30 fs) led to the gradual transition of ordered (laser-induced periodic surface structures; LIPSS) to disordered structures at both micro- and nanoscales on gold [21]. The authors observed that laser-induced nano topography (LIPSS) inhibited bacterial adhesion by reducing the contact points and disrupting the bacterial cells. In addition to their ability to hinder bacterial attachment, such types of superficial modifications could be used to influence the physiochemical properties of materials such as the wettability degree without greatly altering their chemical composition. For example, a study employing fs-laser parameters similar to the ones described in the following sections ($\lambda$ = 800; $\tau$= 130 fs; $\nu$ = 1 kHz) showed that the patterning of micro trenches with roughness on a nanoscale led to a significant increase in hydrophobicity ($\theta$ = 120°~156°) compared to unprocessed surfaces ($\theta$ = 75°) [22]. This process is mainly attributed to the formed topographies that influence the contact between liquids and the modified surface [23].

In this paper, we aim at developing antibacterial surfaces for bone tissue engineering. The novelty behind our approach for achieving this aim includes the use of ultra-short laser pulses for processing biocompatible polymers, which represents a contactless, highly precise, and reproducible method for obtaining surface patterns with defined locations and dimensions. The methodology described in this article involved the production of 3D-printed PCL scaffolds with pre-defined geometry, whose surface was textured via fs-laser modification. The antimicrobial potential of the formed patterns was evaluated by using *S. aureus* and *E. coli*. The effects of the modified scaffolds on the viability, proliferation, and differentiation of osteoblastic cells were further investigated.

## 2. Materials and Methods

### 2.1. Fabrication of Polymeric Scaffolds

PCL-based 3D scaffolds, resembling a woodpile construct, were fabricated via extrusion 3D printing. The standard operating procedure was already described and established by Daskalova et al. [24]. In brief, PCL pellets with $M_n$ = 45 kDa (Sigma-Aldrich, St. Louis, MO, USA) were melted at 70 °C in a cartridge unit and extruded through a 250 µm needle at a pressure of 5 bar and speed of deposition of 95 mm/min. The constructs were fabricated layer by layer with space between separate fibers of 140 µm and a thickness of the layers of 160 µm. The layer deposition angle was set to 90 °C.

### 2.2. Surface Modification by Femtosecond-Laser Machining

The surface of the PCL scaffolds was processed with a Ti:sapphire fs-laser (Quantronix-Integra-C, Hamden, CT, USA) emitting at a central wavelength of 800 nm with a pulse duration ($\tau$) of 130 fs. The repetition rate for ($\nu$) was set at 1 kHz. The number of applied laser pulses (N) was adjusted by the velocity of a vertical translation stage controlled by computer software (LabView). All scaffolds were adjusted on the translation stage (moving in XY directions), which was positioned perpendicular to the direction of the laser beam. The parameters chosen for material processing for all in vitro experiments were as follows: N = 2 or 10; laser fluence (F) 0.08 J/cm$^2$ and distance between two adjacent laser spots ($d_x$) 45 or 35 µm. This decision was based on the findings of a previous investigation into this subject [24]. In addition to these samples, a separate set of PCL scaffolds were modified with the following fs-laser parameters: N = 1/2/5/10; F = 0.08/0.17/0.42/1.23 J/cm$^2$; $d_x$= 65 or 75 µm. These samples were used to assess the morphological changes that a gradual increase in the number of applied laser pulses and laser fluence could induce in the material.

*2.3. Characterization of fs-Laser-Processed Scaffolds*

2.3.1. Assessment of fs Laser Derived Topography Features and Their Effects on the Morphology and Cell Growth Pattern of *S. aureus* and MG-63 Osteoblastic Cell Lines via Microscopic Techniques

Laser-processed matrices were visualized by scanning electron microscopy (SEM) (SU5000 Hitachi High-Tech Europe). Prior to the analyses, the samples were coated with a thin layer of platinum (~4 nm). The images were obtained at 15 kV. The morphological changes on the surface of the materials arising from the increase in applied laser pulses were investigated by a 3D optical surface metrological system Leica DCM3D (Leica Microsystems, Wetzlar, Germany). Two- and three-dimensional topographical images in true colors were obtained by vertical scanning of selected modified or control areas. All images were obtained at 20× magnification with the field of view being 636.61 × 477.25 $\mu m^2$ or at 10× (field of view 1.27 × 0.95 mm$^2$). Based on the acquired images in true colors, the roughness of the modified areas on the material was evaluated in accordance with ISO 4287. For this purpose, the arithmetical mean height (Ra) was taken into account. All data obtained by the 3D optical system was processed via ProfilmOnline software (www.profilmonline.com, accessed on 6 April 2022).

All samples used for in vitro analyses with osteoblast-like and bacterial cells were fixed in 1.5% glutaraldehyde solution (prepared in 25% cacodylate solution, TAAB laboratories equipment Ltd., Aldermaston, England) for 15 min followed by storage in sodium cacodylate solution. They were serially dehydrated in gradually increasing alcohol concentrations (50%, 70%, 90%, 100% ethanol) followed by critical point drying. Samples were sputter-coated with a thin layer of gold-palladium coating and evaluated using SEM (JEOL JSM-700, Tokyo, Japan).

2.3.2. Analysis of Wettability and Surface Energy Changes in Relation to fs-Laser Processing

Wettability studies were performed using a video-based optical contact angle measurement device DSA100 Drop Shape Analyzer (KRÜSS GmbH, Hamburg, Germany). In order to fully evaluate the wetting state and the total surface energy of the control and modified surfaces, three types of liquids were used based on their polarity: distilled water (highly polar), ethylene glycol (medium polarity), and diiodomethane (very low polarity). Static contact angles were measured at room temperature by the sessile drop method on droplets of 2 $\mu$L. The measurements were performed with a minimum of 3 drops per sample type. The drop evolution was followed for a total of 3 min as measurements were taken at each second during the first minute and every minute for the next two. Contact angles and surface energy were calculated by ADVANCE software (KRÜSS GmbH, Hamburg, Germany) fitting the drop profiles to the Young–Laplace equation. Surface energy (SFE) was calculated by the software following the Owens–Wendt–Rabel–Kaelble (OWRK) equation (Equation (1)) [25].

$$\gamma_{12} = \gamma_1 + \gamma_2 - 2\sqrt{(\gamma_1^d \gamma_2^d)} - 2\sqrt{(\gamma_1^p \gamma_2^p)}, \tag{1}$$

*2.4. Degradation Test in Phosphate Buffer Saline*

Fs-laser-irradiated (F = 0.08 J/cm$^2$; N = 2) 3D printed PCL scaffold was immersed in phosphate buffer saline (PBS, pH 7.2, Sigma-Aldrich, St. Louis, MO, USA) for 7 weeks. The pH was read every week with a pH meter with an external sensor (VAT1011, V & A Instrument Co., Ltd., Shanghai, China).

*2.5. In Vitro Cytocompatibility Assessment*

In vitro cytocompatibility assessment was performed on osteoblast-like MG63 (ATCC$^\circledR$CRL-1427™) cells cultured in alpha-MEM medium supplemented with 10% fetal bovine serum (FBS), 100 IU/mL penicillin, 100 $\mu$g/mL streptomycin, and 2.5 $\mu$g/mL amphotericin B (all reagents from Gibco, USA) at 37 °C, 95% humidity and 5% $CO_2$ atmosphere. Cells were seeded over the materials at a density of 2 × 10$^5$ cells/cm$^2$ and cultured for 7 days in basal (as described above) and osteogenic-induced conditions. On induced cultures, MG63 cells were pre-treated with 50 µg/mL ascorbic acid and 10 nM dexamethasone for 48 h. Next, cells were passaged and seeded over the materials in an osteogenic medium containing 50 µg/mL ascorbic acid, 10 nM dexamethasone, and 10 mM beta-glycerophosphate (all from Sigma-Aldrich, St. Louis, MO, USA). Cell cultures were characterized for metabolic activity (Resazurin assay), alkaline phosphatase (ALP) activity, and SEM analysis.

### 2.5.1. Resazurin Assay

Metabolic activity of MG63 cells on PCL samples was assessed through the Resazurin assay on day 1, day 3, and day 7 for both non-induced and induced cell cultures. Briefly, all samples were transferred to a fresh well plate before incubation for 3 h in 10% Resazurin solution (Resazurin sodium salt, Sigma-Aldrich R7017) prepared in a complete medium (alpha-MEM with 10% FBS, 100 IU/mL penicillin, 100 µg/mL streptomycin, and 2.5 µg/mL amphotericin B) at 37 °C. Fluorescence (530 nm excitation/590 nm emission) was measured in a microplate reader (Synergy HT, Biotek, Winooski, VT, USA) with Gen5 1.09 Data Analysis Software.

### 2.5.2. Alkaline Phosphatase (ALP) Activity

The ALP activity of cells was evaluated on day 3 and day 7 of osteoinduction. Cell lysates were prepared in 0.1% Triton X-100 (in distilled water) for 30 min followed by the hydrolysis of p-nitrophenyl phosphate substrate (p-NPP, 25 mM, Sigma-Aldrich, USA) in an alkaline buffer (pH 10.3, 37 °C, 1 h). The reaction was stopped with 5 M NaOH and the product (p-nitrophenol) was measured at 400 nm in a microplate reader (Synergy HT, Biotek, USA). Results were normalized to total protein content, measured using the DCTM Protein Assay (BioRad, Hercules, CA, USA) according to the manufacturer's instructions, and expressed as nanomoles of p-nitrophenol per microgram of protein (nmol/µg protein).

### 2.6. Antimicrobial Activity

Antibacterial activity of all samples was assessed against *Staphylococcus aureus* (ATCC 25923) and *Escherichia coli* (ATCC 25922). PCL samples were incubated with 10,000 CFU/mL of bacterial suspension for 6 h, 24 h, and 48 h in tryptic soy buffer. Biofilm formation/inhibition was assessed using SEM analysis. CFU assay measurements were made after 24 h of incubation.

### 2.7. Statistical Analysis

Results for the biological assays are presented as mean $\pm$ standard deviation of three independent experiments, with three replicas for each experiment. One-way analysis of variance (ANOVA) was used with Bonferroni's post hoc test for data evaluation. Values of $p \leq 0.05$ were considered significant.

## 3. Results

### 3.1. Inducing Morphological Changes in the Surface Profile of the 3D-Printed PCL Scaffolds by fs-Laser Micromachining

The physical effects that the fs-laser treatment exerted on the surface of the scaffolds were visualized by a 3D optical system and SEM. As previously seen, the chosen working parameters F = 0.08 J/cm$^2$ and N = 2 and 10 lead to the formation of surface patterns resembling protrusions (N = 2) and microchannels (N = 10) (Figure 1) [26]. The average height of the protrusions was 12.03 µm, whereas the average depth of the microchannels was 32.87 µm with an average width of 28.7 µm. The surface roughness within the microchannels ranged between 0.14 and 0.8 µm. A measurement of the arithmetical mean height (Ra) across the modified areas was obtained in order to observe the variation of the roughness in relation to the achieved surface morphologies. For N = 2, the Ra value was 1.15 µm and for N = 10 it was found to be 5.1 µm. In comparison, the control sample showed an Ra of 0.08 μm. A study monitoring the attachment and proliferation of MG63 cells on Mg-based metallic glass showed that a surface with a roughness of ~0.22 μm led to the highest cell attachment, whereas a roughness of ~0.24 μm led to an improved calcium deposition [27]. The formed microchannels as a result of the increased number of applied laser pulses, substantially improved the overall roughness of the material, thus providing a surface that would be suitable for cell adhesion and osteogenic differentiation.

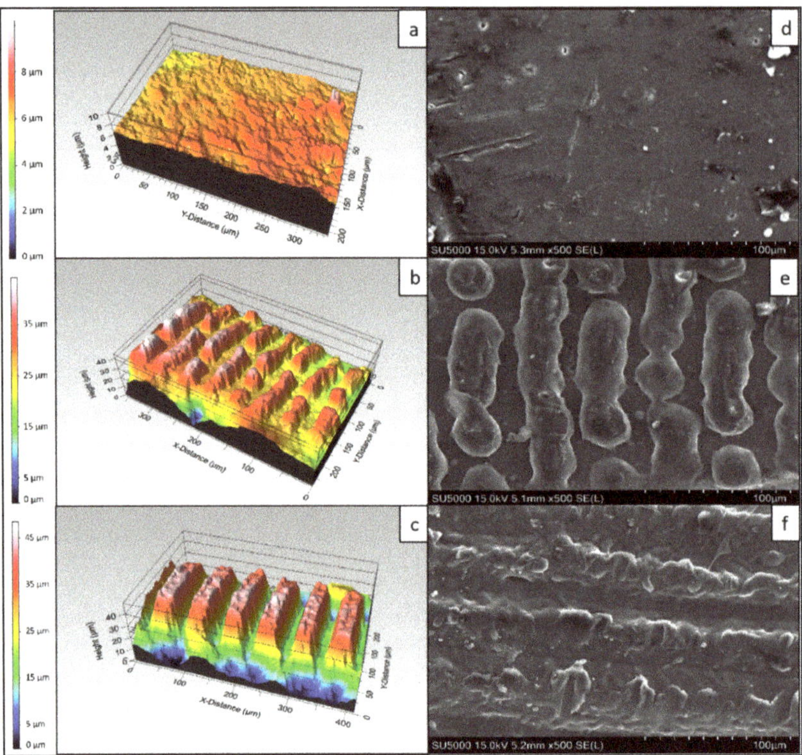

**Figure 1.** 3D and SEM profiles of PCL scaffolds irradiated with an fs-laser. (**a,d**) Control sample; (**b,e**) A fiber of the polymeric scaffold processed with F = 0.08 J/cm$^2$; N = 2; (**c,f**) A fiber of the polymeric scaffold processed with F = 0.08 J/cm$^2$; N = 10.

As an additional assessment, the evolution of the ablation crater on the material in relation to the number of applied laser pulses and laser fluence was investigated. Regardless of the laser fluence used in the experiments, with a lower number of applied pulses (1 or 2) the laser–matter interaction led to a build-up of melted material without any substantial ablation (Figure 2a,b). However, the increase of N to 5 and 10 caused the formation of engraved channels (Figure 2c,d). The morphological observations of the 3D true-color images revealed the transition between the protruding patterns and the microchannels.

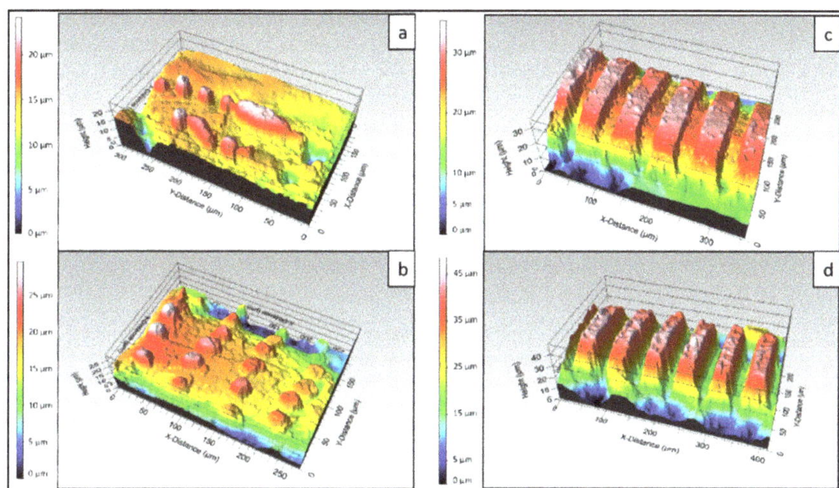

**Figure 2.** 3D profiles of PCL scaffold treated with the same fluence (0.08 J/cm$^2$) and rising applied laser pulses. (**a**) N = 1; (**b**) N = 2; (**c**) N = 5; (**d**) N = 10.

As expected, the highest laser fluence (1.25 J/cm$^2$) formed a laser spot with the largest diameter with both N = 1 and N = 10 (Figure 3). However, the increase in the fluency in the range 0.08–0.42 J/cm$^2$ at N = 1 did not lead to a major change in the dimensions of the physical mark on the material.

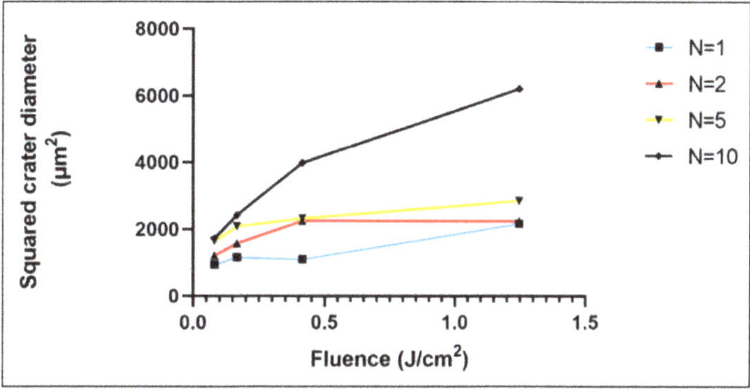


**Figure 3.** Increase in laser crater diameter with rising laser fluence and applied laser pulses.

### 3.2. Hydrophobic Behaviour of fs-Laser-Treated PCL Scaffolds

The water contact angle (WCA) evaluation with a highly polar solvent revealed that the fs-laser treatment with N = 10 led to an increase in the wettability of the polymeric scaffolds; however, the material still retained a contact angle of slightly less than 90° 180 s after the drop deployment (Figures 4 and 5). In contrast, the control samples showed a WCA of 123.3° at t = 1 s and 115.8° at t = 180 s. When comparing the two types of formed surface morphologies, it could be clearly seen that the microchannels had a stronger effect on the improvement in PCL wettability. The reason behind this observation could be that a part of the water droplet spread along the microchannel while entrapping air in the valleys of the channels. Thus, a solid–liquid–air interface could be formed that would not allow the full disruption of the droplet. In this case, the interaction between the liquid and the roughened surface might follow the Cassie–Baxter wetting state [28]. Potentially, the microprotrusions had the same impact on the water droplet, in terms of spreading, as they resembled continuous channels due to their partial fusion. By observing the laser-induced patterns and the wettability results, we could hypothesize that the reduction in the water contact angle values was attributed mainly to the changed surface morphology.

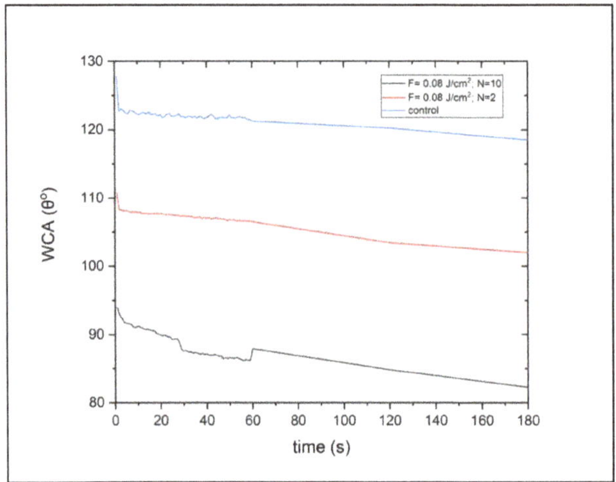


**Figure 4.** Water contact angle evaluation before and after fs-laser treatment. The scaffolds processed with F = 0.08 J/cm$^2$ and N = 10 showed an improved wettability behaviour ($\theta$ = 94° at t = 1 s; $\theta$ = 82.3° at t = 180 s) when compared to the control ones ($\theta$ = 123.3° at t = 1 s; $\theta$ = 115.8° at t = 180 s).

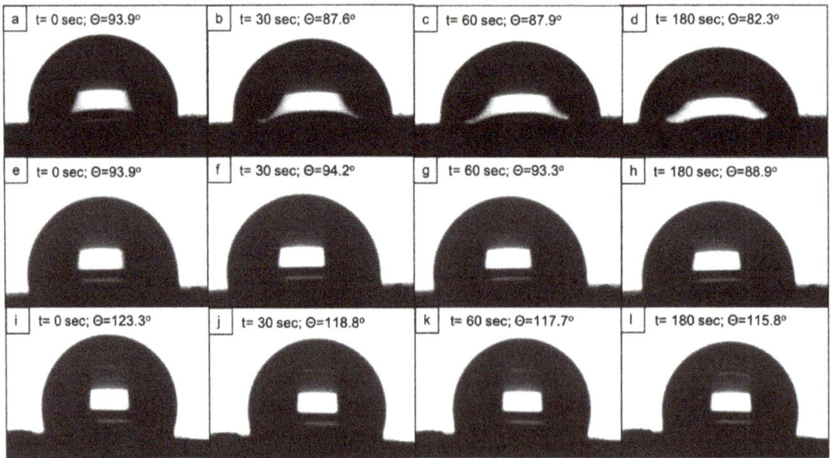


**Figure 5.** Evolution of a water droplet on the surface of fs-laser processed and control scaffolds for a period of 3 min. First-row images (**a–d**) represent a water droplet on scaffolds irradiated with F = 0.08 J/cm$^2$; N = 10; second-row images (**e–h**) represent a water droplet on scaffolds treated with F = 0.08 J/cm$^2$; N = 2; third-row images (**i–l**) depict the hydrophobic nature of the control PCL scaffold.

Contrary to the results with distilled water, ethylene glycol drops fully spread over the surface of both the laser-treated and the control samples approximately 30 s after deposition (Figure 6). In the case of the highly dispersive solution (diiodomethane), the drops of the solvent immediately wetted the whole surface of both the modified and control scaffolds. Hence, a contact angle could not be measured on either of the two types of samples.

Both ethylene glycol and diiodomethane possess a higher dispersive component of their surface energy compared to the polar one, and the same statement has been reported for polycaprolactone, mostly attributed to its nonpolar $CH_2$ groups [29,30]. The complete spread of both solvents on the scaffold's surfaces could be explained by the attraction between the higher dispersive components of surface energies of both the material and the solvents [30]. The authors implied that a decrease in the polarity of a given surface, leaving a higher nonpolar component, would have a repelling effect against a polar solvent such as distilled water but not toward nonpolar solvents.

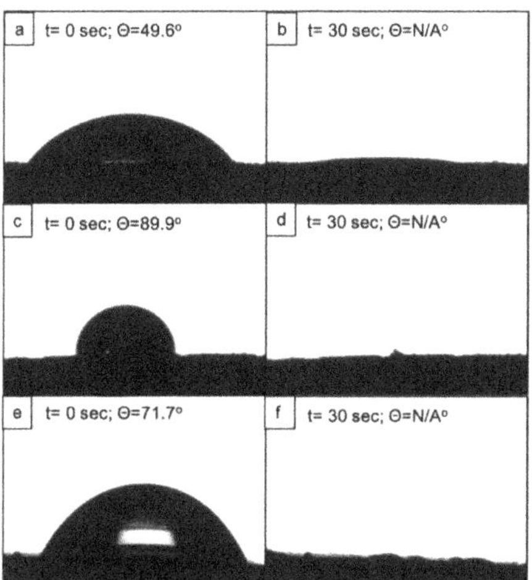


**Figure 6.** Contact angle of ethylene glycol on fs-laser-treated PCL. (**a,b**) Surface with microchannels (F = 0.08 J/cm$^2$; N = 10); (**c,d**) Surface with microprotrusions (F = 0.08 J/cm$^2$; N = 2); (**e,f**) Control scaffold.

Based on the contact angles of distilled water and ethylene glycol, the surface energy of the scaffolds was calculated by the software. Overall, it was noted that the fs-laser irradiation greatly reduced the surface free energy of the material. The analysis of the control samples showed an SFE of 193.8 mN/m. On the contrary, the samples treated with N = 10 exhibited an SFE of 20.8 mN/n and the samples patterned with microprotrusions, 18.3 mN/m. These results were in contradiction to the ones regarding the wettability of the scaffolds. In particular, the higher surface energy would imply that a given surface is hydrophilic and a reduction in the surface energy would result in a shift to a hydrophobic state [30]. However, our findings showed the opposite trend—the laser-treated samples with lower values for SFE were hydrophilic ($\theta$ = 82.3° for the microchannels pattern and $\theta$ = 88.9° for the protrusions pattern), whereas the control scaffolds with a nearly 10-fold higher SFE were found to be hydrophobic ($\theta$ = 115.8°). Furthermore, the SFE was calculated by built-in software taking into consideration the contact angles of water and ethylene glycol. Unlike water, ethylene glycol completely spread over both the treated and control surfaces 30 s after drop deployment, which could have impacted the overall software calculation. Based on these observations and together with the presence of surface laser-induced roughness, it could be concluded that the method used for obtaining SFE values might induce an error in the overall calculation of SFE. Thus, further optimization of the experimental procedure is needed, and using alternative methods for the estimation of surface energy shall be applied.

*3.3. Change in pH of PBS in Accordance with PCL Degradation*

As a preliminary degradation test, the fs-laser-processed PCL was immersed in PBS over 7 weeks and the change in the pH of the solution was monitored. There was an overall reduction trend from a pH of 7.3 at week 0 to a pH of 6.72 at week 7 (Figure 7). The sharp peak at week 3 indicated that a slight increase in the pH values was not considered significant. Up to this point, no morphological analyses during the degradation test have been performed; however, no significant changes are expected. Due to its slow degradation rate (potentially taking up to 4 years), it has been shown that it would take about 12 weeks of PCL in PBS to exhibit initial slight disruptions in the surface morphology [31,32]. Thus, it has been assumed that our preliminary degradation test should not have induced any alterations in the laser-induced surface modifications. Further optimization of the experiment as well as performing it by using simulated body fluid would yield more accurate results.

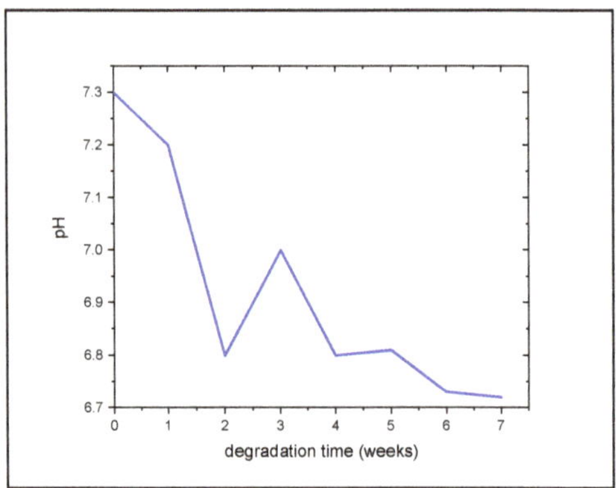

**Figure 7.** Monitoring pH change as a response of fs-laser-treated PCL degradation in PBS over 7 weeks.

### 3.4. Cytocompatibility

In order to assess the cytocompatibility of human osteoblast-like cells on fs-laser-induced microtopographies on PCL, cells were cultured for 11 days on the microchannel-patterned surface along with the non-laser treated controls (Figure 8). The laser-treated samples demonstrated significantly higher cell viability at all points during cell culture compared to the control samples. Further, SEM analysis revealed that cells showed a typical elongated morphology with cytoplasmic extensions and both cell–cell contact and cell–material interaction. Contrastingly, only a few cells with a round and disrupted morphology were seen on the non-laser treated controls. The distinct cell growth pattern and morphological behavior of the laser-treated and untreated samples are clearly seen in Figure 8e, which shows a representative image of a seeded semi-treated PCL sample.

Next, osteoblast induction with osteogenic growth factors (ascorbic acid, dexamethasone, and beta-glycerophosphate) was selectively performed on the two laser-induced topographies— the microchannels and protrusions (Figure 9). The resazurin assay revealed that cells demonstrated higher metabolic activity on protrusions compared to microchannels (statistically significant at day 3, ~20%, $p \leq 0.05$). The same modification also led to a higher production of total protein content by the cells in comparison to the microchannels (not shown). However, the ALP activity (ALP levels normalized to total protein content) was significantly higher in microchannels (at day 7, ~twice, $p \leq 0.05$). This suggests enhanced osteogenic differentiation of cells growing on the microchannels. Nevertheless, SEM analysis revealed cell adhesion and alignment along both microchannels and protrusions (Figure 9). Furthermore, induction of the cells with osteogenic factors led to the deposition of nanovesicles and extracellular matrix proteins that are visible at higher magnification, indicating the osteogenic potential of the cells in both structures.

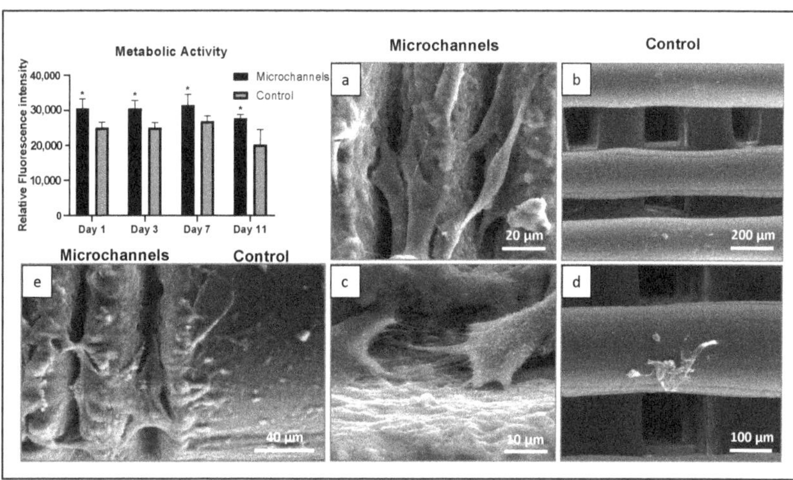

**Figure 8.** Metabolic activity (days 1, 3, 7, and 11) and SEM representative images (day 7) of human osteoblastic cells cultured over laser-induced microchannel topography on PCL (**a,c,e**) and control (**b,d,e**), in basal conditions. Scale bar: 20 μm (**a**); 200 μm (**b**); 10 μm (**c**); 100 μm (**d**), and 40 μm (**e**). * Significant difference from control (unmodified PCL), $p \leq 0.05$.

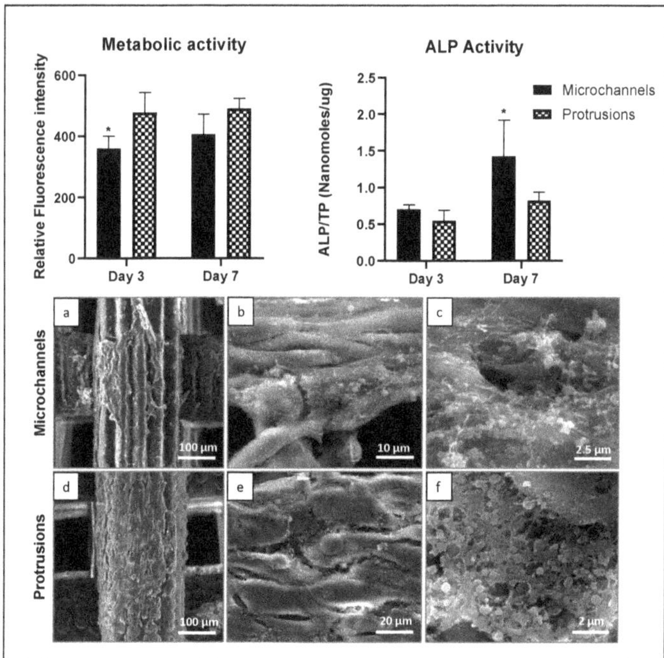

**Figure 9.** Metabolic activity (days 3 and 7), ALP activity (days 3 and 7), and SEM representative images (day 7) of human osteoblastic cells cultured over laser-induced topographies on PCL microchannels (**a–c**) and protrusions (**d–f**), in osteogenic conditions. Scale bar: 100 μm (**a,d**), 10 μm (**b**), 20 μm (**e**), 2.5 μm (**c**), and 2 μm (**f**). * Significant difference from protrusions, $p \leq 0.05$.

### 3.5. Antibacterial Assay

*S. aureus* adhesion was observed on all PCL surfaces (microchannels, protrusions, and the non-laser treated control), but colonization appeared relatively limited. Key factors that could influence bacterial colonization include the physicochemical properties of the material's surface as well as its roughness. Taylor et al. demonstrated that although an increase in surface roughness from 0.04 µm to 1.24 µm stimulated bacterial adhesion, a further rise in roughness beyond 1.86 µm resulted in reduced bacterial retention [33]. Our results are in accordance with these reports and show greater bacterial colonization on protrusions, whereas the surface roughness is below the threshold (Ra = 1.15 µm) compared to the microchannels (Ra = 5 µm) with a higher surface roughness than the threshold. On the control PCL samples, the SEM images suggested a slight increase in colonization from 6 h to 48 h (Figure 10a,b) with the cells maintaining the typical round morphology (Figure 10c) and size of ~ 0.5–1.0 µm in diameter. In contrast, bacterial cell behavior was negatively affected by the PCL laser-treated samples. At 48hrs after incubation, the number of attached cells on the treated samples was clearly lower (Figure 10e,h) than that on the control (Figure 10b). Laser-treated surfaces induced a significant disruption in the cell morphology, which was more evident in the microchannel topography (Figure 10f) compared to the protrusion surface (Figure 10i). Cells lost the typical round morphology, presenting a rough appearance and evident signs of cell lysis (Figure 10f,i). CFU analysis further revealed that the microchannels showed potential antibacterial activity compared to protrusions. The mean size of the surface topographical features has been correlated with the bacterial adhesion and colonization [34]. It has been established that when the size of the micron-scale topographies is lower than the size of bacteria, they prevent bacterial adhesion by limiting the surface area and contact points for the attachment [35]. In our study, the overall surface roughness within the channels was found between 0.14 and 0.8 µm, which is lower than the dimensions of the bacteria types tested in our study, *S. aureus* (0.5–1 µm) and *E. coli* (1–1.5 µm). This explains the greater inhibition of bacterial attachment on microchannels.

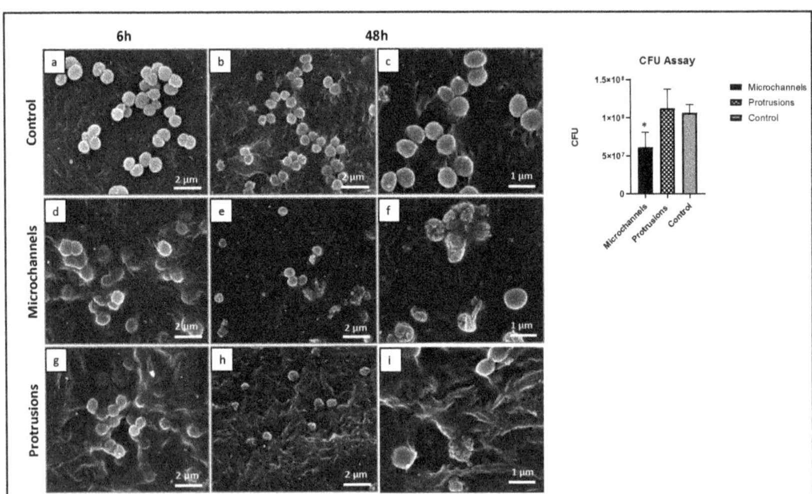


**Figure 10.** Representative SEM images (6 h and 48 h) of PCL surfaces colonized with *S. Aureus*—topographies of the control (**a–c**), laser-induced microchannels (**d–f**), and protrusions (**g–i**),. CFU assay on the three PCL surfaces. Scale bar: 2 µm (**a,b,d,e,g,h**), and 1 µm (**c,f,i**). * Significant difference from control (unmodified PCL), $p \leq 0.05$.

Considering the current visual and analytical results on surface morphology, a reason for this observation could be that the bottom of the microchannels (Figure 10e,f) appeared to have lower roughness due to the laser-induced material redistribution resulting in the partial movement of the surface layers and the subsequent cooling of the material. Thus, there would be less surface area for the bacterial cells to attach to. In comparison, the control sample and the one bearing microprotrusions appeared to have a rougher and more-caressed surface morphology, which could provide for more cells to adhere to. Furthermore, another factor that could influence the adhesion of bacteria is their motility. *S. aureus* is known to be a non-motile bacterium and a lack of motility could affect its successful spread and adhesion to the bottom part of the microchannels. A similar low adhesion of non-motile *P. fluorescence* has been observed by Scheuerman et al. who monitored the attachment of bacteria to microscale grooves on silicon [36].

The behavior of *E. coli* on the PCL samples was clearly different. SEM analysis revealed a significantly higher bacterial coverage compared to that of *S. aureus*. This resulted from the very high growth rate from 6 h to 48 h of incubation. At 6 h, adhesion was higher on the PCL control (Figure 11a) compared to that on the laser-treated surfaces (Figure 11e,i). In these topographies, cell adhesion was mainly observed in the protected bottom of the surface channels (Figure 11e) and on the smooth surface around the protrusions (Figure 11i). The bacteria presented with characteristic rod-like morphology and regular size (~1–2 μm long). During the incubation period, the high cell growth was associated with the formation of an extracellular matrix on all surfaces. Nevertheless, evident differences were seen in the formed biofilm on the untreated (control) and treated PCL surfaces. On the control PCL, the cells were already deeply embedded within a dense matrix at 24 h (Figure 11c) and partially buried at 48 h (Figure 11d). In contrast, the cells growing on the lasered surfaces formed a loose fiber reciprocal connection and were seen as individualized, being unable to produce a continuous sticky matrix (Figure 11g,h,k,l). Furthermore, anomalies in the cell division pattern were noted with the formation of long bacterial chains (Figure 11l).

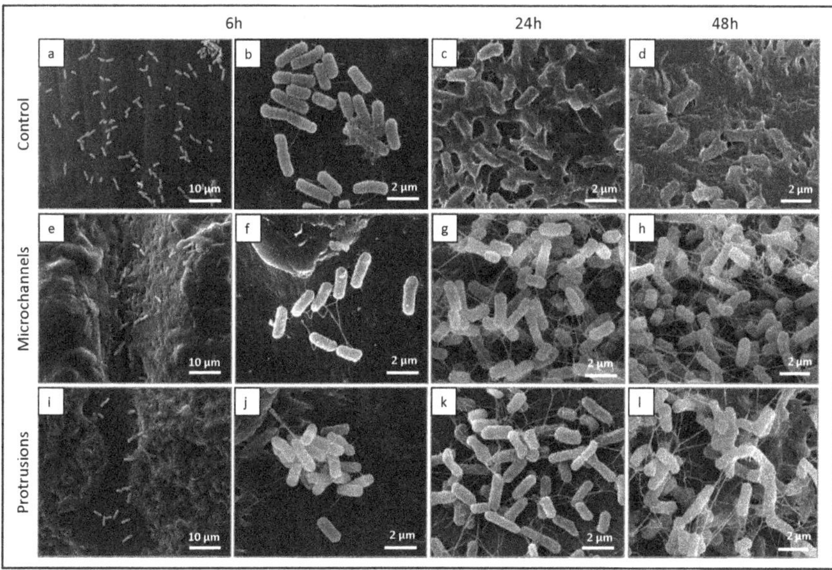

**Figure 11.** Representative SEM images (6 h, 24 h, and 48 h) of PCL surfaces colonized with *E. coli*—topographies of the control (**a–d**), laser-induced microchannels (**e–h**), and protrusions (**i–l**). Scale bar: 10 μm (**a,e,i**), and 2 μm (**b–d,f–h,j–l**).

## 4. Discussion

Cellular attachment and behavior are influenced by both the surface topography as well as the wettability of the scaffolds. Polycaprolactone (PCL) is an FDA-approved, hydrophobic polymer, which has been reported to have poor cell attachment. Our study demonstrated that femtosecond-laser-derived microchannels on the surface of PCL rendered the scaffold more hydrophilic by reducing the water contact angle from 115.8° to 82.3°, thus enhancing cellular adhesion and proliferation. These results are in accordance with previous studies reporting the enhanced laser-induced wettability of PCL and its correlation with cellular adhesion [24,37]. As an addition to its surface hydrophilicity, another important aspect of the behaviour of the PCL scaffold when in contact with the external environment is its stability over time and more specifically the endurance of the surface modifications. Due to the polymeric nature of the PCL, the resulting fs-surface modifications on the 3D constructs would remain solid and permanent as they do not change their morphology over time when stored at room temperature. The colonization of the surface with cells participating in the tissue repair process would take on average between 5 and 15 days, and it has been observed that substantial changes in the surface morphology of PCL in phosphate buffer saline (PBS) occur after 72 weeks. Thus, it could be stated that the laser-induced surface modifications would remain stable long after the initiation of tissue regeneration [32,38].

We generated femtosecond-laser-induced microstructures on PCL such as microchannels and microprotrusions, which promoted the directional and guided growth of osteoblast-like cells. We observed a higher degree of cell attachment and alignment along the microchannels with the enhanced osteogenic potential of cells (ALP activity). We further observed signs of differentiation including extracellular secretions on the microstructured PCL scaffolds indicating the osteogenic potential of the cells over the PCL scaffolds. These results are supported by past studies reporting enhanced attachment and guided cell growth along microchannels for various cell types [39–41]. A possible mechanism of microstructure-induced cell growth is postulated to be the increased surface roughness that supports protein adsorption and thus cellular adhesion [42]. Our analyses showed that the Ra values for the bottom of the laser-derived microchannels ranged between 0.14 and 0.8 μm. Andrukhov et al. explored the effects of microroughness on the osteogenic potential of MG63 cells [43]. This study indicated that titanium surfaces with an Ra of 1 μm enhanced the expression of ALP and osteocalcin, whereas Ra values of more than 2 μm reduced both the cell adhesion and differentiation. Our observations were also in accordance with the findings of Faia–Torres et al., whose study demonstrated that PCL membranes with an Ra of ~0.93 μm improved the levels of ALP expressed by human mesenchymal stem cells [44]. In the future, validation studies employing the proposed experimental methodology can be performed on human mesenchymal stem cells with a detailed analysis of gene expression.

Several cellular studies in the past have been performed on femtosecond-laser-induced microstructures on biodegradable polymers indicating guided cell growth and even differentiation into different lineages in the absence of inducing factors [40,45]. Yeong et al. demonstrated a high degree of alignment and proliferation for C2C12 mouse myoblast cells along the microchannels generated by femtosecond-laser on a poly(L-lactide-co-epsilon-caprolactone) copolymer [39]. Further studies on this co-polymer with human mesenchymal stem cells revealed an upregulation of the expression of myogenic genes [46]. Studies on other biopolymers such as collagen, gelatine, and elastin also demonstrated preferential cell proliferation and migration along laser-irradiated micropatterns [47]. Femtosecond-laser ablation has also been used as an approach for confining the growth of cells. Studies with mouse embryonic cell cultures revealed that fs-laser-ablated PCL-gelatine blends led to the confined growth of cells in the microwells [48]. This strategy has also been exploited for vascular tissue engineering for containing the growth of smooth muscle and endothelial cells along tubular PCL scaffolds [49].

In addition to enhanced cytocompatibility, the results also pointed to the antibacterial potential of the femtosecond-laser PCL surfaces. It is well-established that the adhesion, spreading, and growth of both eukaryotic and prokaryotic cells are greatly affected by surface topography. However, the responses to how they sense topographical features are distinct. Eukaryotic cells have the ability to expand the cytoplasm by adapting their morphology to the underlying surface thus retaining their shape, compared to bacteria that have a characteristic shape and a limited capacity to deform. This hinders bacteria–surface interaction and particular topographical features may further limit bacterial sensing ability, preventing bacterial adherence [50]. Antibacterial activity associated with topographical features is advantageous for its long-term effects as it retains its anti-adhesive effects and/or elicits mechano-inhibitory effects after bacteria attachment, i.e., morphological abnormalities and cell lysis [51]. Femtosecond-laser-induced microtopographies have been reported to influence bacterial cell attachment, colonization, and biofilm formation for both Gram-positive and Gram-negative bacteria [52–54]. Furthermore, the type of laser-induced pattern (such as pillar-like, honeycomb, etc.) has also been demonstrated to influence reduced bacterial attachment and confined entrapment [55–57]. In agreement with this, the present results on PCL surfaces clearly suggest that the femtosecond-laser-induced topographies, in particular the microchannel's patterned surface, had a negative effect on the behavior of *S. aureus*, the most common Gram-positive bacteria associated with biofilm formation in orthopaedic infections. The behavior of *E. coli*, which was occasionally involved in these infections, was also clearly affected as the bacteriostatic effect of the scaffolds was less profound on *E. coli* and they showed reduced sensitivity to microtopography. On the laser-treated surfaces, sessile (adhered) bacteria lost the ability to form a sticky and dense polymeric matrix preventing the establishment of a stable biofilm, which is the most relevant 3D structure for bacteria to evade the host defence mechanisms [58]. Nevertheless, within the observed inhibitory effects, significant differences were noted in the behavior of the two species on the PCL surfaces, namely a high proliferation growth rate of *E. coli* compared to that of *S. aureus*. Previous studies also reported that bacterial responses to underlying surface topography are highly species- and strain-dependent, including in femtosecond-laser-modified surfaces [52,54]. These differences in the antibacterial activity of *S. aureus* and *E. coli* are related to the structure and composition of the cell wall in Gram-positive and negative bacteria. *S. aureus* has a thick, rigid peptidoglycan layer. In contrast, *E. coli*, which is a Gram-negative bacterium, has a thin peptidoglycan layer with an additional outer membrane offering extra resistance to the bacteria. In contrast to the bacteria, the cell membrane of mammalian cells is highly fluidic and can adapt easily to micron-scale topographies allowing cytoplasmic spreading and migration [59]. Cells attach to the substrate using integrin proteins and migrate and then as probing along a surface using filopodia and lamellipodia, which extend to the micron scale. Increased surface roughness provides greater focal adhesion points leading to higher migration and proliferation. This explains the differential responses of bacterial and mammalian cells (including osteoblastic cells) toward surface topography. Overall, the contrasting responses of the bacteria to the laser-processed surfaces are determined by various cell-specific factors. Some of these might include the intrinsic structural, biochemical, and metabolic differences between Gram-positive (*S. aureus*) and Gram-negative (*E. coli*) bacteria that, together with the distinct size and morphology of the two species, determine the specific bacteria/surface interactions during adhesion and subsequent proliferation.

## 5. Conclusions

In this study, we demonstrated a novel approach to patterning the surface of 3D polycaprolactone scaffolds by femtosecond-laser with the aim of developing distinct types of topographies—microchannels and microprotrusions. The parallel microchannels allowed the successful guidance and enhancement of the osteogenic potential of MG63 cells. In combination with the improved cytocompatibility, the same microtopography showed strong antibacterial effects against *S. aureus*. By developing a biodegradable scaffold that has the potential to simultaneously promote bone tissue regeneration while preventing bacterial biofilm formation, we become a step closer to overcoming the current problems in bone tissue engineering.

**Author Contributions:** Writing, E.F.; original draft preparation, E.F.; investigation, E.F.; writing, L.A.; original draft preparation, L.A.; methodology, S.V.; original draft preparation, S.V.; writing, S.V.; methodology, M.H.F.; writing, M.H.F.; supervision, M.H.F.; formal analysis, M.H.F.; investigation, G.M.; formal analysis, G.M.; data curation, M.L.; investigation, M.L.; supervision, I.B.; Conceptualization, A.D.; methodology, A.D.; writing, A.D.; supervision, A.D.; project administration, A.D.; All authors have read and agreed to the published version of the manuscript.

**Funding:** This work was supported by: EUROPEAN UNION'S H2020 research and innovation program under the Marie Sklodowska-Curie Grant Agreement AIMed No. 861138; BULGARIAN NATIONAL SCIENCE FUND (NSF) under grant number No. KP-06-H48/6 (2020–2023), and H2020 FET Open METAFAST Grant Agreement No. 899673.

**Institutional Review Board Statement:** Not applicable.

**Informed Consent Statement:** Not applicable.

**Data Availability Statement:** Not applicable.

**Conflicts of Interest:** The authors declare no conflict of interest.

# References


1. Matassi, F.; Botti, A.; Sirleo, L.; Carulli, C.; Innocenti, M. Porous Metal for Orthopedics Implants. *Clin. Cases Miner. Bone Metab.* **2013**, *10*, 111–115. [PubMed]
2. Hussain, M.; Askari Rizvi, S.H.; Abbas, N.; Sajjad, U.; Shad, M.R.; Badshah, M.A.; Malik, A.I. Recent Developments in Coatings for Orthopedic Metallic Implants. *Coatings* **2021**, *11*, 791. [CrossRef]
3. Ulery, B.D.; Nair, L.S.; Laurencin, C.T. Biomedical Applications of Biodegradable Polymers. *J. Polym. Sci. Part B Polym. Phys.* **2011**, *49*, 832–864. [CrossRef] [PubMed]
4. Silva, M.; Ferreira, F.N.; Alves, N.M.; Paiva, M.C. Biodegradable Polymer Nanocomposites for Ligament/Tendon Tissue Engineering. *J. Nanobiotechnol.* **2020**, *18*, 23. [CrossRef]
5. Woodruff, M.A.; Hutmacher, D.W. The Return of a Forgotten Polymer—Polycaprolactone in the 21st Century. *Prog. Polym. Sci.* **2010**, *35*, 1217–1256. [CrossRef]
6. Jahani, H.; Jalilian, F.A.; Wu, C.; Kaviani, S.; Soleimani, M.; Abbasi, N.; Ou, K.; Hosseinkhani, H. Controlled Surface Morphology and Hydrophilicity of *Polycaprolactone* toward Selective Differentiation of Mesenchymal Stem Cells to Neural like Cells. *J. Biomed. Mater. Res. A* **2015**, *103*, 1875–1881. [CrossRef]
7. Dash, T.K.; Konkimalla, V.B. Poly-ε-Caprolactone Based Formulations for Drug Delivery and Tissue Engineering: A Review. *J. Control. Release* **2012**, *158*, 15–33. [CrossRef]
8. Bartnikowski, M.; Dargaville, T.R.; Ivanovski, S.; Hutmacher, D.W. Degradation Mechanisms of Polycaprolactone in the Context of Chemistry, Geometry and Environment. *Prog. Polym. Sci.* **2019**, *96*, 1–20. [CrossRef]
9. Lu, L.; Zhang, Q.; Wootton, D.M.; Chiou, R.; Li, D.; Lu, B.; Lelkes, P.I.; Zhou, J. Mechanical Study of Polycaprolactone-Hydroxyapatite Porous Scaffolds Created by Porogen-Based Solid Freeform Fabrication Method. *J. Appl. Biomater. Funct. Mater.* **2014**, *12*, 145–154. [CrossRef]
10. Mondrinos, M.; Dembzynski, R.; Lu, L.; Byrapogu, V.; Wootton, D.; Lelkes, P.; Zhou, J. Porogen-Based Solid Freeform Fabrication of Polycaprolactone–Calcium Phosphate Scaffolds for Tissue Engineering. *Biomaterials* **2006**, *27*, 4399–4408. [CrossRef]
11. Wei, S.; Ma, J.-X.; Xu, L.; Gu, X.-S.; Ma, X.-L. Biodegradable Materials for Bone Defect Repair. *Mil. Med. Res.* **2020**, *7*, 54. [CrossRef] [PubMed]
12. Malhotra, R.; Dhawan, B.; Garg, B.; Shankar, V.; Nag, T.C. A Comparison of Bacterial Adhesion and Biofilm Formation on Commonly Used Orthopaedic Metal Implant Materials: An In Vitro Study. *Indian J. Orthop.* **2019**, *53*, 148–153. [CrossRef] [PubMed]
13. Veerachamy, S.; Yarlagadda, T.; Manivasagam, G.; Yarlagadda, P.K. Bacterial Adherence and Biofilm Formation on Medical Implants: A Review. *Proc. Inst. Mech. Eng.* **2014**, *228*, 1083–1099. [CrossRef] [PubMed]
14. Kreve, S.; Reis, A.C.D. Bacterial Adhesion to Biomaterials: What Regulates This Attachment? A Review. *Jpn. Dent. Sci. Rev.* **2021**, *57*, 85–96. [CrossRef]
15. Anselme, K.; Davidson, P.; Popa, A.M.; Giazzon, M.; Liley, M.; Ploux, L. The Interaction of Cells and Bacteria with Surfaces Structured at the Nanometre Scale. *Acta Biomater.* **2010**, *6*, 3824–3846. [CrossRef]
16. Vorobyev, A.Y.; Guo, C. Direct Femtosecond Laser Surface Nano/Microstructuring and Its Applications: Direct Femtosecond Laser Surface Nano/Microstructuring and Its Applications. *Laser Photonics Rev.* **2013**, *7*, 385–407. [CrossRef]
17. Krüger, J.; Kautek, W. Ultrashort Pulse Laser Interaction with Dielectrics and Polymers. In *Polymers and Light*; Lippert, T.K., Ed.; Advances in Polymer Science; Springer: Berlin/Heidelberg, Germany, 2004; Volume 168, pp. 247–290; ISBN 978-3-540-40471-2.
18. Toyserkani, E.; Rasti, N. Ultrashort Pulsed Laser Surface Texturing. In *Laser Surface Engineering*; Elsevier: Amsterdam, The Netherlands, 2015; pp. 441–453; ISBN 978-1-78242-074-3.

19. Siddiquie, R.Y.; Gaddam, A.; Agrawal, A.; Dimov, S.S.; Joshi, S.S. Anti-Biofouling Properties of Femtosecond Laser-Induced Submicron Topographies on Elastomeric Surfaces. *Langmuir* **2020**, *36*, 5349–5358. [CrossRef]
20. Chen, C.; Enrico, A.; Pettersson, T.; Ek, M.; Herland, A.; Niklaus, F.; Stemme, G.; Wågberg, L. Bactericidal Surfaces Prepared by Femtosecond Laser Patterning and Layer-by-Layer Polyelectrolyte Coating. *J. Colloid Interface Sci.* **2020**, *575*, 286–297. [CrossRef]
21. Jalil, S.A.; Akram, M.; Bhat, J.A.; Hayes, J.J.; Singh, S.C.; ElKabbash, M.; Guo, C. Creating Superhydrophobic and Antibacterial Surfaces on Gold by Femtosecond Laser Pulses. *Appl. Surf. Sci.* **2020**, *506*, 144952. [CrossRef]
22. Martínez-Calderon, M.; Rodríguez, A.; Dias-Ponte, A.; Morant-Miñana, M.C.; Gómez-Aranzadi, M.; Olaizola, S.M. Femtosecond Laser Fabrication of Highly Hydrophobic Stainless Steel Surface with Hierarchical Structures Fabricated by Combining Ordered Microstructures and LIPSS. *Appl. Surf. Sci.* **2016**, *374*, 81–89. [CrossRef]
23. McHale, G.; Newton, M.I.; Shirtcliffe, N.J. Dynamic Wetting and Spreading and the Role of Topography. *J. Phys. Condens. Matter* **2009**, *21*, 464122. [CrossRef] [PubMed]
24. Daskalova, A.; Ostrowska, B.; Zhelyazkova, A.; Święszkowski, W.; Trifonov, A.; Declercq, H.; Nathala, C.; Szlazak, K.; Lojkowski, M.; Husinsky, W.; et al. Improving Osteoblasts Cells Proliferation via Femtosecond Laser Surface Modification of 3D-Printed Poly-ε-Caprolactone Scaffolds for Bone Tissue Engineering Applications. *Appl. Phys. A* **2018**, *124*, 413. [CrossRef]
25. Rožić, M.; Šegota, N.; Vukoje, M.; Kulčar, R.; Šegota, S. Description of Thermochromic Offset Prints Morphologies Depending on Printing Substrate. *Appl. Sci.* **2020**, *10*, 8095. [CrossRef]
26. Daskalova, A.; Filipov, E.; Angelova, L.; Stefanov, R.; Tatchev, D.; Avdeev, G.; Sotelo, L.; Christiansen, S.; Sarau, G.; Leuchs, G.; et al. Ultra-Short Laser Surface Properties Optimization of Biocompatibility Characteristics of 3D Poly-ε-Caprolactone and Hydroxyapatite Composite Scaffolds. *Materials* **2021**, *14*, 7513. [CrossRef] [PubMed]
27. Wong, P.-C.; Song, S.-M.; Tsai, P.-H.; Nien, Y.-Y.; Jang, J.S.-C.; Cheng, C.-K.; Chen, C.-H. Relationship between the Surface Roughness of Biodegradable Mg-Based Bulk Metallic Glass and the Osteogenetic Ability of MG63 Osteoblast-Like Cells. *Materials* **2020**, *13*, 1188. [CrossRef]
28. Roach, P.; Shirtcliffe, N.J.; Newton, M.I. Progess in Superhydrophobic Surface Development. *Soft Matter* **2008**, *4*, 224–240. [CrossRef]
29. Lins, L.; Bugatti, V.; Livi, S.; Gorrasi, G. Ionic Liquid as Surfactant Agent of Hydrotalcite: Influence on the Final Properties of Polycaprolactone Matrix. *Polymers* **2018**, *10*, 44. [CrossRef]
30. Song, K.; Lee, J.; Choi, S.-O.; Kim, J. Interaction of Surface Energy Components between Solid and Liquid on Wettability, and Its Application to Textile Anti-Wetting Finish. *Polymers* **2019**, *11*, 498. [CrossRef]
31. Arakawa, C.K.; DeForest, C.A. Polymer Design and Development. In *Biology and Engineering of Stem Cell Niches*; Elsevier: Amsterdam, The Netherlands, 2017; pp. 295–314. ISBN 978-0-12-802734-9.
32. Rangel, A.; Nguyen, T.N.; Egles, C.; Migonney, V. Different Real-time Degradation Scenarios of Functionalized Poly(E-caprolactone) for Biomedical Applications. *J. Appl. Polym. Sci.* **2021**, *138*, 50479. [CrossRef]
33. Taylor, R.L.; Verran, J.; Lees, G.C.; Ward, A.J.P. The Influence of Substratum Topography on Bacterial Adhesion to Polymethyl Methacrylate. *J. Mater. Sci. Mater. Med.* **1998**, *9*, 17–22. [CrossRef]
34. Wang, Y.; Subbiahdoss, G.; Swartjes, J.; van der Mei, H.C.; Busscher, H.J.; Libera, M. Length-Scale Mediated Differential Adhesion of Mammalian Cells and Microbes. *Adv. Funct. Mater.* **2011**, *21*, 3916–3923. [CrossRef]
35. Lutey, A.H.A.; Gemini, L.; Romoli, L.; Lazzini, G.; Fuso, F.; Faucon, M.; Kling, R. Towards Laser-Textured Antibacterial Surfaces. *Sci. Rep.* **2018**, *8*, 10112. [CrossRef] [PubMed]
36. Scheuerman, T.R.; Camper, A.K.; Hamilton, M.A. Effects of Substratum Topography on Bacterial Adhesion. *J. Colloid Interface Sci.* **1998**, *208*, 23–33. [CrossRef]
37. Toosi, S.F.; Moradi, S.; Hatzikiriakos, S.G. Fabrication of Micro/Nano Patterns on Polymeric Substrates Using Laser Ablation Methods to Control Wettability Behaviour: A Critical Review. *Rev. Adhes. Adhes.* **2017**, *5*, 55–78. [CrossRef]
38. Felgueiras, H.P.; Antunes, J.C.; Martins, M.C.L.; Barbosa, M.A. Fundamentals of Protein and Cell Interactions in Biomaterials. In *Peptides and Proteins as Biomaterials for Tissue Regeneration and Repair*; Elsevier: Amsterdam, The Netherlands, 2018; pp. 1–27; ISBN 978-0-08-100803-4.
39. Yeong, W.Y.; Yu, H.; Lim, K.P.; Ng, K.L.G.; Boey, Y.C.F.; Subbu, V.S.; Tan, L.P. Multiscale Topological Guidance for Cell Alignment via Direct Laser Writing on Biodegradable Polymer. *Tissue Eng. Part C Methods* **2010**, *16*, 1011–1021. [CrossRef] [PubMed]
40. Li, H.; Wong, Y.S.; Wen, F.; Ng, K.W.; Ng, G.K.L.; Venkatraman, S.S.; Boey, F.Y.C.; Tan, L.P. Human Mesenchymal Stem-Cell Behaviour On Direct Laser Micropatterned Electrospun Scaffolds with Hierarchical Structures. *Macromol. Biosci.* **2013**, *13*, 299–310. [CrossRef]
41. Ortiz, R.; Moreno-Flores, S.; Quintana, I.; Vivanco, M.; Sarasua, J.R.; Toca-Herrera, J.L. Ultra-Fast Laser Microprocessing of Medical Polymers for Cell Engineering Applications. *Mater. Sci. Eng. C* **2014**, *37*, 241–250. [CrossRef]
42. Ji, Y.; Zhang, H.; Ru, J.; Wang, F.; Xu, M.; Zhou, Q.; Stanikzai, H.; Yerlan, I.; Xu, Z.; Niu, Y.; et al. Creating Micro-Submicro Structure and Grafting Hydroxyl Group on PEEK by Femtosecond Laser and Hydroxylation to Synergistically Activate Cellular Response. *Mater. Des.* **2021**, *199*, 109413. [CrossRef]
43. Andrukhov, O.; Huber, R.; Shi, B.; Berner, S.; Rausch-Fan, X.; Moritz, A.; Spencer, N.D.; Schedle, A. Proliferation, Behavior, and Differentiation of Osteoblasts on Surfaces of Different Microroughness. *Dent. Mater.* **2016**, *32*, 1374–1384. [CrossRef]

44. Faia-Torres, A.B.; Charnley, M.; Goren, T.; Guimond-Lischer, S.; Rottmar, M.; Maniura-Weber, K.; Spencer, N.D.; Reis, R.L.; Textor, M.; Neves, N.M. Osteogenic Differentiation of Human Mesenchymal Stem Cells in the Absence of Osteogenic Supplements: A Surface-Roughness Gradient Study. *Acta Biomater.* **2015**, *28*, 64–75. [CrossRef]

45. Lee, B.L.-P.; Jeon, H.; Wang, A.; Yan, Z.; Yu, J.; Grigoropoulos, C.; Li, S. Femtosecond Laser Ablation Enhances Cell Infiltration into Three-Dimensional Electrospun Scaffolds. *Acta Biomater.* **2012**, *8*, 2648–2658. [CrossRef] [PubMed]

46. Li, H.; Wen, F.; Wong, Y.S.; Boey, F.Y.C.; Subbu, V.S.; Leong, D.T.; Ng, K.W.; Ng, G.K.L.; Tan, L.P. Direct Laser Machining-Induced Topographic Pattern Promotes up-Regulation of Myogenic Markers in Human Mesenchymal Stem Cells. *Acta Biomater.* **2012**, *8*, 531–539. [CrossRef] [PubMed]

47. Daskalova, A.; Nathala, C.S.R.; Bliznakova, I.; Stoyanova, E.; Zhelyazkova, A.; Ganz, T.; Lueftenegger, S.; Husinsky, W. Controlling the Porosity of Collagen, Gelatin and Elastin Biomaterials by Ultrashort Laser Pulses. *Appl. Surf. Sci.* **2014**, *292*, 367–377. [CrossRef]

48. Lim, Y.C.; Johnson, J.; Fei, Z.; Wu, Y.; Farson, D.F.; Lannutti, J.J.; Choi, H.W.; Lee, L.J. Micropatterning and Characterization of Electrospun Poly(ε-Caprolactone)/Gelatin Nanofiber Tissue Scaffolds by Femtosecond Laser Ablation for Tissue Engineering Applications. *Biotechnol. Bioeng.* **2011**, *108*, 116–126. [CrossRef] [PubMed]

49. Lee, C.H.; Lim, Y.C.; Farson, D.F.; Powell, H.M.; Lannutti, J.J. Vascular Wall Engineering Via Femtosecond Laser Ablation: Scaffolds with Self-Containing Smooth Muscle Cell Populations. *Ann. Biomed. Eng.* **2011**, *39*, 3031–3041. [CrossRef] [PubMed]

50. Li, W.; Thian, E.S.; Wang, M.; Wang, Z.; Ren, L. Surface Design for Antibacterial Materials: From Fundamentals to Advanced Strategies. *Adv. Sci.* **2021**, *8*, 2100368. [CrossRef]

51. Wu, S.; Zhang, B.; Liu, Y.; Suo, X.; Li, H. Influence of Surface Topography on Bacterial Adhesion: A Review. *Biointerphases* **2018**, *13*, 060801. [CrossRef]

52. Wu, X.; Ao, H.; He, Z.; Wang, Q.; Peng, Z. Surface Modification of Titanium by Femtosecond Laser in Reducing Bacterial Colonization. *Coatings* **2022**, *12*, 414. [CrossRef]

53. Shaikh, S.; Kedia, S.; Singh, D.; Subramanian, M.; Sinha, S. Surface Texturing of Ti6Al4V Alloy Using Femtosecond Laser for Superior Antibacterial Performance. *J. Laser Appl.* **2019**, *31*, 022011. [CrossRef]

54. Schwibbert, K.; Menzel, F.; Epperlein, N.; Bonse, J.; Krüger, J. Bacterial Adhesion on Femtosecond Laser-Modified Polyethylene. *Materials* **2019**, *12*, 3107. [CrossRef]

55. Valle, J.; Burgui, S.; Langheinrich, D.; Gil, C.; Solano, C.; Toledo-Arana, A.; Helbig, R.; Lasagni, A.; Lasa, I. Evaluation of Surface Microtopography Engineered by Direct Laser Interference for Bacterial Anti-Biofouling. *Macromol. Biosci.* **2015**, *15*, 1060–1069. [CrossRef] [PubMed]

56. Yang, M.; Ding, Y.; Ge, X.; Leng, Y. Control of Bacterial Adhesion and Growth on Honeycomb-like Patterned Surfaces. *Colloids Surf. B Biointerfaces* **2015**, *135*, 549–555. [CrossRef] [PubMed]

57. Ge, X.; Leng, Y.; Lu, X.; Ren, F.; Wang, K.; Ding, Y.; Yang, M. Bacterial Responses to Periodic Micropillar Array: Bacterial Responses to Periodic Micropillar Array. *J. Biomed. Mater. Res. A* **2015**, *103*, 384–396. [CrossRef] [PubMed]

58. Arciola, C.R.; Campoccia, D.; Montanaro, L. Implant Infections: Adhesion, Biofilm Formation and Immune Evasion. *Nat. Rev. Microbiol.* **2018**, *16*, 397–409. [CrossRef] [PubMed]

59. Pacha-Olivenza, M.Á.; Tejero, R.; Fernández-Calderón, M.C.; Anitua, E.; Troya, M.; González-Martín, M.L. Relevance of Topographic Parameters on the Adhesion and Proliferation of Human Gingival Fibroblasts and Oral Bacterial Strains. *BioMed Res. Int.* **2019**, *2019*, 8456342. [CrossRef] [PubMed]

 *polymers*

*Article*

# Nanostructured Polyacrylamide Hydrogels with Improved Mechanical Properties and Antimicrobial Behavior

Elena Olăreț [1], Ștefan Ioan Voicu [1,2], Ruxandra Oprea [2], Florin Miculescu [3], Livia Butac [2], Izabela-Cristina Stancu [1] and Andrada Serafim [1,*]

[1] Advanced Polymer Materials Group, University Politehnica of Bucharest, 011061 Bucharest, Romania; elena.olaret@upb.ro (E.O.); stefan.voicu@upb.ro (Ș.I.V.); izabela.stancu@upb.ro (I.-C.S.)

[2] Faculty of Chemical Engineering and Biotechnologies, University Politehnica of Bucharest, 011061 Bucharest, Romania; ruxandra_aela@yahoo.com (R.O.); livia_butac@hotmail.com (L.B.)

[3] Faculty of Materials Science and Engineering, University Politehnica of Bucharest, 060042 Bucharest, Romania; florin.miculescu@upb.ro

* Correspondence: andrada.serafim0810@upb.ro

**Abstract:** This work proposes a simple method to obtain nanostructured hydrogels with improved mechanical characteristics and relevant antibacterial behavior for applications in articular cartilage regeneration and repair. Low amounts of silver-decorated carbon-nanotubes (Ag@CNTs) were used as reinforcing agents of the semi-interpenetrating polymer network, consisting of linear polyacrylamide (PAAm) embedded in a PAAm-methylene-bis-acrylamide (MBA) hydrogel. The rational design of the materials considered a specific purpose for each employed species: (1) the classical PAAm-MBA network provides the backbone of the materials; (2) the linear PAAm (i) aids the dispersion of the nanospecies, ensuring the systems' homogeneity and (ii) enhances the mechanical properties of the materials with regard to resilience at repeated compressions and ultimate compression stress, as shown by the specific mechanical tests; and (3) the Ag@CNTs (i) reinforce the materials, making them more robust, and (ii) imprint antimicrobial characteristics on the obtained scaffolds. The tests also showed that the obtained materials are stable, exhibiting little degradation after 4 weeks of incubation in phosphate-buffered saline. Furthermore, as revealed by micro-computed tomography, the morphometric features of the scaffolds are adequate for applications in the field of articular tissue regeneration and repair.

**Keywords:** nanostructured polyacrylamide; silver-decorated carbon nanotubes; mechanical properties; antibacterial activity

**Citation:** Olăreț, E.; Voicu, Ș.I.; Oprea, R.; Miculescu, F.; Butac, L.; Stancu, I.-C.; Serafim, A. Nanostructured Polyacrylamide Hydrogels with Improved Mechanical Properties and Antimicrobial Behavior. *Polymers* **2022**, *14*, 2320. https://doi.org/10.3390/polym14122320

Academic Editor: Alberto Romero García

Received: 13 May 2022
Accepted: 6 June 2022
Published: 8 June 2022

**Publisher's Note:** MDPI stays neutral with regard to jurisdictional claims in published maps and institutional affiliations.



## 1. Introduction

Hydrogels are used for the fabrication of scaffolds with biomedical applications due to their ability to uptake large amounts of water and their resemblance to the extracellular matrix, but, usually, their low mechanical resistance to effort limits their use as tissue substitutes. To address this issue, several approaches have been proposed, including the synthesis of interpenetrated networks (IPNs) [1,2], reinforcement with micro- or nanospecies [3], or the use of specific combinations of two or more components [4]. Owing its broad usefulness to properties such as chemical inertia, tailorable mechanical properties, high swelling degree, and optical transparency, polyacrylamide (PAAm) has been intensively investigated for various applications in the biomedical field, ranging from the fabrication of contact lenses [5] to bone tissue engineering [6,7]. Although they exhibit adjustable elasticity, PAAm hydrogels are usually purely elastic, displaying little or no dissipation of deformation energy, unlike natural soft tissue, which is viscoelastic [8]. Recent studies have exploited routes of obtaining viscoelastic PAAm hydrogels, with mechanical properties, which show a closer resemblance to native soft tissue [8,9]. A more usual approach to modifying the mechanical properties of hydrogels is represented by nanostructuring. As a natural consequence of using fillers in a polymeric matrix, the mechanical properties are also altered, usually leading to more mechanically robust material.

Nanostructuring is being increasingly researched in various domains, including the development of new materials with biomedical applications. In this respect, carbon nanostructures have gained a lot of attention due to their mechanical, electronic, and biological properties [10]. Available in various structural forms—0D (fullerenes, carbon dots), 1D (single-wall carbon nanotubes (SWCNT) or multi-wall carbon nanotubes (MWCNT)), 2D (graphene), or 3D (nanodiamonds)—carbon nanostructures offer promising potential for the development of several types of materials, with medical applications ranging from biosensors [11–13] to drug delivery systems [14–16] or structures for tissue engineering [17–19]. Among the investigated nanospecies, carbon nanotubes (CNTs) represent an appealing candidate, due to their high conductivity, high strength, biocompatibility and ease of functionalization [20]. However, as most nanospecies, their use is restricted, due to their low dispersibility, especially in aqueous media [21]. Nonetheless, the presence of functional groups [22,23] or appropriate ultrasonication treatment [24–26] has led to improved results and ultimately to materials with superior characteristics. The viscosity of the polymeric matrix [27] plays an important role in obtaining a homogeneous distribution of the nanospecies in a CNT-based composite. In highly viscous matrices, the mobility of the nanospecies is restricted, thus preventing the agglomeration and sedimentation of the nanoscale filler.

Moreover, several research studies have demonstrated that CNTs also possess strong inhibitory effects against both Gram-negative and Gram-positive microorganisms when placed in direct contact, either through interactions in aqueous solutions [28] or as surface coatings [29,30]. CNTs inhibit the bacterial growth, either due to their geometry and small dimensions, which permit the breakage of the cellular membrane resulting in cell's death [31], or by disrupting the equilibrium between the production of free radicals and antioxidant defenses, thus resulting in oxidative stress followed by cell death [32,33]. Due to their notable properties, CNTs' compatibility with the live tissue has been scrutinized, and intensive research has been performed to this end [34]. Due to their small dimensions and structural similarity to asbestos, CNT inhalation has been investigated, and their effect on the respiratory tract has been documented as by some studies as harmful [35–38]. There are studies that suggest that CNTs' cytotoxicity and overall effect on cells are strongly influenced by a series of factors, such as CNT type, surface chemistry and concentration, and exposure time [39]. However, when used in the fabrication of composites for tissue engineering, CNTs' behavior differs considerably from that of inhaled CNTs [40]. When embedded in a biocompatible matrix, the toxicity of these nanospecies is hampered, but they are able to provide composites with remarkable structural reinforcement or electro-conductive properties, which are adequate for bone [41] or nerve regeneration and repair [42], respectively. Furthermore, incorporating CNTs into polymeric matrices imprints a certain degree of antimicrobial activity on the material, depending on the ratio, type, and features of the used nanospecies, treatment time, and the characteristics of the microorganism [30,43,44]. To further increase the antibacterial activity of CNTs, various molecules are adsorbed on their surface (e.g., silver [45,46], peptides [47], enzymes [48], etc.), and the so-obtained decorated CNTs were further used to obtain composites with the ability to inhibit the activity of various microorganisms. The design of materials that, in addition to being able to repair or replace a certain tissue, also possess antimicrobial characteristics [49] or release in a controlled manner certain active species [50] is of great interest to material scientists, and extensive efforts have been made in this direction.

The present study proposes a one-pot synthesis of nanostructured viscoelastic materials with applications in the low-load bearing articular tissue regeneration and repair. Additionally, this study shows that the incorporation into the monomer-cross-linker system of the corresponding linear polymer leads to an elastic, high-performance material that is able to withstand repeated compressions. The microarchitectural features of lyophilized materials were investigated through micro-computed tomography, and the morphometric parameters were correlated with the samples' composition. The stability of the samples was investigated through the incubation in acellular media for 28 days. Furthermore, this work also investigates the antimicrobial behavior of the composites when various ratios of silver decorated CNTs are used for nanostructuring.

## 2. Materials and Methods

The monomer (acrylamide, AAm), crosslinker (*N,N*-Methylenebis(acrylamide), MBA), and carbon nanospecies (carboxyl-functionalized multiwall carbon nanotubes, CNTs) were purchased from Sigma. Polyacrylamide (PAAm), with an average molecular weight of $3.76 \times 10^5$ g/mol, was obtained in the laboratory through the free radical photopolymerization of the monomer in an aqueous solution in the presence of Irgacure 2959 (Sigma, St. Louis, MO, USA), purified through dialysis against distilled water (dH$_2$O) using dialysis membranes (SpectraPor, MWCO 12–14 kDa, Repligen, CA, USA). The polymer's synthesis and the method employed for the assessment of its viscosity and average molecular weight are described in Appendix A. Ammonium persulphate (APS, Sigma) and trietanolamine (TEA, Sigma) were used as the redox initiating system of the AAm-MBA system. Phosphate-buffered saline (PBS 0.01 M, pH 7.4, Sigma-Aldrich, St. Louis, MO, USA) was prepared according to manufacturer's instructions. For the antimicrobial tests, *Staphylococcus aureus* (ATCC 6538TM) and *Escherichia coli* (ATCC 10536TM) prepared according to the manufacturer's recommendations were used as references. Casein soyabean digest broth (TSB, Sigma) was used as dispersant for the lyophilized bacteria, while casein soyabean digest agar (TSA, Sigma) was used as culture medium.

### 2.1. Synthesis

Two series of samples were synthesized to highlight the effect of embedding linear PAAm in the classical 3D polymeric network of PAAm-MBA. For simplicity, the two series were further denominated AC (acrylamide-carbon nanotubes) and PAC (polyacrylamide-acrylamide-carbon nanotubes). The two series were prepared following the same protocol, as described below for the AC series.

Firstly, various amounts of CNTs were ultrasonicated in dH$_2$O for 90 min, as described in Table 1. The adequate amounts of monomer (acrylamide—AAm, 10% wt./vol. in the final solution) and crosslinker (*N,N*-Methylenebis(acrylamide)—MBA, AAm:MBA = 100:2 wt./wt.) were added under vigorous stirring at room temperature. After the complete dissolution of all reagents, the initiators—ammonium persulfate (APS) and triethanolamine (TEA)—were added, and the hydrogel precursor was poured into molds. After two hours at 37 °C, all hydrogels (except the samples used for establishing the polymerization efficiency) were removed from the molds and washed with large amounts of dH$_2$O for 24 h at 40 °C under gentle stirring. In the case of PAC series, the PAAm was added in the polymerization mixture maintaining a ratio of AAm:PAAm = 100:25 (wt./wt.) immediately after the dissolution of the AAm and MBA.

The PAC series was further used to establish the minimum ratio of silver that would imprint a relevant antimicrobial effect. To this end, the same synthesis protocol was followed, but the CNTs were decorated with silver ions (Ag) though the direct dispersion of the nanospecies in AgNO$_3$ aqueous solution, maintaining a CNT:Ag ratio of 100:1 (wt./wt.). For simplicity, this series will be denominated Ag@PAC below.

**Table 1.** Synthesized materials' denomination and composition of the polymerization mixture.

| Series | Composition * | AAm:PAAm, wt./wt. | CNT:AAm, wt./wt. | CNT:Ag, wt./wt. |
|---|---|---|---|---|
| AC | AC-0 | - | 0:100 | - |
| | AC-0.125 | - | 0.125:100 | - |
| | AC-0.25 | - | 0.25:100 | - |
| | AC-0.5 | - | 0.5:100 | - |
| | AC-1 | - | 1:100 | - |
| PAC | PAC-0 | | 0:100 | |
| | PAC-0.125 | | 0.125:100 | - |
| | PAC-0.25 | 100:25 | 0.25:100 | - |
| | PAC-0.5 | | 0.5:100 | - |
| | PAC-1 | | 1:100 | - |
| Ag@PAC | Ag@PAC-0 | | 0:100 | |
| | Ag@PAC-0.125 | | 0.125:100 | |
| | Ag@PAC-0.25 | 100:25 | 0.25:100 | 100:1 |
| | Ag@PAC-0.5 | | 0.5:100 | |
| | Ag@PAC-1 | | 1:100 | |

* Constant parameters: AAm concentration—10% wt./vol. in the final solution; AAm:MBA = 100:2 wt./wt.

## 2.2. Polymerization Efficiency

The polymerization efficiency was assessed though gel fraction (*GF*, %), as described in [2]. Immediately after synthesis, samples of each composition were dried in the oven at 37 °C for 24 h and weighted ($w_i$). Then, they were thoroughly washed with dH$_2$O for 48 h, at 37 °C under mild stirring, dried, and weighted again ($w_f$). *GF* was determined using Equation (1):

$$GF,\ \% = \frac{w_f - w_i}{w_i} \times 100 \tag{1}$$

where $w_i$ is the initial mass of the dried samples as they result from the crosslinking process and $w_f$ is the mass of the dried samples after extraction in dH$_2$O at 37 °C for 48 h.

## 2.3. Water Uptake Ability

The swelling ability of the hydrogels was assessed following the protocol described in [51]. Briefly, samples with a diameter of 6 ± 0.2 mm and a height of 10 ± 0.5 mm were incubated in plastic tubes containing 15 mL of dH$_2$O and placed in the oven for 24 h, at 37 °C. The hydration equilibrium was considered achieved when three identical consecutive measurements performed at relevant timepoints (2 h) were obtained. The tests were performed in triplicate for each composition. The swelling degree (*SD*, %) was determined gravimetrically using Equation (2):

$$SD,\ \% = \frac{w_f - w_0}{w_0} \times 100 \tag{2}$$

where $w_0$ denotes the initial mass of the samples prior to incubation and $w_f$ is the mass of the equilibrium-hydrated samples. Before the test, the samples were purified in large amounts of water for 24 h and subsequently dried in the oven, at 37 °C, for 24 h.

## 2.4. Investigation of the Morphometric Parameters

The morpho-structural characteristics of the synthesized hydrogels were assessed using micro-computed tomography (micro-CT) performed on lyophilized samples. To this end, samples of both series were hydrated at equilibrium and subsequently lyophilized using Martin Christ equipment (−80 °C, 48 h, under vacuum). After freeze drying, the materials were kept at room temperature in sealed containers to avoid humidity. No changes in the shape and dimensions of the samples were noticed during storage. A SkyScan 1272 high-resolution X-Ray micro-tomograph (Bruker MicroCT, Belgium) was used to assess the morphometric parameters of the porous materials. The projections were recorded at a camera binning of $1 \times 1$ with a voltage of 50 kV and an emission current of 130 µA. All samples were scanned at a pixel size of 1.25 µm with 4 average frames at every $0.1°$ angle step. The projections were reconstructed using the NRecon software (version 1.7.1.6, Bruker, Kontich, Belgium), and the obtained cross-sections were further used to obtain 3D images of the scanned samples using CTVox software (version 3.3.or1403, Bruker, Kontich, Belgium). The porosity measurements were performed maintaining the same volume of interest for all samples, using the CTAnalyzer software (version 1.18.4.0+, Bruker, Kontich, Belgium).

### 2.5. Mechanical Properties

The mechanical characteristics of AC and PAC series were assessed through uniaxial compression tests performed using a CT3 Texture analyzer (Brookfield Engineering Laboratories Inc., Middleboro, MA, USA) equipped with a 4500 g cell load and a TA4/1000 compression accessory. Cylindrical samples of each composition (d = $10 \pm 0.5$ mm, h = $6 \pm 0.2$ mm), hydrated to equilibrium, were placed on the bottom plate of the equipment, and the upper plate was lowered up to a certain deformation. Ten cyclic loading-unloading tests were performed up to a deformation of 50% on the same sample immediately after the initial loading at a crosshead speed of 0.1 mm/s to determine the elasticity and deformation of the materials during subsequent exposure to effort. The stress–strain curve was plotted using the dedicated software (TexturePro CT V1.8 Build 31). The stress was read in the linear region at 2% deformation on the loading curve of the first cycle and the compression modulus ($E'$, kPa) was computed using Equation (3). Hysteresis was also computed using Equation (4) at 40% deformation in the first and last loading–unloading cycle. Single loading tests were performed with a crosshead speed of 1 mm/s to determine the ultimate compression stress the material can withstand before breakage.

$$E', kPa = \frac{\sigma}{\varepsilon} = \frac{\frac{F}{A}}{\frac{\Delta L}{L}} \tag{3}$$

where

$E'$ = compression modulus, (kPa)
$\sigma$ = applied compressive stress, (kPa)
$\varepsilon$ = strain
$F$ = applied compressive force, (kN)
$A$ = samples' area, (m$^2$)
$\Delta L$ = compressed length, (m)
$L$ = original length of the sample, (m)

$$H, \% = \frac{\sigma_{up} - \sigma_d}{\sigma_{max}} \times 100 \tag{4}$$

where

$H$ = hysteresis, (%)
$\sigma_{up}$ = the value of stress read on the loading curve, (kPa)
$\sigma_d$ = the value of stress read on the unloading curve, (kPa)
$\sigma_{max}$ = the maximum value of stress, (kPa)

### 2.6. Stability in Acellular Medium

Hydrogels' stability was investigated in acellular conditions using as incubation medium phosphate-buffered saline (PBS), as described in [52]. Briefly, three samples of each composition (d = $10 \pm 0.5$ mm, h = $6 \pm 0.2$ mm) were kept for 4 weeks in individual test tubes containing 15 mL of PBS, at a constant temperature (37 °C) on an orbital rotator

(IKA KS 4000) at 50 rpm. The incubation medium was refreshed every two days throughout the experiment. The remaining mass (*RM*, %) was calculated using Equation (5):

$$RM, \% = \frac{w_f}{w_i} \times 100 \tag{5}$$

where $w_i$ is the weight of dried specimen before incubation in PBS and $w_f$ is the weight of dried specimen after incubation.

### 2.7. Antimicrobial Activity

The antibacterial activity of the compositions was determined through indirect contact, as described in [53]. Briefly, bacterial suspensions of *E. coli* and *S. aureus* (concentration of colonies of $10^3$ CFU/mL) were placed in contact with samples of each composition. The positive control was evaluated by bacterial suspension incubation in polypropylene sterile tubes, in the same conditions as analyzed samples (without contact with any other material). Following 24 h of incubation at 30–35 °C, PBS was added over the samples, and the aliquot was transferred to Petri dishes and spread onto fresh agar plates to acquire images and count the colonies. The samples were incubated for 3 days at 30–35 °C. The antimicrobial activity was computed using Equation (6):

$$AA, \% = \frac{N_i - N_c}{N_i} \times 100 \tag{6}$$

where

$AA$ represents the antibacterial activity;
$N_i$ represent the initial number of bacterial colonies (1000 CFU/mL);
$N_c$ represents the obtained number of bacterial colonies.

### 2.8. Statistical Analyses

All experiments were conducted in triplicate ($n = 3$), and results are expressed as mean $\pm$ standard deviation. Statistical relevance was performed using GraphPad Prism Software 6.0 (GraphPad Software Inc., San Diego, CA, USA), one-way ANOVA method, Bonferroni post-test, and differences were considered statistically significant for $p < 0.05$.

## 3. Results and Discussions

The present paper discusses the possibility of modulating the mechanical performances of PAAm hydrogels using two different approaches simultaneously: on one hand, the nanostructuring of the polymeric matrix with CNTs, and on the other hand, the addition of the corresponding linear polymer in the hydrogel precursor mixture. While the first route would lead to stiffer materials, it is expected that the second one would imprint a certain elasticity on the scaffolds. In addition to tailoring the mechanical properties of the classical MBA-cross-linked PAAm hydrogels through the addition of linear PAAm and CNTs nanostructuring, the study also aimed at establishing the lowest Ag:CNTs ratio that imprints a relevant antimicrobial behavior on the obtained scaffolds.

The one-pot synthesis of the materials consisted of two preparatory steps: (1) first, the CNTs dispersion/Ag decoration, performed through the sonochemical method, followed by (2) the addition of the polymer matrix components (monomer, cross-linker, and linear polymer) and polymerization initiators (Figure 1A). For the Ag@PAC series, during the first step, the cavitation bubbles generated by the ultrasonical field enabled the decoration of the CNTs [54] with Ag ions from the AgNO$_3$ aqueous solution. A notable aspect regarding the designed materials is represented by the low ratio of both CNTs in the system (a maximum of 1% weight ratio with respect to the monomer) and low ratio of metallic ions (only 1% Ag weight ratio with respect to the CNTs).

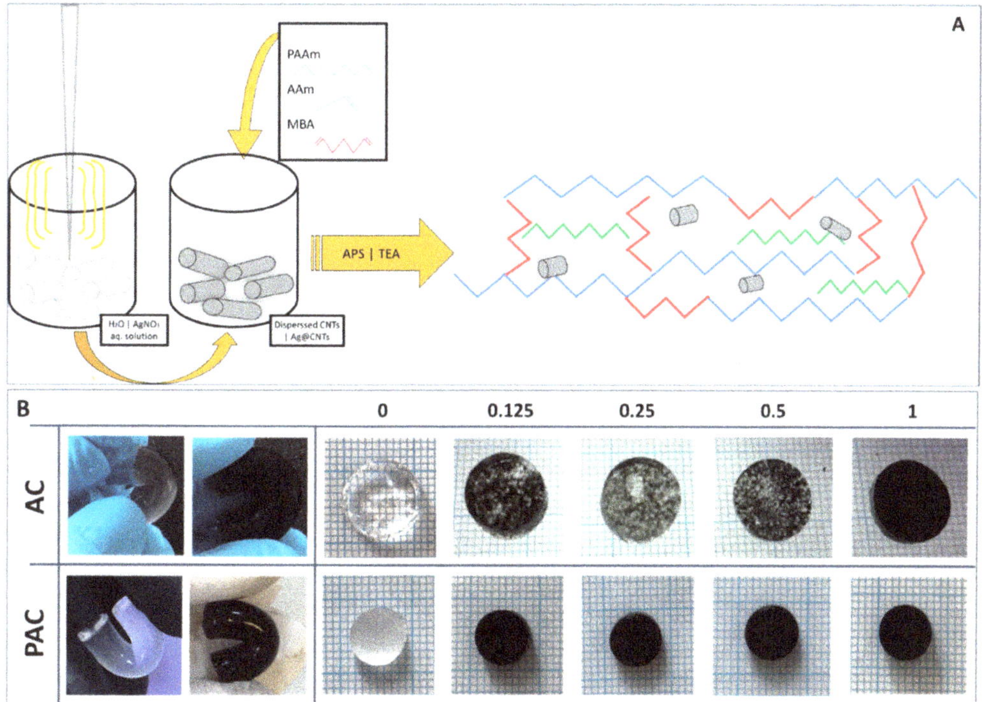


**Figure 1.** Panel (**A**): schematical depiction of the synthesis process of the polyacrylamide-based hydrogels reinforced with Ag-decorated CNTs; Panel (**B**): digital images of the obtained hydrogels at hydration equilibrium.

The resulting hydrogels are elastic and smooth. Figure 1, panel B, presents aggregates of CNTs observed in the majority of AC compositions. The addition of the linear polymer leads to hydrogels with a homogeneous appearance, which may be assigned to an improved dispersion of the CNTs in the polymeric matrix (PAC series in Figure 1B). These data support the hypothesis that PAAm acts as dispersion stabilizer.

PAAm was previously used to enhance the elasticity of PAAm-MBA hydrogels in order to better resemble natural tissues [8,9]. To the best of our knowledge, this paper is the first to report the employment of this synthesis method to obtain composite materials.

### 3.1. Polymerization Efficiency

The polymerization efficiency of the materials was determined through gel fraction analysis. Values above 84% were registered for the PAC series and above 90% for the AC compositions (Figure 2A). The values registered for PAC series are slightly lower than the ones registered for the classically synthesized AC series. This behavior may be attributed to the high content of linear polymer added in the system (25% with respect to the monomer), which was not completely trapped in the 3D network of the PAAm-MBA hydrogel. Considering that the linear PAAm amounts for about 20% (wt./wt.) of the total solid content of the polymerization mixture, the GF values confirm the presence of the linear polymer in the final purified material. Additional details regarding the stability of the purified scaffolds are presented in "Section 3.5. Stability in simulated physiologic conditions"

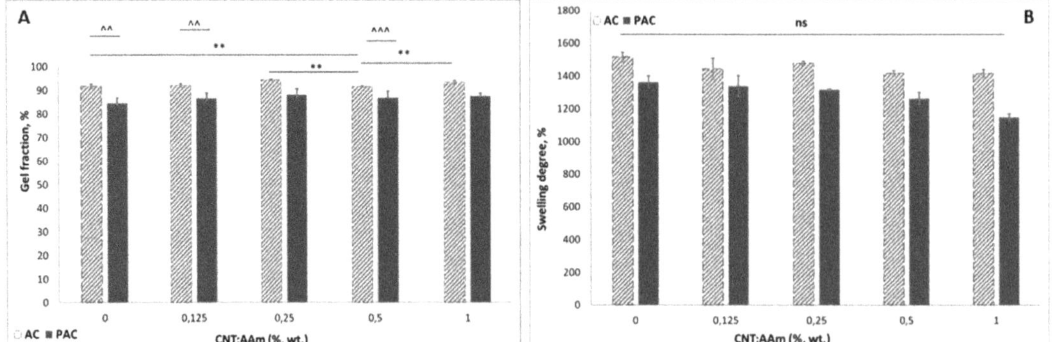

**Figure 2.** Gel fraction (**A**) and swelling degree (**B**) values computed for the AC (patterned columns) and PAC (full color columns) series. Statistical significance: ^^, ** $p < 0.01$, ^^^ $p < 0.001$, ns—not statistically significant.

### 3.2. Water Uptake Ability

The swelling in $dH_2O$ was assessed through the classical gravimetric method, as described in the methods section. Although at low CNT fillings (0.125 and 0.25% CNTs with respect to the AAm content) there are no notable differences between the neat hydrogel and the composites, at filling ratios above 0.5% CNTs, a slight decrease in the SD value can be observed in both AC and PAC series (Figure 2B). This behavior can be attributed to the presence of CNTs above a critical filling ratio; the presence of the CNTs in the network's spaces obstructs water accumulation. For example, the SD value of AC-0 is $1515 \pm 30\%$, while for AC-1, the SD reaches $1148 \pm 23\%$. Furthermore, some differences might also be noticed between the SD values of the two series. The SD decrease in the PAC series when compared to the AC series can also be assigned to an increase in the total solid content in the composition. Although not statistically significant, the addition of CNTs impacts the PAC water affinity when compared to the AC series: SD of AC-1 is 6.5% lower than AC-0, while the SD of PAC-1 is almost 16% lower than the neat semi-IPN hydrogel. This behavior shows that the water absorption was additionally reduced by the presence of the CNTs in the PAC network.

### 3.3. Architectural Characteristics

The morphometric parameters of the hydrogels were assessed on porous samples resulting from freeze drying, using a micro-CT scanner at a resolution of 1.25 µm. The images were visualized as 3D objects, and the pores were color-coded yellow to red according to their dimensions, while the walls were kept on a grey scale. Structure separation was quantitatively evaluated, and the results were graphically represented as bar charts (Figures 3 and 4). The morphometric parameters (Table 2) were obtained for the same volume of interest (VOI) and number of layers, maintaining the upper and lower grey threshold constant for all samples. The graphical representation presented in Figures 3 and 4 shows the percentage of pores space within different range values in the VOI.

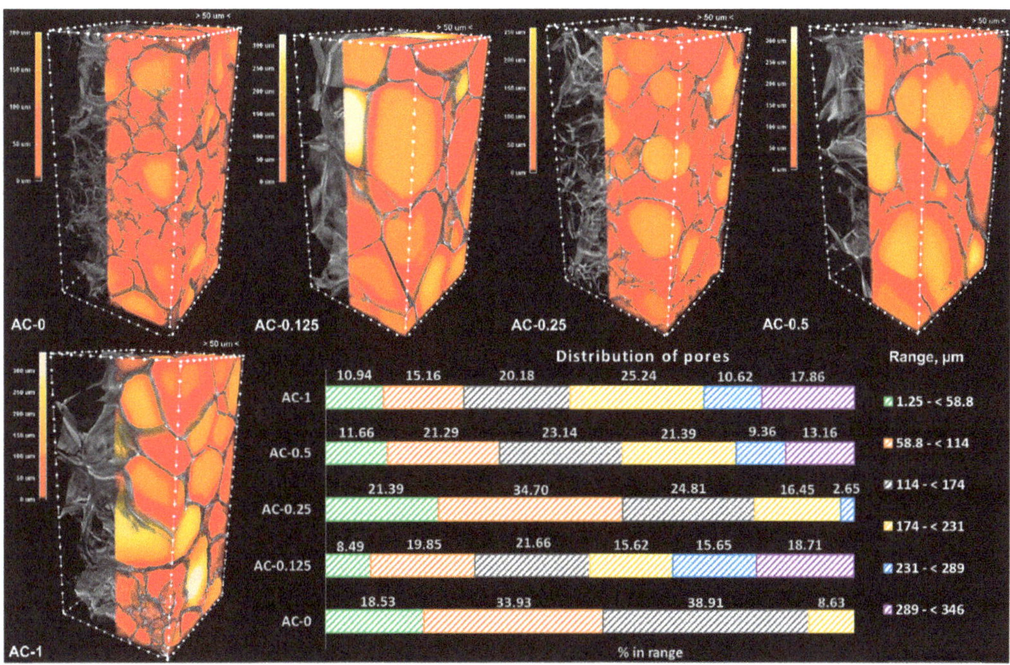

**Figure 3.** Three-dimensional images of the AC samples and graphical representation of the porosity measurements.

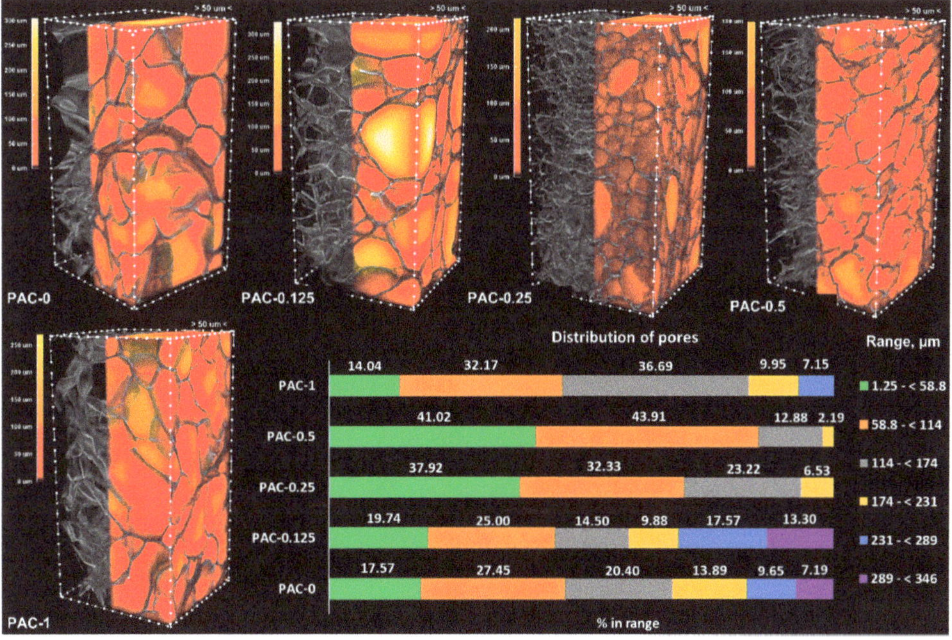

**Figure 4.** Three-dimensional images of the PAC samples and graphical representation of the porosity measurements.

**Table 2.** Morphometric parameters registered through micro-CT.

| Sample | Specific Surface Area, mm$^{-1}$ | Total Porosity, % | Open Porosity, % | Closed Porosity, % |
|---|---|---|---|---|
| AC-0 | 437.10 | 92.584 | 92.584 | 0.000 |
| AC-0.125 | 289.33 | 91.985 | 91.983 | 0.020 |
| AC-0.25 | 456.90 | 92.818 | 92.818 | 0.000 |
| AC-0.5 | 384.21 | 94.607 | 94.607 | 0.000 |
| AC-1 | 310.30 | 91.438 | 91.440 | 0.024 |
| PAC-0 | 335.98 | 89.112 | 89.109 | 0.027 |
| PAC-0.125 | 345.72 | 89.094 | 89.094 | 0.000 |
| PAC-0.25 | 382.09 | 86.189 | 86.188 | 0.002 |
| PAC-0.5 | 441.45 | 88.894 | 88.894 | 0.000 |
| PAC-1 | 384.99 | 91.782 | 91.782 | 0.005 |

Both series have a total porosity of around 90%, consisting mostly of opened pores. The addition of the linear PAAm in the system leads to an overall decrease in porosity (for example, in the case of the neat hydrogels, the total porosity decreases from 92.6% for AC-0 to 89.1% for PAC-0), but the addition of the CNTs cannot be correlated with a clear trend of the morphometric parameters. However, the fact that the CNTs were added in low amounts must not be overlooked (the maximum amount of CNTs is 1% with respect to the monomer, the equivalent of 0.1% (wt.) in the final system). Moreover, in the case of the AC series, the results registered for the specific surface area could not be correlated with the composition, probably due to the poor distribution of the CNTs within the polymer matrix.

As depicted in Figure 3, the control sample (AC-0) is lacking pores greater than 231 µm, with most of them (over 52%) being in the range of 1.25–114 µm. The addition of nanoparticles in the system leads to an increase in the pores' dimensions and to the appearance of larger ones, with dimensions in the interval 231–346 µm; the sample with the highest content of CNTs has the majority of pores (53.72%) in the interval 174–346 µm.

For the PAC series, the results indicated an opposite trend, with the neat hydrogel exhibiting larger pores when compared to the nanocomposites. This behavior indicates that the addition of the linear PAAm in the system has a greater influence when compared to the addition of CNTs, especially since the PAAm content is much higher (25-fold when compared to the maximum load of CNTs). As presented in Figure 4, the samples with the lowest pore dimensions are PAC-0.5 and PAC-0.25, which present only pores lower than 231 µm, with the vast majority (over 85% for PAC-0.5 and 70% for PAC-0.125) lower than 114 µm.

Materials designed for articular cartilage tissue regeneration and repair require open porosity, which is useful for scaffolds' colonization with chondrocytes. These are specialized cells found in the articular cartilage with dimensions in the range of 7–30 µm [55]. However, studies have shown that increasing both porosity and pore size leads to improved cell viability [56,57]. As presented in Table 2, both series of materials exhibited great open porosity, but series PAC, especially compositions PAC-0.25 and PAC-0.5, have a large number of pores in the relevant porosity interval, considerably higher than their AC counterparts (patterned green for AC in Figure 3 and green for PAC in Figure 4).

The surface morphology of the samples was also investigated through scanning electron microscopy, using a FEI XL30 equipment. No noticeable differences were observed between series AC and PAC. The registered images showed that all samples have heterogeneously distributed pores, with smooth surfaces. Relevant images of the scaffolds are presented in Supplementary Figure S3.

### 3.4. Mechanical Behavior

The fatigue resistance of the materials was evaluated through cyclic loading–unloading tests performed in a quasi-static manner at a low compression speed (0.1 mm/s). As depicted in Figure 5, all samples exhibited good elasticity, as the stress decreased gradually after load removal. There were notable differences between the AC and PAC samples. Firstly, in the case of the AC series, hysteresis may be observed even at low strain values (strain 0.5), especially above a 0.5% filling ratio (wt., CNT:AAm), while in the case of PAC samples, hysteresis may be observed at slightly higher values. Even more, the hysteresis of the PAC samples is lower when compared to the values computed for the AC series (Table 3). In the first compression cycle, there are no notable differences between the hysteresis values of the neat hydrogels of the two series (11.55% for AC-0 and 11.37% for PAC). The hysteresis of the neat hydrogels is significantly altered by the addition of nanoparticles in both series. For the AC series, the hysteresis of the low-loading composites (AC-0.125 and AC-0.25) is smaller, while for the compositions with 0.5% and 1% filling ratios (AC-0.5 and AC-1, respectively), it is higher than the neat hydrogel. AC-0.5 and AC-1 behave quite differently under stress compared to the rest of the compositions of the same series, even at 10% deformation, as depicted in Figure 5. The plotted strain–stress curves show that when AC-0.5 and AC-1 are considerably stiffer, requiring increased stress to reach the same deformation as the rest of the hydrogels, the stress–strain curves registered for the other two composites are similar to the ones registered for the neat hydrogel. This behavior indicates that the minimum ratio of CNTs that has a significant impact on the mechanical properties of the AC hydrogels is 0.5% (wt.) CNT:AAm.

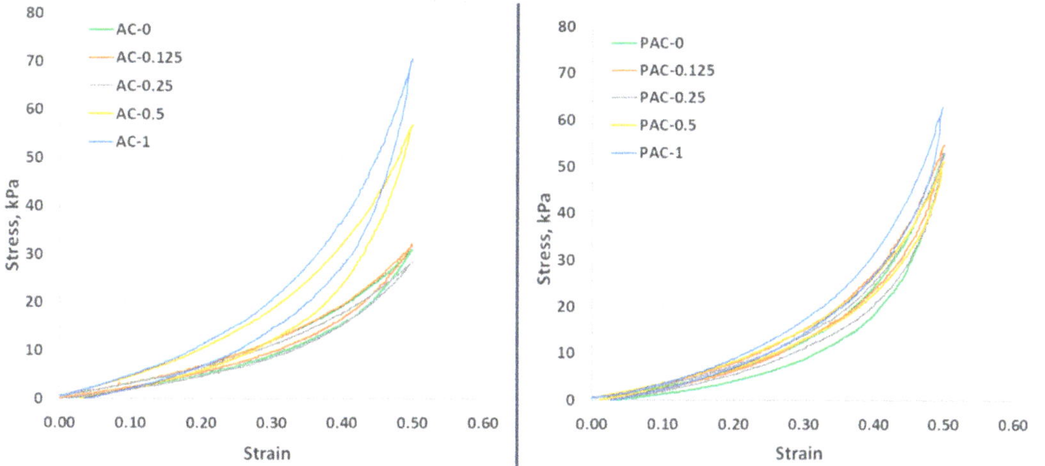

**Figure 5.** First cycle of loading–unloading compressions for the AC (**left**) and PAC (**right**) series.

**Table 3.** Mechanical parameters of the AC and PAC hydrogels.

| Sample | Hysteresis, % 1st Compression Cycle | Hysteresis, % 10th Compression Cycle | Ultimate Compression Stress, kPa | $E'$ at 2% Deformation, kPa |
|---|---|---|---|---|
| AC-0 | 11.55 | 12.15 | 212.83 ± 63.08 | 46.81 ± 3.44 |
| AC-0.125 | 8.42 | 9.49 | 142.49 ± 70.6 | 54.37 ± 8.00 |
| AC-0.25 | 8.62 | 10.06 | 150.32 ± 12.75 | 51.10 ± 4.15 |
| AC-0.5 | 14.76 | 17.49 | 285.66 ± 73.4 | 50.66 ± 5.60 |
| AC-1 | 12.66 | 12.68 | 357.53 ± 26.94 | 52.31 ± 5.00 |
| PAC-0 | 11.37 | 13.25 | * | 38.95 ± 2.79 |
| PAC-0.125 | 5.95 | 6.97 | * | 39.63 ± 6.17 |
| PAC-0.25 | 9.36 | 9.80 | * | 39.68 ± 2.96 |
| PAC 0.5 | 6.71 | 8.66 | 447.67 ± 75.97 | 42.44 ± 3.00 |
| PAC-1 | 7.37 | 7.65 | 630.81 ± 88.65 | 42.93 ± 0.69 |

* no breakage within the equipment compression limits.

In the case of the PAC series, the highest hysteresis was registered for the neat hydrogel, while the presence of CNTs lowered the hysteresis for all composite materials. The behavior under repeated compressions is also very similar for all compositions; slight differences may be observed only for PAC-1, indicating that the reinforcing effect of the nanoparticles is hindered by the presence of the linear polymer. Compared to the AC series, in which the water around the polymers chains plays the role of lubricant [58] and helps the chains to adjust their position under stress, in the PAC compositions, the linear PAAm also contributes to the polymer chains' rearrangement during compression, resulting in smaller relaxation times and therefore smaller hysteresis. Continuing the compression cycles leads to an increased hysteresis (the values of the 1st and 10th cycles are presented in Table 3). We attribute the hysteresis cycle to the relaxation time of the materials: a better dispersion of CNTs due to the presence in the initial system of PAAm leads to stiffer materials, as shown by both a reduced hysteresis and a higher breaking stress. However, there is some accumulation of irreversible deformation, as shown by the increase in hysteresis from the 1st to the 10th cycle, but also by the lower stress needed to achieve the same deformation in the 10th cycle as compared with the first (Supplementary Figure S1). For both series, the smallest differences in terms of hysteresis values were registered for the samples with the highest CNTs content: in the case of AC-0, the hysteresis increases from the first to the last compression test with only 0.2%, while in the case of PAC-1, the hysteresis increases with 3.65%.

It is important that the CNTs are homogenously distributed in the polymeric matrix. This is one of the main challenges when working with nanoparticles, and it is very hard to accomplish in low viscosity media [27], such as the precursor in the AC series. The presence of the PAAm in the initial reaction system leads to a significant increase in the viscosity, thus reducing the sedimentation/aggregation tendency of the CNTs to levels low enough to allow the polymerization/cross-linking to be completed before non-homogeneity occurs.

The ability of the materials to withstand changes in length when subjected to compressive loads was assessed through uniaxial compression tests performed at a test speed of 1 mm/s. The ultimate compression stress is considered the stress value reached by the material when it fails completely under the applied load. As presented in Table 3, compositions AC-0.125 and AC-0.25 break at considerably lower stress values compared to the neat hydrogel or the two other composite materials of the same series. This inability to withstand deformation is attributed to the poor dispersion of the nanoparticles. Although agglomerations can also be noticed in AC-0.5, the CNTs seem to have a reinforcing effect on the hydrogel, thus leading to slightly higher values of the ultimate compression stress compared to the neat composition. In this series, the only composition that exhibits a significantly higher resistance to stress compared to the control sample (AC-0) is the one with the highest filling ratio of nanoparticles (AC-1). The addition of the linear polymer (series PAC) leads to a considerably different behavior when compressed to failure. The samples with no or small amounts of CNTs (PAM-0, PAM-0.125, and PAM-0.25) reach a deformation of 85% without breaking, confirming the plasticizer effect of the linear polymer embedded in the 3D network. At higher filling ratios (PAC-0.5 and PAC-1), the effect of the nanoparticle reinforcing becomes visible, and although the materials can withstand higher deformations compared to the other compositions of the series (around 80% deformation), they break when high stress is applied (almost 450 kPa for PAC-0.5 and 630 kPa for PAC-1). The stress–strain curves are presented in Supplementary Figure S2. The compression modulus of the samples was also computed from the stress–strain curves at a deformation of 2%. The results (Table 3) indicate that at small deformations, the addition of nanoparticles does not have a significant impact on the samples' elasticity. Numerous studies present the possibility to tune the mechanical properties of composites using various CNTs loading ratios [59–63]. For example, in the system MWCNT-poly(vinyl alcohol) (PVA), the addition of 3% wt. CNTs led to an increase in Young's modulus with about 250% when compared to the neat PVA hydrogel [63]. In a similar CNTs-PAAm system, the addition of 5% wt. carboxyl-functionalized MWCNT increases Young's modulus 3-fold when compared to a

1% wt. loading [60]. However, there are also studies that present the changes in the composites' mechanical properties at the addition of low filling ratios. Lan et al. [41] showed that the addition of 0.25% wt. CNT in PVA hydrogel leads to approximately 4-fold increase of the elasticity modulus.

Our study indicates that in the AC series, the addition of small amounts of CNTs (0.125 and 0.25% wt. with respect to AAm) leads to the stiffest materials, but this behavior can be attributed to the poor dispersibility of the nanoparticles in the hydrogel precursor. At 1% (wt.) CNTs:AAm (AC-1), the elasticity modulus increases with about 10% compared to the neat hydrogel. The same increase may be observed for samples PAC-0.5 and PAC-1 compared to the PAC control sample (PAC-0). The higher values of E' indicate that the addition of PAAm in the classical PAAm-MBA system leads to more elastic materials.

### 3.5. Stability in Simulated Physiological Conditions

The stability of purified samples was investigated following incubation for 28 days in PBS, at 37 °C. The remaining mass (RM, %) was computed, and the results are depicted in Figure 6 as bar charts. The AC series is highly stable, with no mass loss. Figure 6 indicates that the presence of linear PAAm reduces the stability of the purified scaffolds, with the strongest effect (statistically significant) recorded for the sample without nanoparticles (PAC-0 registered the highest mass loss, around 14%). The lower values of the RM for the PAC series may be attributed to the solubilization of the linear polymer, immobilized in the 3D polymeric network. The lower values of mass loss (below 10%) recorded for the CNTs-loaded samples indicate that the presence of the nanospecies stabilizes the compositions.

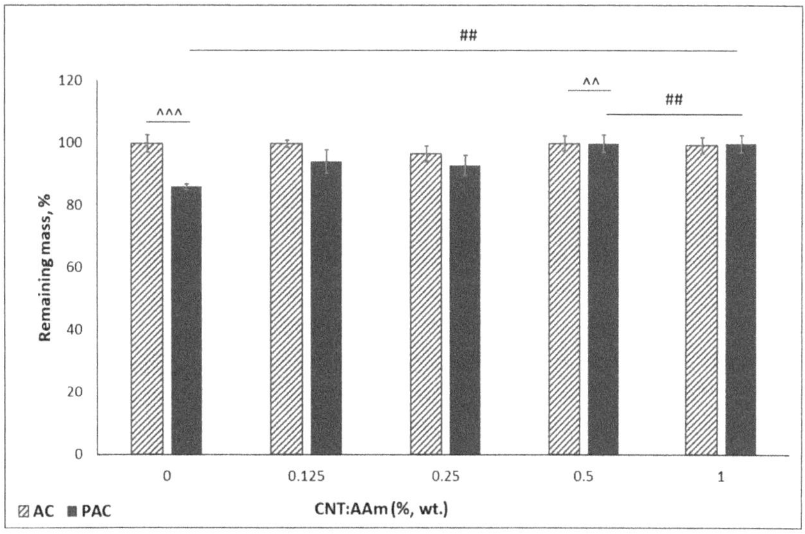


**Figure 6.** Hydrogels stability after 28 days incubation in PBS. Statistical significance: ^^, ## $p < 0.01$, ^^^ $p < 0.001$.

### 3.6. The Antimicrobial Activity

Considering the previously described results indicating that the addition of PAAm to the classical AAm-MBA precursor system leads not only to a visibly improved dispersion of the CNTs but also to more resilient materials to repeated compressions, only the PAC series was subjected to anti-microbial tests.

Due to the well-known antibacterial spectrum of Ag [46], great attention has been paid to the use of Ag-decorated CNTs as efficient antimicrobial agents, but most studies concentrate on the laborious and time-consuming preparation methods, such as the Tollens process [45] or wet impregnation followed by thermal treatment [64]. As the antimicrobial performance of the materials highly depends on the Ag content, several studies discuss the minimum loading of Ag on the CNTs surface. Seo et al. reported that a minimum of 30 µg/mL Ag-MWCNT with a Ag:MWCNT ratio of 2:1 balances both the adequate antibacterial effect and the slightest cytotoxicity when placed in direct contact with both bacteria and cells [65]. The study by Hamouda et al. concludes that 6% represents an optimal Ag loading on the MWCNT surface for efficient bactericidal performances [64]. The antibacterial mechanism of the silver ions has been largely discussed in the reported literature; briefly, the ability of Ag to kill or disrupt the activity of various bacteria is based on the generation of reactive oxygen species, which leads to the membrane disruption [66,67]. Another very important issue to consider is represented by the appropriate amount of silver that would exhibit a relevant antimicrobial effect while being cytotoxic, which differs depending on the application. According to Poon et al., concentrations of silver nitrate higher than $33.3 \times 10^{-4}$% are toxic towards monocultured fibroblasts, while concentrations higher than $60 \times 10^{-4}$% are needed to exhibit toxic behavior towards cells cultured in three-dimensional collagen gel [68]. Khansa et al. recently reviewed the use of silver in wound therapy and concluded that for silver-loaded wound dressing to be both effective and safe, it should have a concentration of silver in the range of 30–60 ppm [69]. Silver has also been employed in various concentrations in nanostructuring composite systems. Three-dimensionally printed bioceramic scaffolds modified with silver-decorated graphite oxide have shown, in addition to osteogenic activity, excellent antimicrobial behavior at concentrations over 3.5 ppm [70].

The antimicrobial activity of the PAC materials was assessed using silver decorated CNTs (Ag@CNTs) as fillers, while maintaining the synthesis protocol, as previously described. The ratio Ag:CNT was maintained 1:100 in all samples, resulting in a maximum of $78.4 \times 10^{-4}$% wt./wt. in PAC-1 and half that amount in PAC-0.5. The registered results (Figure 7 and Table 4) show an efficiency of over 95% for Ag@PAC-0.5 and Ag@PAC-1 when tested against the Gram-negative bacteria (*E. coli*). Better results were obtained in the case of Gram-positive bacteria (*S. aureus*), where a Ag@CNT loading ratio of only 0.25% (wt.) with respect to AAm was sufficient for a 50% efficiency. Higher Ag@CNT loading led to higher antimicrobial efficiencies of 98% (for Ag@PAC-0.5) and 100% (for Ag@PAC-1).

**Table 4.** Silver content (% wt./wt. with respect to the total solid content) and antibacterial efficiency of the PAC and Ag@PAC series against Gram-negative (*E. coli*) and Gram-positive (*S. aureus*) samples.

| Sample | Silver Content, % wt./wt. | Antibacterial Activity, % | |
|---|---|---|---|
| | | *E. coli* | *S. aureus* |
| control | - | 5 | 5 |
| PAC-0.125 | - | 5 | 5 |
| PAC-0.25 | - | 15 | 5 |
| PAC-0.5 | - | 10 | 5 |
| PAC-1 | - | 10 | 10 |
| Ag@PAC-0 | 0 | 25 | 15 |
| Ag@PAC-0.125 | $9.8 \times 10^{-4}$ | 25 | 25 |
| Ag@PAC-0.25 | $19.6 \times 10^{-4}$ | 35 | 50 |
| Ag@PAC-0.5 | $39.2 \times 10^{-4}$ | 95 | 98 |
| Ag@PAC-1 | $78.4 \times 10^{-4}$ | 99 | 100 |

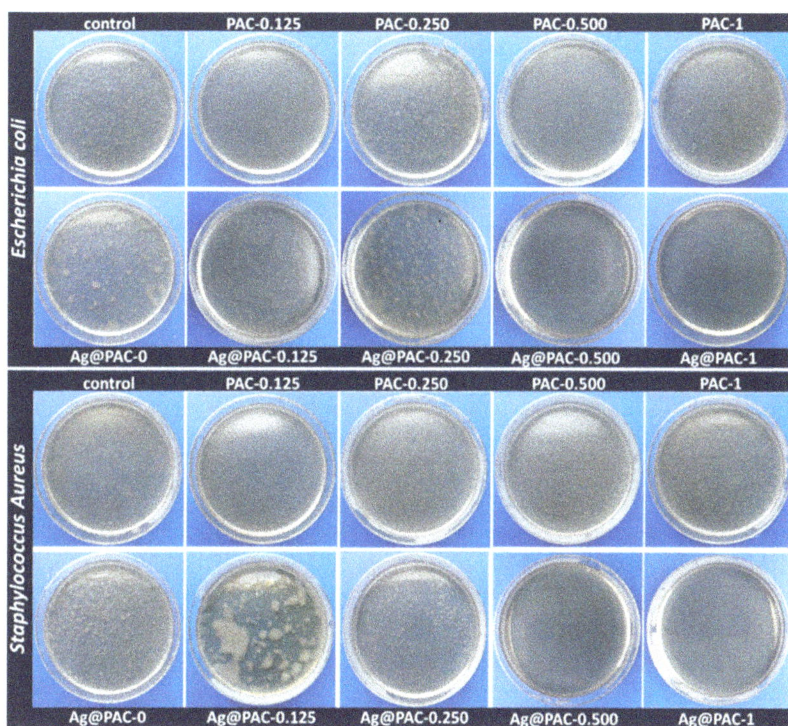

**Figure 7.** Antimicrobial efficiency of the Ag@PAC series compared to PAC counterparts.

## 4. Conclusions

The main challenge in using hydrogels in applications related to tissue regeneration is represented by their inadequate mechanical properties; although hydrogels are usually soft, most of them are not able to withstand high values of stress. The present paper aims to design nanocomposites that exhibit both elasticity and toughness by simultaneously using two different approaches: (1) the embedding of the linear PAAm in the 3D network of the corresponding monomer and cross-linker, aiming to improve CNTs' dispersion in the precursor and scaffolds' elasticity, and (2) the use of low ratios of nanoparticles as fillers, with the aim of providing toughness to the so obtained nanostructured system. The addition of the linear polymer (PAAm) in the classical PAAm-MBA system has a great contribution to the dispersibility of the CNTs in the precursor, as it can be easily observed even from the synthesis stage (Figure 1). Our results suggest that adding linear PAAm might contribute to chains' rearrangement and increase their resistance under load. Furthermore, the PAC materials have greater resistance to compression compared to their AC counterparts. At a low CNT-filling ratio (less than 0.5% CNT:AAm, wt.), the ultimate compression stress could not be reached, while the values registered for PAC-0.5 and PAC-1 were considerably higher when compared to AC-0.5 and AC-1, respectively. This behavior confirms that the addition of the linear PAAm leads to more resistant materials under load, while the nanoparticles reinforce the materials, leading to stiffer compositions. As indicated by micro-computed tomography, there are also some differences with regard to the architectural features of the lyophilized samples. The PAC series exhibit a slightly lower total porosity, and the pores' dimensions were mostly smaller compared to the AC series. The porosity of both series shows that these compositions are suitable for applications in the field of articular cartilage regeneration and repair. In addition, the paper demonstrates that the decoration of CNTs with low amounts of Ag leads to materials with excellent antimicrobial properties. Our results indicate that these materials have an efficiency of over 95% for both Gram-negative (*Escherichia coli*) and Gram-positive (*Staphylococcus aureus*) bacteria at a Ag loading of only $39.2 \times 10^{-4}\%$ (wt./wt.) in the final precursor dispersion. The silver content in Ag@PAC-0.5 is within the safety range established by Khansa et al. [69], while in Ag@PAC-1, the silver content is a bit higher. Our findings suggest that the obtained silver-decorated nanocomposites represent effective materials with enhanced mechanical properties and excellent antimicrobial activity which may find potential applications, particularly tissue regeneration and repair.

**Supplementary Materials:** The following supporting information can be downloaded at https://www.mdpi.com/article/10.3390/polym14122320/s1. Supplementary Figure S1: Hysteresis curves for all hydrogels; Supplementary Figure S2: strain–stress curves of the AC and PAC series, respectively; Supplementary Figure S3: Relevant SEM images of the PAC freeze-dried samples, indicating the smooth surface of the pores' walls.

**Author Contributions:** Conceptualization, A.S. and Ș.I.V.; methodology, A.S. and Ș.I.V.; validation, A.S., E.O., Ș.I.V., L.B. and I.-C.S.; formal analysis, A.S., E.O. and R.O.; investigation, A.S., E.O., F.M., L.B. and R.O.; resources, A.S.; writing—original draft preparation, A.S., E.O., L.B. and R.O.; writing—review and editing, A.S., Ș.I.V. and I.-C.S.; supervision, A.S.; project administration, A.S.; funding acquisition, A.S. All authors have read and agreed to the published version of the manuscript.

**Funding:** This work was supported by a grant of the Romanian Ministry of Education and Research, CNCS—UEFISCDI, project number PN-III-P1-1.1-TE-2019-1161, within PNCDI III.

**Data Availability Statement:** The data presented in this study are available on request from the corresponding author.

**Conflicts of Interest:** The authors declare no conflict of interest.

### Appendix A

Appendix A contains experimental details regarding the determination of the average molecular weight of the synthesized polyacrylamide used in obtaining the PAC and Ag@PAC series.

*Determination of the Average Molecular Weight of the Synthesized Polyacrylamide (PAAm)*

An Ubbelohde capillary viscosimeter was employed to determine the relative, specific and reduced viscosity of the synthesized PAAm, using the protocol described in [71]. Based on these data, the average molecular weight was computed, using Mark-Houwink equation (Equation (A1)), expressed as

$$[\eta] = K \cdot M_V^\alpha \tag{A1}$$

where $\eta$ represents the intrinsic viscosity [dL/g]

$K$ and $\alpha$ constant parameters for polymer-solvent solutions (k = $6.31 \times 10^{-5}$ dL/g; $\alpha$ = 0.8 for polyacrylamide-water, at 30 °C [72]).

$M_w$ represents the average molecular weight [Da].

Briefly, several PAAm aqueous solutions were prepared, with concentrations ranging from 0.05 g/dL to 0.8 g/dL were prepared. The solutions and the solvent were filtered and subsequently introduced in the Ubbelohde viscosimeter, placed in a water bath with a precise control of the temperature, at 30 °C. The flowing time of each solution is measured, and the average is used to determine the relative viscosity following Equation (A2).

$$\eta_{rel} = \frac{\tau_{solution}}{\tau_{solvent}} \tag{A2}$$

$$\eta_{sp} = \eta_{rel} - 1 \tag{A3}$$

To obtain the intrinsic viscosity a graph $\frac{\eta_{sp}}{c} = f(c)$ was plotted (Figure A1) and the intrinsic viscosity was read by extrapolating the concentration to zero.

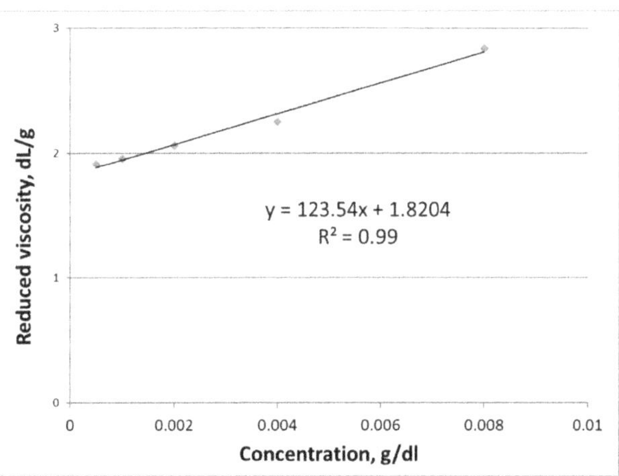


**Figure A1.** Reduced viscosity of the PAAm aqueous solutions as function of concentration.

By substituting the obtained value for the intrinsic viscosity (1.8204 dL/g) in Mark-Houwink equation (Equation (A1)), the obtained average molecular weight of the synthesized PAAm is $3.76 \times 10^5$ Da.

# References


1. Olăreț, E.; Bălănucă, B.; Onaș, A.M.; Ghițman, J.; Iovu, H.; Stancu, I.-C.; Serafim, A. Double-Cross-Linked Networks Based on Methacryloyl Mucin. *Polymers* **2021**, *13*, 1706. [CrossRef] [PubMed]
2. Serafim, A.; Tucureanu, C.; Petre, D.G.; Dragusin, D.M.; Salageanu, A.; Van Vlierberghe, S.; Dubruel, P.; Stancu, I.C. One-pot synthesis of superabsorbent hybrid hydrogels based on methacrylamide gelatin and polyacrylamide. Effortless control of hydrogel properties through composition design. *New J. Chem.* **2014**, *38*, 3112–3126. [CrossRef]
3. Lee, W.; Cha, S. Improvement of Mechanical and Self-Healing Properties for Polymethacrylate Derivatives Containing Maleimide Modified Graphene Oxide. *Polymers* **2020**, *12*, 603. [CrossRef] [PubMed]
4. Eastwood, E.; Viswanathan, S.; O'Brien, C.P.; Kumar, D.; Dadmun, M.D. Methods to improve the properties of polymer mixtures: Optimizing intermolecular interactions and compatibilization. *Polymer* **2005**, *46*, 3957–3970. [CrossRef]
5. Sokolyuk, A.; Kokosha, N.; Ulberg, Z.; Ovcharenko, F. A Method for the Production of a Soft Contact Lens. Patent WO 94/13717, 23 June 1994.
6. Fang, J.; Li, P.; Lu, X.; Fang, L.; Lü, X.; Ren, F. A strong, tough, and osteoconductive hydroxyapatite mineralized polyacrylamide/dextran hydrogel for bone tissue regeneration. *Acta Biomater.* **2019**, *88*, 503–513. [CrossRef]
7. Zhao, B.; He, J.; Wang, F.; Xing, R.; Sun, B.; Zhou, Y. Polyacrylamide-Sodium Alginate Hydrogel Releasing Oxygen and Vitamin C Promotes Bone Regeneration in Rat Skull Defects. *Front. Mater.* **2021**, *8*, 758599. [CrossRef]
8. Charrier, E.E.; Pogoda, K.; Li, R.; Park, C.Y.; Fredberg, J.J.; Janmey, P.A. A novel method to make viscoelastic polyacrylamide gels for cell culture and traction force microscopy. *APL Bioeng.* **2020**, *4*, 036104. [CrossRef]
9. Charrier, E.E.; Pogoda, K.; Wells, R.G.; Janmey, P.A. Control of cell morphology and differentiation by substrates with independently tunable elasticity and viscous dissipation. *Nat. Commun.* **2018**, *9*, 1–13. [CrossRef]
10. Alshehri, R.; Ilyas, A.M.; Hasan, A.; Arnaout, A.; Ahmed, F.; Memic, A. Carbon Nanotubes in Biomedical Applications: Factors, Mechanisms, and Remedies of Toxicity. *J. Med. Chem.* **2016**, *59*, 8149–8167. [CrossRef]
11. Chiticaru, E.A.; Pilan, L.; Damian, C.-M.; Vasile, E.; Burns, J.S.; Ioniță, M. Influence of Graphene Oxide Concentration when Fabricating an Electrochemical Biosensor for DNA Detection. *Biosensors* **2019**, *9*, 113. [CrossRef]
12. Becheru, D.F.; Vlăsceanu, G.M.; Banciu, A.; Vasile, E.; Ioniță, M.; Burns, J.S. Optical Graphene-Based Biosensor for Nucleic Acid Detection; Influence of Graphene Functionalization and Ionic Strength. *Int. J. Mol. Sci.* **2018**, *19*, 3230. [CrossRef] [PubMed]
13. Jiang, Z.; Feng, B.; Xu, J.; Qing, T.; Zhang, P.; Qing, Z. Graphene biosensors for bacterial and viral pathogens. *Biosens. Bioelectron.* **2020**, *166*, 112471. [CrossRef] [PubMed]
14. de Carvalho Lima, E.N.; Piqueira, J.R.C.; Maria, D.A. Advances in Carbon Nanotubes for Malignant Melanoma: A Chance for Treatment. *Mol. Diagn. Ther.* **2018**, *22*, 703–715. [CrossRef] [PubMed]
15. Zhao, C.; Song, X.; Liu, Y.; Fu, Y.; Ye, L.; Wang, N.; Wang, F.; Li, L.; Mohammadniaei, M.; Zhang, M.; et al. Synthesis of graphene quantum dots and their applications in drug delivery. *J. Nanobiotechnol.* **2020**, *18*, 142. [CrossRef] [PubMed]
16. Liu, J.; Cui, L.; Losic, D. Graphene and graphene oxide as new nanocarriers for drug delivery applications. *Acta Biomater.* **2013**, *9*, 9243–9257. [CrossRef] [PubMed]

17. Ghitman, J.; Biru, E.I.; Cojocaru, E.; Pircalabioru, G.G.; Vasile, E.; Iovu, H. Design of new bioinspired GO-COOH decorated alginate/gelatin hybrid scaffolds with nanofibrous architecture: Structural, mechanical and biological investigations. *RSC Adv.* **2021**, *11*, 13653–13665. [CrossRef]

18. Olăreţ, E.; Drăgușin, D.-M.; Serafim, A.; Lungu, A.; Șelaru, A.; Dobranici, A.; Dinescu, S.; Costache, M.; Boerașu, I.; Vasile, B.Ș.; et al. Electrospinning Fabrication and Cytocompatibility Investigation of Nanodiamond Particles-Gelatin Fibrous Tubular Scaffolds for Nerve Regeneration. *Polymers* **2021**, *13*, 407. [CrossRef]

19. Cojocaru, E.; Ghitman, J.; Biru, E.I.; Pircalabioru, G.G.; Vasile, E.; Iovu, H. Synthesis and Characterization of Electrospun Composite Scaffolds Based on Chitosan-Carboxylated Graphene Oxide with Potential Biomedical Applications. *Materials* **2021**, *14*, 2535. [CrossRef]

20. Zare, H.; Ahmadi, S.; Ghasemi, A.; Ghanbari, M.; Rabiee, N.; Bagherzadeh, M.; Karimi, M.; Webster, T.J.; Hamblin, M.R.; Mostafavi, E. Carbon nanotubes: Smart drug/gene delivery carriers. *Int. J. Nanomed.* **2021**, *16*, 1681–1706. [CrossRef]

21. Kharissova, O.V.; Kharisov, B.I.; de Casas Ortiz, E.G. Dispersion of carbon nanotubes in water and non-aqueous solvents. *RSC Adv.* **2013**, *3*, 24812–24852. [CrossRef]

22. Zhu, J.; Kim, J.; Peng, H.; Margrave, J.L.; Khabashesku, V.N.; Barrera, E. V Improving the Dispersion and Integration of Single-Walled Carbon Nanotubes in Epoxy Composites through Functionalization. *Nano Lett.* **2003**, *3*, 1107–1113. [CrossRef]

23. do Amaral Montanheiro, T.L.; Cristóvan, F.H.; Machado, J.P.B.; Tada, D.B.; Durán, N.; Lemes, A.P. Effect of MWCNT functionalization on thermal and electrical properties of PHBV/MWCNT nanocomposites. *J. Mater. Res.* **2015**, *30*, 55–65. [CrossRef]

24. Yang, K.; Yi, Z.; Jing, Q.; Yue, R.; Jiang, W.; Lin, D. Sonication-assisted dispersion of carbon nanotubes in aqueous solutions of the anionic surfactant SDBS: The role of sonication energy. *Chin. Sci. Bull.* **2013**, *58*, 2082–2090. [CrossRef]

25. Voicu, S.I.; Pandele, M.A.; Vasile, E.; Rughinis, R.; Crica, L.; Pilan, L.; Ionita, M. The impact of sonication time through polysulfone-graphene oxide composite films properties. *Dig. J. Nanomater. Biostruct.* **2013**, *8*, 1389–1394.

26. Ma, P.C.; Siddiqui, N.A.; Marom, G.; Kim, J.K. Dispersion and functionalization of carbon nanotubes for polymer-based nanocomposites: A review. *Compos. Part A Appl. Sci. Manuf.* **2010**, *41*, 1345–1367. [CrossRef]

27. Socher, R.; Krause, B.; Müller, M.T.; Boldt, R.; Pötschke, P. The influence of matrix viscosity on MWCNT dispersion and electrical properties in different thermoplastic nanocomposites. *Polymer* **2012**, *53*, 495–504. [CrossRef]

28. Kang, S.; Pinault, M.; Pfefferle, L.D.; Elimelech, M. Single-walled carbon nanotubes exhibit strong antimicrobial activity. *Langmuir* **2007**, *23*, 8670–8673. [CrossRef]

29. Hirschfeld, J.; Akinoglu, E.M.; Wirtz, D.C.; Hoerauf, A.; Bekeredjian-Ding, I.; Jepsen, S.; Haddouti, E.-M.; Limmer, A.; Giersig, M. Long-term release of antibiotics by carbon nanotube-coated titanium alloy surfaces diminish biofilm formation by Staphylococcus epidermidis. *Nanomed. Nanotechnol. Biol. Med.* **2017**, *13*, 1587–1593. [CrossRef]

30. Teixeira-Santos, R.; Gomes, M.; Gomes, L.C.; Mergulhão, F.J. Antimicrobial and anti-adhesive properties of carbon nanotube-based surfaces for medical applications: A systematic review. *iScience* **2021**, *24*, 102001. [CrossRef]

31. Kang, S.; Herzberg, M.; Rodrigues, D.F.; Elimelech, M. Antibacterial effects of carbon nanotubes: Size does matter! *Langmuir* **2008**, *24*, 6409–6413. [CrossRef]

32. Shvedova, A.A.; Castranova, V.; Kisin, E.R.; Schwegler-Berry, D.; Murray, A.R.; Gandelsman, V.Z.; Maynard, A.; Baron, P. Exposure to carbon nanotube material: Assessment of nanotube cytotoxicity using human keratinocyte cells. *J. Toxicol. Environ. Health A* **2003**, *66*, 1909–1926. [CrossRef] [PubMed]

33. Nel, A.; Xia, T.; Mädler, L.; Li, N. Toxic potential of materials at the nanolevel. *Science* **2006**, *311*, 622–627. [CrossRef] [PubMed]

34. Aoki, K.; Saito, N. Biocompatibility and carcinogenicity of carbon nanotubes as biomaterials. *Nanomaterials* **2020**, *10*, 264. [CrossRef] [PubMed]

35. Lam, C.-W.; James, J.T.; McCluskey, R.; Hunter, R.L. Pulmonary toxicity of single-wall carbon nanotubes in mice 7 and 90 days after intratracheal instillation. *Toxicol. Sci.* **2004**, *77*, 126–134. [CrossRef] [PubMed]

36. Catalán, J.; Siivola, K.M.; Nymark, P.; Lindberg, H.; Suhonen, S.; Järventaus, H.; Koivisto, A.J.; Moreno, C.; Vanhala, E.; Wolff, H.; et al. In vitro and in vivo genotoxic effects of straight versus tangled multi-walled carbon nanotubes. *Nanotoxicology* **2016**, *10*, 794–806. [CrossRef]

37. Nagai, H.; Okazaki, Y.; Chew, S.H.; Misawa, N.; Yamashita, Y.; Akatsuka, S.; Ishihara, T.; Yamashita, K.; Yoshikawa, Y.; Yasui, H.; et al. Diameter and rigidity of multiwalled carbon nanotubes are critical factors in mesothelial injury and carcinogenesis. *Proc. Natl. Acad. Sci. USA* **2011**, *108*, E1330–E1338. [CrossRef]

38. Xu, J.; Futakuchi, M.; Shimizu, H.; Alexander, D.B.; Yanagihara, K.; Fukamachi, K.; Suzui, M.; Kanno, J.; Hirose, A.; Ogata, A.; et al. Multi-walled carbon nanotubes translocate into the pleural cavity and induce visceral mesothelial proliferation in rats. *Cancer Sci.* **2012**, *103*, 2045–2050. [CrossRef]

39. Ju, L.; Wu, W.; Yu, M.; Lou, J.; Wu, H.; Yin, X.; Jia, Z.; Xiao, Y.; Zhu, L.; Yang, J. Different Cellular Response of Human Mesothelial Cell MeT-5A to Short-Term and Long-Term Multiwalled Carbon Nanotubes Exposure. *Biomed Res. Int.* **2017**, *2017*, 2747215. [CrossRef]

40. Saito, N.; Haniu, H.; Usui, Y.; Aoki, K.; Hara, K.; Takanashi, S.; Shimizu, M.; Narita, N.; Okamoto, M.; Kobayashi, S.; et al. Safe clinical use of carbon nanotubes as innovative biomaterials. *Chem. Rev.* **2014**, *114*, 6040–6079. [CrossRef]

41. Lan, W.; Zhang, X.; Xu, M.; Zhao, L.; Huang, D.; Wei, X.; Chen, W. Carbon nanotube reinforced polyvinyl alcohol/biphasic calcium phosphate scaffold for bone tissue engineering. *RSC Adv.* **2019**, *9*, 38998–39010. [CrossRef]

42. Kunisaki, A.; Kodama, A.; Ishikawa, M.; Ueda, T.; Lima, M.D.; Kondo, T.; Adachi, N. Carbon-nanotube yarns induce axonal regeneration in peripheral nerve defect. *Sci. Rep.* **2021**, *11*, 19562. [CrossRef] [PubMed]
43. Saleemi, M.A.; Fouladi, M.H.; Yong, P.V.C.; Wong, E.H. Elucidation of Antimicrobial Activity of Non-Covalently Dispersed Carbon Nanotubes. *Materials* **2020**, *13*, 1676. [CrossRef] [PubMed]
44. Al-Jumaili, A.; Alancherry, S.; Bazaka, K.; Jacob, M.V. Review on the antimicrobial properties of Carbon nanostructures. *Materials* **2017**, *10*, 1066. [CrossRef] [PubMed]
45. Dinh, N.X.; Van Quy, N.; Huy, T.Q.; Le, A.T. Decoration of silver nanoparticles on multiwalled carbon nanotubes: Antibacterial mechanism and ultrastructural analysis. *J. Nanomater.* **2015**, *2015*, 63. [CrossRef]
46. Rangari, V.K.; Mohammad, G.M.; Jeelani, S.; Hundley, A.; Komal, V.; Singh, S.R.; Pillai, S. Synthesis of Ag / CNT hybrid nanoparticles and fabrication of their Nylon-6 polymer nanocomposite fibers for antimicrobial. *Nanotechnology* **2010**, *21*, 095102. [CrossRef] [PubMed]
47. Chaudhari, A.A.; Joshi, S.; Vig, K.; Sahu, R.; Dixit, S.; Baganizi, R.; Dennis, V.A.; Singh, S.R.; Pillai, S. A three-dimensional human skin model to evaluate the inhibition of Staphylococcus aureus by antimicrobial peptide-functionalized silver carbon nanotubes. *J. Biomater. Appl.* **2019**, *33*, 924–934. [CrossRef]
48. Nepal, D.; Balasubramanian, S.; Simonian, A.L.; Davis, V.A. Strong Antimicrobial Coatings: Single-Walled Carbon Nanotubes Armored with Biopolymers. *Nano Lett.* **2008**, *8*, 1896–1901. [CrossRef]
49. Zhang, Y.; Li, S.; Xu, Y.; Shi, X.; Zhang, M.; Huang, Y.; Liang, Y.; Chen, Y.; Ji, W.; Kim, J.R.; et al. Engineering of hollow polymeric nanosphere-supported imidazolium-based ionic liquids with enhanced antimicrobial activities. *Nano Res.* **2022**, *15*, 5556–5568. [CrossRef]
50. Zhang, Y.; Song, W.; Lu, Y.; Xu, Y.; Wang, C.; Yu, D.-G.; Kim, I. Recent Advances in Poly($\alpha$-L-glutamic acid)-Based Nanomaterials for Drug Delivery. *Biomolecues* **2022**, *12*, 636. [CrossRef]
51. Leu Alexa, R.; Iovu, H.; Ghitman, J.; Serafim, A.; Stavarache, C.; Marin, M.-M.; Ianchis, R. 3D-Printed Gelatin Methacryloyl-Based Scaffolds with Potential Application in Tissue Engineering. *Polymers* **2021**, *13*, 727. [CrossRef]
52. Serafim, A.; Olaret, E.; Cecoltan, S.; Butac, L.M.; Balanuca, B.; Vasile, E.; Ghica, M.; Stancu, I.C. Bicomponent hydrogels based on methacryloyl derivatives of gelatin and mucin with potential wound dressing applications. *Mater. Plast.* **2018**, *55*, 68–74. [CrossRef]
53. Fan, S.; Chen, K.; Yuan, W.; Zhang, D.; Yang, S.; Lan, P.; Song, L.; Shao, H.; Zhang, Y. Biomaterial-Based Scaffolds as Antibacterial Suture Materials. *ACS Biomater. Sci. Eng.* **2020**, *6*, 3154–3161. [CrossRef] [PubMed]
54. Muhulet, A.; Tuncel, C.; Miculescu, F.; Pandele, A.M.; Bobirica, C.; Orbeci, C.; Bobirica, L.; Palla-Papavlu, A.; Voicu, S.I. Synthesis and characterization of polysulfone–TiO2 decorated MWCNT composite membranes by sonochemical method. *Appl. Phys. A Mater. Sci. Process.* **2020**, *126*, 1–9. [CrossRef]
55. Şenol, M.S.; Özer, H. Chapter 7—Architecture of cartilage tissue and its adaptation to pathological conditions. In *Comparative Kinesiology of the Human Body. Normal and Pathological Conditions*; Angin, S., Şimşek, I.E., Eds.; Academic Press: Cambridge, MA, USA, 2020; pp. 91–100. ISBN 978-0-12-812162-7.
56. Nava, M.M.; Draghi, L.; Giordano, C.; Pietrabissa, R. The effect of scaffold pore size in cartilage tissue engineering. *J. Appl. Biomater. Funct. Mater.* **2016**, *14*, e223–e229. [CrossRef] [PubMed]
57. Moradi, A.; Pramanik, S.; Ataollahi, F.; Abdul Khalil, A.; Kamarul, T.; Pingguan-Murphy, B. A comparison study of different physical treatments on cartilage matrix derived porous scaffolds for tissue engineering applications. *Sci. Technol. Adv. Mater.* **2014**, *15*, 06500. [CrossRef] [PubMed]
58. Lei, J.; Zhou, Z.; Liu, Z. Side Chains and the Insufficient Lubrication of Water in Polyacrylamide Hydrogel—A New Insight. *Polymers* **2019**, *11*, 1845. [CrossRef] [PubMed]
59. Wang, W.; Zhu, Y.; Liao, S.; Li, J. Carbon nanotubes reinforced composites for biomedical applications. *Biomed Res. Int.* **2014**, *2014*, 518609. [CrossRef] [PubMed]
60. Kausar, A.; Shi, T. Properties of Polyacrylamide and Functional Multi-walled Carbon Nanotube Composite. *Am. J. Nanosci. Nanotechnol. Res.* **2016**, *4*, 1–8.
61. Huang, Y.; Zheng, Y.; Song, W.; Ma, Y.; Wu, J.; Fan, L. Poly(vinyl pyrrolidone) wrapped multi-walled carbon nanotube/poly(vinyl alcohol) composite hydrogels. *Compos. Part A Appl. Sci. Manuf.* **2011**, *42*, 1398–1405. [CrossRef]
62. Mihajlovic, M.; Mihajlovic, M.; Dankers, P.Y.W.; Masereeuw, R.; Sijbesma, R.P. Carbon Nanotube Reinforced Supramolecular Hydrogels for Bioapplications. *Macromol. Biosci.* **2019**, *19*, 1800173. [CrossRef]
63. Bin, Y.; Mine, M.; Koganemaru, A.; Jiang, X.; Matsuo, M. Morphology and mechanical and electrical properties of oriented PVA–VGCF and PVA–MWNT composites. *Polymer* **2006**, *47*, 1308–1317. [CrossRef]
64. Hamouda, H.I.; Abdel-Ghafar, H.M.; Mahmoud, M.H.H. Multi-walled carbon nanotubes decorated with silver nanoparticles for antimicrobial applications. *J. Environ. Chem. Eng.* **2021**, *9*, 105034. [CrossRef]
65. Seo, Y.; Hwang, J.; Kim, J.; Jeong, Y.; Hwang, M.P.; Choi, J. Antibacterial activity and cytotoxicity of multi-walled carbon nanotubes decorated with silver nanoparticles. *Int. J. Nanomed.* **2014**, *9*, 4621–4629. [CrossRef]
66. Kędziora, A.; Wieczorek, R.; Speruda, M.; Matolínová, I.; Goszczyński, T.M.; Litwin, I.; Matolín, V.; Bugla-Płoskońska, G. Comparison of Antibacterial Mode of Action of Silver Ions and Silver Nanoformulations With Different Physico-Chemical Properties: Experimental and Computational Studies. *Front. Microbiol.* **2021**, *12*, 1–12. [CrossRef] [PubMed]

67. Bonilla-Gameros, L.; Chevallier, P.; Sarkissian, A.; Mantovani, D. Silver-based antibacterial strategies for healthcare-associated infections: Processes, challenges, and regulations. An integrated review. *Nanomed. Nanotechnol. Biol. Med.* **2020**, *24*, 102142. [CrossRef]
68. Poon, V.K.M.; Burd, A. In vitro cytotoxity of silver: Implication for clinical wound care. *Burns* **2004**, *30*, 140–147. [CrossRef]
69. Khansa, I.; Schoenbrunner, A.R.; Kraft, C.T.; Janis, J.E. Silver in Wound Care—Friend or Foe?: A Comprehensive Review. *Plast. Reconstr. Surg. Glob. Open* **2019**, *7*, 1–10. [CrossRef]
70. Zhang, Y.; Zhai, D.; Xu, M.; Yao, Q.; Zhu, H.; Chang, J.; Wu, C. 3D-printed bioceramic scaffolds with antibacterial and osteogenic activity. *Biofabrication* **2017**, *9*, 025037. [CrossRef]
71. Zeynali, M.E.; Rabbii, A. Alkaline hydrolysis of polyacrylamide and study on poly(acrylamide-co-sodium acrylate) properties. *Iran. Polym. J. (Engl. Ed.)* **2002**, *11*, 269–275.
72. American Polymer Standards Corporation. Available online: http://www.ampolymer.com/Mark-Houwink.html (accessed on 1 May 2022).

*Review*

# A Comprehensive Review of Biopolymer Fabrication in Additive Manufacturing Processing for 3D-Tissue-Engineering Scaffolds

Nurulhuda Arifin [1], Izman Sudin [2], Nor Hasrul Akhmal Ngadiman [2,*] and Mohamad Shaiful Ashrul Ishak [3]

[1] Quality Engineering, Malaysian Institute of Industrial Technology, Universiti Kuala Lumpur (UniKL), Persiaran Sinaran Ilmu, Bandar Seri Alam 81750, Johor, Malaysia; nurulhuda@unikl.edu.my

[2] School of Mechanical Engineering, Faculty of Engineering, Universiti Teknologi Malaysia, 81310 UTM Skudai, Johor Bahru 81310, Johor, Malaysia; izman@utm.edu.my

[3] Faculty of Mechanical Engineering Technology, Universiti Malaysia Perlis, Kampus Pauh Putra, Arau 02600, Perlis, Malaysia; mshaiful@unimap.edu.my

* Correspondence: norhasrul@utm.my

**Abstract:** The selection of a scaffold-fabrication method becomes challenging due to the variety in manufacturing methods, biomaterials and technical requirements. The design and development of tissue engineering scaffolds depend upon the porosity, which provides interconnected pores, suitable mechanical strength, and the internal scaffold architecture. The technology of the additive manufacturing (AM) method via photo-polymerization 3D printing is reported to have the capability to fabricate high resolution and finely controlled dimensions of a scaffold. This technology is also easy to operate, low cost and enables fast printing, compared to traditional methods and other additive manufacturing techniques. This article aims to review the potential of the photo-polymerization 3D-printing technique in the fabrication of tissue engineering scaffolds. This review paper also highlights the comprehensive comparative study between photo-polymerization 3D printing with other scaffold fabrication techniques. Various parameter settings that influence mechanical properties, biocompatibility and porosity behavior are also discussed in detail.

**Keywords:** additive manufacturing; tissue engineering; biomaterials; scaffold; 3D printing

**Citation:** Arifin, N.; Sudin, I.; Ngadiman, N.H.A.; Ishak, M.S.A. A Comprehensive Review of Biopolymer Fabrication in Additive Manufacturing Processing for 3D-Tissue-Engineering Scaffolds. *Polymers* **2022**, *14*, 2119. https://doi.org/10.3390/polym14102119

Academic Editors: Andrada Serafim and Stefan Ioan Voicu

Received: 26 March 2022
Accepted: 1 May 2022
Published: 23 May 2022

**Publisher's Note:** MDPI stays neutral with regard to jurisdictional claims in published maps and institutional affiliations.

## 1. Introduction

The failure of organs or tissues due to trauma or ageing is a primary concern in healthcare, as they are costly and devastating problems. Nowadays, technology transplantation from one individual into another has faced a significant challenge: to access enough tissue and organs for all patients. In addition, a problem exists with the immune system, which has a higher tendency to produce chronic rejection and destruction over time. These constraints have generated a need for a new solution to provide needed tissue. This has led to the development of tissue engineering (TE), which aims to create biological substitutes to repair or replace the failing organs and tissues [1].

Tissue engineering has gained more attention in the past decade, owing to its success in enabling tissue regeneration. The tissue-engineering field applies the knowledge of engineering, life, and clinical sciences toward solving the critical problems of tissue loss and organ failure. Tissue engineering also aims to produce patient-specific biological substitutes, to circumvent the limitations of existing clinical treatments for damaged tissue or organs. These limitations include the shortage of donor organs, chronic rejection, and cell morbidity [2].

Tissue-engineering scaffold technology provides a temporary template from which to develop biological substitutes that restore, maintain, or improve tissue function or a whole damaged organ [3]. Tissue-engineering technology is unique in that it can establish three-dimensional environments for propagated cells and specific signaling molecules that can mimic native tissue environments. Typically, three groups of biomaterials—ceramics, synthetic polymers, and natural polymers—are used in the fabrication of tissue-engineering scaffolds. The scaffolds can be natural, synthetic or a hybrid of both. One example of a both natural and synthetic biomaterial is amphiphilic conetwork (APCN), which is useful for the controlled release of both hydrophilic and hydrophobic properties. Amphiphilic conetwork (APCN) gels are made up of hydrophilic and hydrophobic polymer chains that are covalently connected. The majority of APCN gels are made by the radical polymerization of telechelic macro-monomers having at least two polymerizable groups and a low molecular weight monomer [4–6]. End linking well-defined polymer chains with di-functional monomers produces APCN gels with a regulated structure as shown in Figure 1 [4].

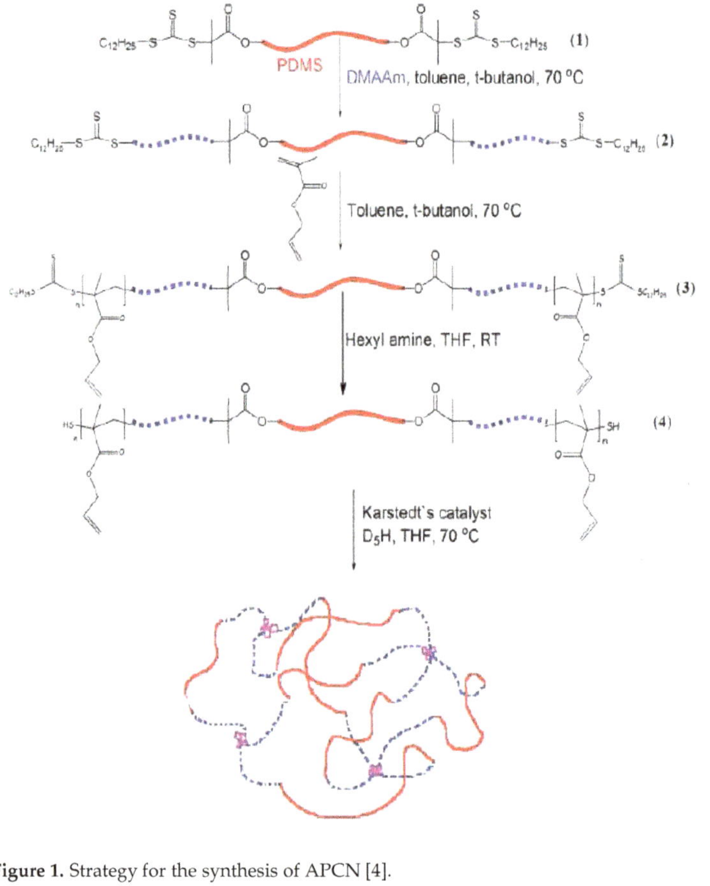

**Figure 1.** Strategy for the synthesis of APCN [4].

On the other hands, biomaterials used in tissue engineering also can be categorized according to their origin by category: natural polymers (collagen, chitosan, hyaluronic acid, elastin and gelatin), synthetic polymers (poly(lactic acid) (PLA) and poly(glycolic acid) (PGA), polycaprolactone (PCL)), ceramics (HA, TCP and biphasic calcium phosphates) and metals (magnesium and nickel alloy) [7–10].

Tissue-engineering techniques have the potential to create tissues and organs. They involve the in-vitro seeding and attachment of human cells onto a scaffold. The cells then proliferate, migrate and differentiate into the specific cell type to repair tissue [11,12]. Therefore, the choice of scaffold is crucial to enable the cells to behave in the required manner, to produce tissues and organs of the desired shape and size. A useful tissue-engineering scaffold should fulfil the biological and mechanical requirements of the target tissue. The scaffold should: have a suitable microstructure to promote cell proliferation, be contained within an open-pore geometry with a highly porous surface that enables cell ingrowth, have a proper surface morphology and be made from biomaterials with a predictable rate of degradation with a nontoxic degraded material [2].

During the development of the scaffold, the primary aim is to imitate the structural and mechanical properties of bone as close as possible. Thus, the scaffold fabrication technique should be flexible, to build scaffold architectures with biomimetic designs. Generally, conventional methods are used to construct tissue-engineering scaffolds. There are various traditional methods used to construct tissue-engineering scaffolds, including the molding technique, solvent casting and particulate leaching, gas foaming, and electrospinning. Although a lot of conventional fabrication methods can be used to produce scaffolds, unfortunately, each of these methods has their own limitations as they are not able to precisely control the internal topology and architecture [13,14]. To the best of our knowledge, none of the traditional methods is satisfactory to produce scaffolds with a control dimension architecture, porosity and faced the difficulty for mimicking the biological function of natural tissue [13–16].

As an alternative to conventional scaffold-fabrication methods, additive manufacturing techniques have recently been developed in tissue engineering, such as a rapid prototype by which a 3D scaffold is fabricated by laying down multiple, precisely formed layers in series [17]. Subia et al. (2010) claimed that the rapid-prototype technique (RP) has drawn tremendous attention with its potential to overcome most of the limitations faced by conventional techniques for the fabrication of 3D scaffolds [18].

This review provides an overview of the advantages and limitations offered by the additive manufacturing process (AM), specifically in the photo-polymerization 3D printing technique compared to other conventional methods. The overview includes their advantages and limitations regarding mechanical properties and the internal architecture porosity of fabricated scaffolds. The potential of the photo-polymerization 3D printing technique in the fabrication of tissue-engineering scaffold hydrogels is also discussed detail.

## 2. Concept of TE Scaffold

Over two decades, many works have been carried out to develop potentially applicable scaffolds for tissue engineering. The scaffolds are designed in three dimensions (3D), with a porous solid structure to perform some or all of the following functions: (i) promote cell–biomaterial interactions, cell adhesion, and ECM deposition; (ii) permit sufficient transport of gases, nutrients, and regulatory factors to allow cell survival, proliferation, and differentiation; (iii) biodegrade at a controllable rate that approximates the rate of tissue regeneration under the culture conditions of interest; (iv) provoke a minimal degree of inflammation or toxicity in vivo; and (vi) contain the porous interconnected structure that is necessary to allow the spread of waste products from the scaffolding [19,20].

Scaffolds can be tough, as a mimic of a physiologic environment that serves to promote proper cell proliferation, differentiation and organization; all cell migration and interaction is often greatly influenced by the local environment [21]. In tissue engineering, researchers have designated the substitution of a native ECM as a "scaffold", "template", or "artificial matrix". The scaffold provides a three-dimensional (3D) ECM analogue that functions as a template required for the infiltration and proliferation of cells into the targeted functional tissue or organ [22,23]. The concept of a tissue engineering scaffold involves the in-vitro seeding and attachment of human cells onto a scaffold. These cells then proliferate, migrate and differentiate into the specific tissue [11] and recover damaged tissues [12].

One of the more promising approaches in tissue engineering is to grow cells on a biodegradable scaffold that mimics the function of the natural extracellular matrix, providing a temporary template for the growth of target tissues [15]. The extracellular matrix (ECM) is the optimal support for tissue engineering, as it provides the perfect chemical composition, surface topology and physical properties experienced by cells in vivo [24]. The use of ECM derived from decellularized tissue is increasingly frequent in regenerative medicine and tissue-engineering strategies, with new applications including the use of three-dimensional ECM scaffolds [24]. One of the principal methods behind tissue engineering involves growing the relevant cells in vitro into the required three-dimensional (3D) organ or tissue. However, cells are difficult to grow in the favored 3D orientations and formed anatomical shape of the tissue. Instead, they randomly migrate to form a two-dimensional (2D) layer of cells. However, 3D tissues are required, and this is achieved by seeding the cells onto porous matrices to enhance the cells' attachment and colonization [13,14,25].

*Requirement of TE Scaffold*

Several requirements are identified as crucial for the production of tissue engineering scaffolds. Most of the researchers have summarized an ideal scaffold as having the following characteristics: (i) a scaffold should possess interconnecting pores of an appropriate scale to favor tissue integration and vascularization [26,27], (ii) a scaffold should be made from material with controlled biodegradability so that tissue will eventually replace the scaffold [15,26,28], (iii) have appropriate surface chemistry to favor cellular attachment, differentiation and proliferation [29], (iv) possess adequate mechanical properties to match the intended site of implantation and handling [28,30], (v) should not induce any adverse response [28], (vi) be easily fabricated into a variety of shapes and sizes [15,28], and (vii) must facilitate the ingrowth of tissue and possibly allow for the inclusion of seeded cells, proteins and/or genes to accelerate tissue regeneration [26,28]. All of these highlighted properties of a tissue engineering scaffold are important, to ensure the ability of the scaffold to be metabolized by the body, allowing it to be gradually replaced by new cells to form functional tissues.

In addition, the criteria for choosing materials as biomaterials in biomedical applications are based on their material chemistry, molecular weight, solubility, shape and composition, hydrophilicity/hydrophobicity, degradation of water absorption, and erosion mechanism [31]. The scaffold should have the mechanical strength needed for implantation and an appropriate strength that can influence the biostability of implants. The porosity and pore size of a supporting 3D scaffold is vital for tissue regeneration [27,32]. A large surface area also favors cell attachment and growth. Other than that, highly porous scaffolds are desirable for the diffusion of nutrients and waste products from the implant [33]. Hydrophilicity is also an essential factor need to consider. It is because hydrophilicity will enhance cell growth and proliferation of 3D scaffolds, as discussed previously [15]. Therefore, due to the important character and behavior of mechanical and porosity scaffolds, a nanofiber material is most suitable for nano-based scaffolding systems with the appropriate mechanical integrity, pore size, and hydrophilic property that will provide an excellent potential for tissue engineering scaffolds.

## 3. Fabrication of 3D TE Scaffolds

Currently, there are two broad categories of scaffold fabrication methods which are the conventional and advanced processing methods (additive manufacturing). The fabrication of tissue engineering scaffolds commonly involves traditional techniques such as (i) solvent casting, (ii) particulate leaching, (iii) electrospinning, (iv) phase separation, (v) extrusion deposition, (vi) pressing, (vii) freeze drying, and (viii) gas foaming [34–37]. Even though these methods have been extensively studied and optimized, they still have a lot of limitations.

There are critiques concerning the practicality of conventional methods. These methods were identified as techniques incapable of precisely controlling pore size, pore geometry, pore interconnectivity, and the spatial distribution of pores to allow construction of internal channels within the scaffolds, as argued by Zhu and Che (2013) [17]. In addition, several of these techniques are contingent upon using organic solvents with inherent biocompatibility when using a toxic solvent that may be toxic to the cells if they are not wholly and adequately removed [1,34].

The revolutionary technology of rapid prototyping in the additive manufacturing process offers potential and opportunities for manufacturing to fabricate 3D materials with optimized properties and multi-functionality. RP is also called the solid free-form technique or additive manufacturing (AM). This technique is a more advanced technique for scaffold fabrication. It is a computer-controlled fabrication technique that can rapidly produce a 3D object by using the layer manufacturing method. The RP technique generally comprises the design of a scaffold model by using computer-aided design (CAD) software [18,36].

There are numerous benefits offered by this rapid-prototype technology, such as ease of use, reliability, cost-effectiveness, and the diversity of the compatible materials [33]. Rapid-prototype techniques also hold much promise over conventional methods in terms of part consistency, design repeatability and the control of scaffold architecture such as pore size, porosity, surface area and the external shape of the scaffold architecture. The control over scaffold architecture is particularly important, as a TE scaffold mimics the original environment organ to regenerate damaged tissues [1] successfully.

The comparative study between fabrications techniques of 3D tissue engineering scaffolds is discussed detail in Table 1. From the summarized reviews on various fabrication techniques of 3D tissue engineering scaffolds, the rapid-prototyping method has lots of advantages to fabricate an excellent profile of biocompatibility properties with a more accurate scaffold architecture. This technique also promises the capability of mimicking the extracellular matrix (ECM) in the human body. Even though pure-polymer products built by rapid prototyping lack in strength, the technique of rapid-prototype polymer composites has solved these problems by combining the matrix and reinforcements to achieve high mechanical performance and excellent functionality.

**Table 1.** Advantages and disadvantages of various fabrication techniques of 3D tissue engineering scaffolds.

| Fabrication Techniques | Advantages | Disadvantages | Ref. |
|---|---|---|---|
| Solvent-casting and particulate-leaching | ▪ Simple process<br>▪ Inexpensive<br>▪ Control porosity | ▪ Limited size<br>▪ Low reproducibility<br>▪ Limited feature control<br>▪ Thickness < 4 mm<br>▪ Inefficient<br>▪ Poor mechanical properties | [17,18,38–42] |
| Gas foaming | ▪ Control porosity<br>▪ Organic process | ▪ Poor mechanical properties<br>▪ Imperfect pore<br>▪ Distinct structure<br>▪ Non-porous external surface | [17,42] |
| Phase separation | ▪ Can combine with other fabrication technique<br>▪ Control porosity<br>▪ High porosity | ▪ Complicated process<br>▪ Difficult control porosity<br>▪ Non-uniform porosity | [18,43–45] |
| Freeze drying | ▪ Easy process<br>▪ Homogenous<br>▪ porosity<br>▪ Durable<br>▪ Flexible | ▪ Small pore size<br>▪ Longer processing time<br>▪ Lower porosity | [39,42,46,47] |
| Fibre bonding | ▪ High surface to<br>▪ volume ratio<br>▪ High porosity<br>▪ Easy process | ▪ Poor mechanical properties<br>▪ limited applications<br>▪ Difficult control porosity<br>▪ Lack of solvent<br>▪ Complicated to set process parameters | [18,46] |
| Electro-spinning | ▪ Low cost<br>▪ Flexible process<br>▪ Simple process<br>▪ Easy to find solvent<br>▪ Smooth fiber produced | ▪ Low productivity<br>▪ Clogging problem<br>▪ Fragile fibers produced<br>▪ High-density nanofiber | [48–52] |

**Table 1.** *Cont.*

| Fabrication Techniques | Advantages | Disadvantages | Ref. |
|---|---|---|---|
| Additive manufacturing (AM): rapid prototyping | ▪ High accuracy<br>▪ High resolution<br>▪ Versatile scaffolds<br>▪ Homogenous cell distribution<br>▪ Interconnected pores<br>▪ Mimicking ECM<br>▪ Fast and easy process<br>▪ Custom made<br>▪ High reproducible<br>▪ No contamination<br>▪ Produces high cells density<br>▪ Conducted at room temperature<br>▪ Multi-color printing scaffolds<br>▪ Automated process<br>▪ Print scaffold with cells | ▪ Lack of strength<br>▪ Limited raw materials | [53–61] |

### 3.1. Additive Manufacturing in TE Scaffold Fabrication

Additive manufacturing, also known as rapid prototyping or 3D-printing technologies in tissue engineering, has been growing in recent years. As pointed out by Bose et al. (2013), among the different technology options for tissue engineering scaffolds, the rapid-prototyping technique is becoming popular due to the ability to print porous scaffolds with a designed shape, controlled chemistry and interconnected porosity [62]. Wang et al. (2017) claimed that rapid prototyping can fabricate complex composite structures of tissue engineering scaffolds without the typical waste, compared to traditional methods. The size and geometry of composites can also be precisely controlled with the help of computer-aided design in the rapid-prototyping process [62]. Thus, tissue engineering scaffolds fabricated by rapid prototyping will have a higher performance.

Rapid prototyping is an additive manufacturing technique that builds the objects piece by piece, using only the material that will become part of the object and avoiding loss of material in the process. The technology of the 3D printer is tied to that of computer-aided design—CADs software—where the design of the object is made [55]. It is based on 2D cross-sectional data obtained from slicing a computer-aided design (CAD) model of the object [37]. Rapid-prototyping technologies can be divided into five main groups according to the working principles used to produce 3D objects. The main rapid-prototyping technologies developed within few years are (i) the photo-polymerization technique, (ii) fused deposition modeling (FDM), (iii) selective laser sintering (SLS), (iv) 3D printing (3DP), and (v) bioprinting (3D plotting or direct writing) [63].

Table 2 summarizes a detailed review of advantages and disadvantages for each rapid-prototyping technique that is useful for tissue engineering scaffold fabrication. Although there are a lot of different advanced technology approaches to fabricating scaffolds' 3D structures, each of them has its limitations. Some of the techniques have a limitation when trying to mimic the biological function of natural tissue, due to the difficulty of finely controlling the scaffold architecture, dimensions, and porosity [64]. Chia and Wu (2015) claimed that the selection of a fabrication technique depends upon the materials of interest, machine limitations, and the specific requirements of the final scaffold. Other than that, design architecture is important for the structural, nutrient transport and cell–matrix interaction conditions of tissue engineering scaffolds [58].

**Table 2.** Typical advanced manufacturing process for 3D tissue engineering scaffolds.

| Fabrication Technique | Advantages | Disadvantages | Ref. |
|---|---|---|---|
| Photo-polymerization technique: stereolithography (SLA)/digital light processing (DLP)) | Rapid response rate High-form precision Allows fabrication of internal pore scaffold Produces strong construction of complex tissue geometries scaffolds Cells can be incorporated Offers better dimensional dimensions Flexibility in design Higher accuracy and resolution compared to SLS/FDM/3D printing Resolution up to 100 nm Wide variety of application Able to create complex forms with the internal architecture Easily removes un-polymerized resin Versatile design | Requires photo-reactive biodegradable polymer in the process Ultraviolet irradiation used Produces layered stratification that may disable cell contact between layers Limited number of biocompatible resins because few bio-compatible polymers are stable under exposure to laser light Not suitable for high production rates due to the slow printing process | [65–71] |
| Fused deposition modelling (FDM) | Does not need any solvents and preserves flexibility in material handling and processing Highly controllable porosity Good mechanical properties Offers sufficient dimensional accuracy | Thermoplastic material used must have good melting viscosity properties Biomaterial used must be available in filament form Limited shape complexity for biological scaffolding materials Inability to incorporate living cells due to the high processing temperature during extrusion Insufficient surface | [36,58,68,72–74] |
| Selective laser sintering (SLS) | Able to produce complex shapes High mechanical strength Powder bed provides support for complex structures Fine resolution | Laser intensity can induce polymer degradation Limitation on materials (must be shrinkage and heat resistant) Trapped non-sintered material Poor control over surface topography High porosity Expensive and time consuming High-temperature process required | [73–76] |

### 3.1.1. Photo-polymerization 3D Printing

There is a critique concerning the importance of considering the porosity and architecture of an engineering scaffold. As pointed out by Annabi et al. (2010), the porosity of a scaffold plays an important role in directing tissue formation and function. The authors suggested a substantial amount of scaffold porosity is often necessary to allow for homogeneous cell distribution and interconnection throughout engineered tissues. In addition, increased porosity can have a beneficial effect on the diffusion of nutrients and oxygen [77]. Based on a review study, Mondschein et al. (2017) suggested that photo-polymerization via the stereolithography (SLA) technique will allow a greater control of the tissue scaffold's dimensions and features, compared to other additive-manufacturing techniques. The SLA technique can also precisely control the architecture and features of the scaffold, which offers a great benefit to regenerative medicine, whether to construct repeatable identical scaffolds or to fabricate patient-specific templates [76].

In the SLA approach, resolution is inversely related to print speed. Parts with a feature size down to 10 µm can be formed with z-axis print speeds of 25–1000 mm/min, which take several hours with conventional SLA techniques [78]. To overcome the limitation of SLA printing speed efficiency, another trend of the photo cross-linking 3D printing field is the emerging use of digital-light-projection (DLP) technology [79]. Surface patterned exposure from digital-light-projection (DLP) sources, and the high-power LED sources used, allows any selected portion of the entire x/y workspace to be exposed simultaneously to dynamic writing with a condensed laser beam. Even though high laser scanning velocities are employed in the SLA approach, the ability to simultaneously photo cure all portions of a given slice with the DLP technique significantly speeds up the cycle times between layers [78–80]. Thus, it allows control of spot-to-spot (lateral) and interlayer (vertical) binding and improves the resolution of printed parts [81].

### 3.1.2. Stereolithography (SLA)-Based 3D Printing

Stereolithography-based 3D printing was developed by 3D systems in 1986 and is the first commercially available solid freeform (SFF) technique [82]. SLA is a particularly versatile manufacturing technique for the freedom of designing structures. In the biomedical field, the development of SLA technology has led to the fabrication of mold-assisted implant fabrication, aids for complicated surgery, and fabrication of tissue engineering scaffolds [83,84]. The manufacturing of 3D objects by SLA is based on the spatially controlled solidification of a liquid resin by photo-polymerization [85]. SLA allows the fabrication of parts from a computer-aided design (CAD) file. The CAD file describes the geometry and size of the parts to be built. This designed structure is (virtually) sliced into layers of the thickness that is to be used in the layer-by-layer fabrication process [82,86].

The SLA technique based on free-radical photo-polymerization is an efficient method for converting a liquid prepolymer resin into a solid polymer network under light exposure [87]. In free-radical chain-growth photo-polymerization, a photoinitiator absorbs light either in the UV or visible light range, which excites the photoinitiator molecules into a high-energy radical state. The initiator radicals interact with precursor molecules, forming the primary radicals that initiate the polymerization reaction. Chain-growth polymerization then propagates until a complex crosslinked network is formed in a process called photo crosslinking [88].

The SLA process can be divided into two major categories, which are the projection and laser-scanning types. The scanning-type stereolithography process uses a UV laser beam to scan and cure the surface of the resin layer by layer. On the other hand, in the projection-type stereolithography process, a digital light projection (DLP) is utilized to project a whole cross-sectional area of mask projection on the resin surface [81]. The scanning-type stereolithography apparatus consists of a bath to be filled with a liquid photocurable resin, a laser source (commonly, UV light), a system that controls the XY-movement of the light beam, and a fabrication platform that permits movement in the vertical plane [79]. In the bath configuration, the UV beam traces a 2D cross-section onto a base submerged in a tank of liquid photoactive resin that polymerizes upon illumination. After completion of the 2D cross-section, the UV beam begins the addition of the next layer, which is polymerized on top of the previous layer. In between layers, a blade loaded with resin levels the surface of the resin to ensure a uniform layer of liquid prior to another round of UV light exposure. This process is repeated, slice by slice, until the 3D object is completed [89]. After the structure is completed, the un-polymerized liquid resin is removed by draining and post-curing converts any unreacted groups.

### 3.1.3. SLA versus DLP 3D Printing for TE Scaffold Fabrication

Digital light processing (DLP) is identical to SLA except for the light source: a projector is used to cure an entire layer at a time (Table 3). Instead of using a UV laser, a DLP projector is used to project the entire cross-sectional layer of the 3D structure [90]. The photosensitive resin is exposed to the light through patterns on a digital mirror device. The exposed parts are cured, and one layer is finished. Then, the platform raises a layer, and the next exposure starts [91], projecting the image of a layer of the part to be built onto the photosensitive resin allows one to fabricate one layer at once, a fact responsible for the high building speed at a significantly lower cost of equipment achieved with these machines [92,93]. For the fabrication of 3D parts, the CAD model was sliced, and every slice was projected onto the bottom layer of the resin tank by a micro-mirror array. Here, the first layer of the light-sensitive resin cured in a few seconds. The polymer adhered to the z-stage, which was then moved upwards [94].

There are two types of projection-type stereolithography DLP process, namely, free surface and constrained surface. In free surface (top-down stereolithography), the layer is cured on the photopolymer from the surface towards the bottom of the vat. On the other hand, in constrained-surface (bottom-up) stereolithography, the layer is cured through the bottom of the vat, so that the printed structure does not adhere to the substrate. It causes the curing of liquid resin to be sealed from the oxygen-rich environment. By eliminating the oxygen inhibition effect, the liquid photopolymer resin can be cured faster, which offers an advantage over the free surface-based system [95].

According to a study conducted by Low et al. (2017), Groth et al. (2014), and A. Woesz (2008), the DLP system has the same general advantages and disadvantages as the SLA method [90,92,96]. In principle, the main advantages of DLP over SLA in the context of scaffold fabrication are that DLP 3D printing does not use a laser, which reduces the system costs significantly. It also has a higher build speed due to the exposure of one layer at one time. A detailed study of the specifications for SLA and DLP 3D printing are summarized in Table 3.

**Table 3.** Comparison between laser-scanning-type SLA and projection-type SLA (DLP) 3D printing.

| Technique | Resolution µm | Light Source | Advantages | Disadvantages | Ref. |
|---|---|---|---|---|---|
| Laser-scanning-type SLA | 200–300 | UV | NA | • Slow | |
| Projection-type SLA (DLP) | 15–100 | Projector | • Higher speed than SLA<br>• Low cost | • Lower light intensity | [90,92,96,97] |

Compared with other advanced manufacturing techniques, SLA shows its superiority in its high resolution. The higher the resolution at which a part can be built, the smaller will be its maximum size. In order to achieve a high resolution, SLA requires a high level of control over the layer thickness being crosslinked. In the SLA technique, control of the thickness of the layer is crucial. The study of curing depths in photo-polymerization is an important aspect of the curing process because it affects the final dimensions of the cured sample. Therefore, it is important to optimize the cure for these systems.

The smallest feature size that can be produced depends on the resin and setting parameters. Melchiorri et al. (2016) examined techniques and materials developed for the DLP printing of vascular tissue engineering scaffolds utilizing poly (propylene fumarate)

(PPF). The researcher found that the mechanical properties of the 3D-printed structure relied largely on the amount of post-printing time to radiation, which increased polymer cross-linking [98]. In line with this, Valentincic et al., (2017) proved that the setting parameter of DLP is important to produce the optimum result. The researcher argued that there are three main printing parameters in DLP printing that needs to take consideration which is curing time, layer thickness and time between the consecutive exposures. It was recorded in their study that an increased exposure time significantly increased illumination intensity on the whole of the projection surface [99].

Although the principles of projection or laser-scanning processes are similar, the effects of process parameters on curing the polymeric-photo material can be quite different. The light in the projection-type process and the UV-scanning type can be different in energy densities due to various control parameters such as curing time and scanning speed, which will correspond to the varying degrees of polymerization. Therefore, it is essential to determine the critical energy density by the UV projector or laser, to form a solid network.

### 3.1.4. Influence Process Parameter

A previous study by Chong et al. (2016) and Tureyen et al. (2015) agreed that the most challenging in photo-polymerization 3D printing is to control curing issues. The researchers reported that light intensity has a significant role in the curing depth of the resin. The curing depth and width of printed parts can be controlled by adjusting the curing time or laser-scanning speed, respectively: the energy density is increased by extending the curing time and lowering laser-scanning speeds. At the same energy density, it is shown that the projection-type SLA process obtains a larger curing depth than the laser-scanning type due to the difference in intensity between both the systems [81,100]. Table 4 summarizes a list of recent references on the research work performed to produce TE scaffolds by using photo-polymerization processes. In view of Table 4, the influences of the resin used and parameter setting are also studied.

**Table 4.** The influences of resin selection and parameter setting for photo-polymerization 3D printing.

| Input | | Responses | | | |
|---|---|---|---|---|---|
| | | **Mechanical Properties** | **Bio-Compatibility** | **Porosity** | **Thickness Diameter** |
| Resin used | | [82,85,101] | [64,79,102] | [85,103,104] | [67,81,105–107] |
| Resin viscosity | | [67,108] | [108] | [65,108,109] | [81,109] |
| Parameter setting | Curing time | [98,99,110] | | | |
| | Power light source | | | | [81,100] |
| | Resolution | | | | [81,110] |
| | Layer of thickness | [91,102] | | | |
| | Scan speed (velocity) | | | | [81,100] |

Based on the summarized review of previous research in Table 4, it is shown that different input setting parameters and the type of resin used will give a different response on the cured thickness of solidified resin, mechanical properties, biocompatibility, and the porosity of the scaffold. Wang et al. (1996) found that the curing degree is approximately proportional to the intensity of the light source, scanning speed and type of resin used. The final degree of the cure of a photo-polymerization-based 3D-printing prototype is determined by the combination of all of these factors. Researchers declared that if the intensity is increased, the curing degree will be increased. This is because, by using a high intensity power source, the resins will have more cross-linking. When the scan is fast, the exposure energy in a unit area is less; thus, the curing degree will be low [105].

In the same context, some researchers have studied the curing process of photo-polymerization resin for SLA 3D printing. Most of the research revealed that the curing phase in SLA is important for further solidification and, thus, causes an enhancement in the prototypes' mechanical properties. Thermal and the heating effects during the curing process also led to the existence of shrinkage and distortion within the structure of a cured resin [105,111]. As compared to the laser-scanning SLA process, projection-based SLA (DLP) has a lower light intensity and resulted in less polymerization, causing the printed parts to be deformed due to inhomogeneous curing [97]. However, to the best of our knowledge, there are only a limited numbers of studies performed on the investigation effect of curing time via DLP.

## 4. Materials for Photo-Polymerization 3D Printing TE Scaffold

The selection of biomaterials plays a crucial role in tissue engineering. The materials should obtain interactions with cells to enhance cellular attachment, proliferation, and new-tissue formation. However, the limitation of biocompatible resins with suitable SLA processing has often been considered as the main disadvantage of this method. Resins utilized in this process should be a liquid that solidifies quickly on illumination with light. The resin used must not only exhibit fast photo-crosslinking kinetics but also enhance adhesion and cell proliferation and has proper mechanical properties after crosslinking [68]. Other than that, the SLA process requires resin that has a melting temperature™ below room temperature and a glass transition temperature (Tg) low enough to maintain the polymer in a liquid-like state, to allow chain mobility. The low viscosity of resin used is also important to obtain optimum cure rates, thus decreasing overall construction times [64].

There are differences between polymers and resins as shown in Table 5. It can be summarized that polymers have large molecules with repeating structural units of monomers, while resins are an organic material that naturally forms in plants. Resins have low molecular weights, whereas polymers have large molecular weights. Additionally, resin is a viscous liquid that can be clear or dark brown in color, whereas polymers can be solid or liquid.

**Table 5.** Differences between polymer and resin [85,101,108].

| Item | Polymer | Resin |
| --- | --- | --- |
| Definition | Repeating structure unit of monomers | Organic material form in plant |
| Properties | Large molecular weight | Small molecular weight |
| Nature | Can be solid or liquid | Solid or highly viscous liquid |

In the photo-polymerization process, the process is driven by a chemical reaction that produces free radicals when exposed to specific wavelengths of light. The problem with this photo-polymerization process is that the created free radicals can cause damage to the cell membrane, protein and nucleic acids. Therefore, it is important to find a cytocompatibility photo-initiator resin for the SLA 3D printing method [68]. One of the remaining big basic issues in polymer science is controlling comonomer sequences in synthetic polymerization techniques. Modern synthetic processes, in fact, do not allow for accurate control of polymer microstructures. This is particularly true for radical chain-growth polymerizations, which frequently result in statistical comonomer inclusion in polymer chains as illustrated in Figure 2 [88].

The first resins developed for use in SLA are based on low-molecular-weight poly-acrylate and epoxide or viny ester-based resin [112], which form glassy networks upon photo-initiated polymerization and cross-linking [82]. When preparing biomedical implants, the use of epoxy- or acrylate-based resins is limited. The advantages of these materials include several useful properties, such as low viscosity, high photosensitivity, controllable mechanical properties, and relative insensitivity to temperature and humidity changes. However, its disadvantages are still noticeable. These materials are usually not biocompatible or biodegradable [113] and have poor dimensional stability and high-volume shrinkage during the post-curing process [114].

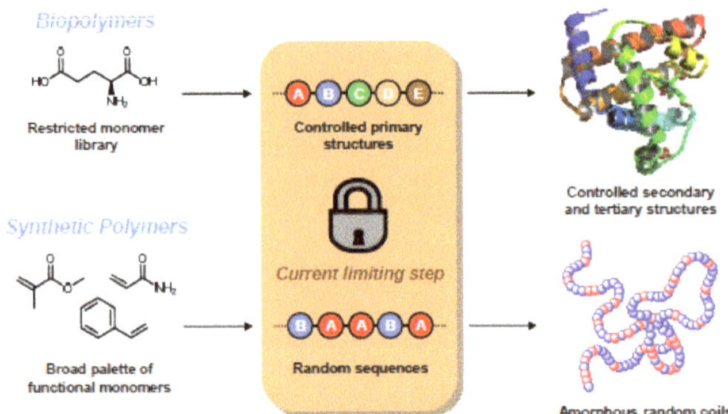


**Figure 2.** A schematic representation of controlled comonomer sequences: a key-step toward highly organized polymer-based materials [88].

In biomedical applications, the photosensitive resin used should integrate without having a toxic effect on the living system, and be biodegradable and bioactive for biomedical applications [112]. Current research is tremendously focused on the development of light-curable and highly biocompatible resin for SLA, such as poly(ε-caprolactone) (PCL) based materials, poly(propylene fumarate) (PPF), poly(D,L-lactide) (PDLLA) resins, and polyethylene glycol (PEG)-based resins, polyglycolic acid (PGA), and polylactic-co-glycolic acid (PLGA) [115–119] which are resins that are usually composed with a photoinitiator, polymerizable oligomers, and a reactive or non-reactive diluent and additives [120].

The amorphousness of poly(D, L-lac-tide) (PDLLA) has successfully been applied in resorbable bone-fixation devices clinically [121]. A PDLLA-based resin was developed using ethyl lactate with a non-reactive diluent such as methyl methacrylate, butanedimethacrylate and N-vinyl-2-pyrroli (NVP) [119]. PDLLA resin material has a glass transition temperature of approximately 55 °C and an elasticity modulus up to 3 GPa, making it one of the biodegradable polymers with mechanical properties that closely similar to the E-modulus of bone (3 to 30 GPa) [122]. PDLLA-based materials by SLA would optimize structures for bone-tissue engineering with regard to mechanical properties and cell seeding [123]. Other than that, Poly(D,L-lactide) (PDLLA) also has been successfully applied in resorbable bone-fixation devices clinically [124] and is well-suited for bone tissue engineering [125]. However, polylactides have occasionally been found to undergo rapid bulk degradation. Degradation products of these materials reduce local pH, accelerating the polyester degradation rate and leading to a localized acidosis and inflammation [126,127].

In 2007, Lee et al. successfully synthesized and modified poly(propylene fumarate) (PPF) by adding diethyl fumarate (PPF/DEF) resin using the SLA process. PPF is a biodegradable and UV-curable material, as an SLA resin. Since synthetic PPF has a high viscosity, it cannot be used in SLA systems directly. Therefore, DEF was added to reduce the viscosity, and photoinitiator dimethoxy phenyl acetophenon (DMPA) [102] or bisacrylphosphrine oxide (BAPO) [118] was used to initiate the UV polymerization of PPF/DEF. The researcher proved the mechanical properties and cell adhesion of the PPF/DEF scaffold has good potential as a bone scaffold for tissue engineering and the finding in this research showed that the measured mechanical properties of the PPF/DEF scaffold were similar to those of human trabecular bone, which proved that the possibility of the PPF/DEF scaffold as a bone scaffold [102].

Another one of the suitable resins is a photocrosslinkable poly (e-caprolactone) (PCL), which has been studied by Eloma et al. (2011). Poly (e-caprolactone) (PCL) is a highly biocompatible elastic polymer with a low melting temperature. The researcher developed a PCL resin and applied it using SLA, without any additional solvents required during the structure-preparation process. Results recorded the photo-crosslinkable and highly interconnected pore network, and biodegradable PCL resin for the solvent-free fabrication of tissue engineering scaffolds by stereolithography with no observable material shrinkage in 3-D scaffolds produced [118].

Even though there are many types of synthetic polymeric biomaterial which are noteworthy in use, most researchers declared that poly (ethylene) glycol (PEG) resin was widely used in biomedicine because of its excellent biocompatibility and hydrophilicity efficiency, making it appropriate for biomedical applications [110,128–135]. Due to its features as non-toxic, non-immunogenic and being readily removed from the body, PEG synthetic hydrogel polymer is widely used in tissue regeneration. The hydrogel-based scaffolds provide an environment with a high water content, enabling high cell-encapsulation densities [13,79,93]. These hydrogels are permeable to oxygen, nutrients and other water-soluble metabolites and have a smooth consistency that makes them soft-tissue-like [127]. Further, because PEG hydrogels are water-soluble, their chains can be easily modified by photoreactive and cross linkable groups such as acrylates or methacrylates with the high crosslinked hydrophilic polymer network that is recommended for use in a diversity of biomedical applications. These hydrogels exhibit 1–100 kPa of mechanical strength [112] and offer flexible, tunable mechanical properties and are soft due to being intermolecular crosslinked to form a stability similar to tissue of the body.

The list of photocurable resin material SLA that have been studied in the development of scaffolds is compiled in Table 6. Based on the review, it can be expected that the different selection of resin materials used in a photo-polymerization 3D printing TE scaffold greatly influences the mechanical properties of the scaffold produced.

*Commercial Available Materials and Global Market in the Future Prospective*

Biomaterials used in tissue engineering can generally be categorized according to their origin by category: natural polymers, synthetic polymers, ceramics and metals. Each of these biomaterials has particular benefits and disadvantages. The metals group is not a suitable option for applications in tissue engineering scaffolds as they are not biodegradable, and their processability is very limited. Ceramic scaffolds have high mechanical stiffness (Young's modulus), low elasticity and a brittle surface. They are highly biocompatible due to their chemical and structural likeness to the bone-mineral phase. However, their clinical applications for tissue engineering are restricted due to their fragility, implantation difficulty [64]. For these reasons, polymeric biomaterials have become increasingly common, due to their biodegradable and biocompatible properties. The group of polymer biomaterials is very efficient because they can be produced with a tailor-made design and their degradation characteristics can be controlled by varying the polymer itself or the individual polymer structure [103].

Polymeric scaffolds play very significant roles in tissue engineering as they are intended to bring cells together and regulate their function to enhance tissue growth [41]. Due to their distinctive characteristics such as a high surface-to-volume ratio, good in biodegradability and mechanical properties, polymeric scaffolds draw tremendous attention. They also offer various advantages of bio-compatibility, chemistry versatility and biological characteristics that are important for tissue engineering and organ substitution [31].

**Table 6.** Mechanical properties of various biomaterial resins for photo-polymerization 3D TE scaffolds.

| Resin Materials | Filler/ Additive | Ratio | Photo Initiator | Ratio | Diluent | Ratio | Young's Modulus (GPa) | Tensile Strength (MPa) | Porosity | Ref. |
|---|---|---|---|---|---|---|---|---|---|---|
| Poly(D,L-lactide) (PDLLA) | Fumaric acid mono ethyl ester (FAME) | | Lucirin-TPO | 5 wt% | N-vinyl-2-pyrrolidone (NVP) | 35 wt% NVP | n/a | n/a | 76% | [116] |
| | | | | 6 wt% | | 0 wt% NVP | Dry: 0.01 | Dry: 1.30 | | |
| | | | | | | 30 wt% NVP | Dry: 1.50 ± 0.1 Wet: 0.80 ± 0.1 | Dry: 42.0 ± 4 Wet: 20.0 ± 3 | | |
| | | | | | | 40 wt% NVP | Dry: 1.80 ± 0.1 Wet: 0.80 ± 0.1 | Dry: 34.0 ± 10 Wet: 19.0 ± 1 | | |
| Poly (propylene fumarate) (PPF) | hydroxyl apatite (HA) | 7 wt% | Bisacylph osphine oxide (BAPO) | 1 wt% | Diethyl fumarate (DEF) | 75 wt% | 0.02–0.2 | 20–70 | 95% | [108] |
| | | | | | | 30 wt% | n/a | n/a | 330 μm to 360 μm | [134] |
| | | 0 wt% | | 1 wt% | | 30 wt % | n/a 0.026 ± 0.001 | n/a 0.6 ± 0.2 | 65% | [136] |
| ethylene | nano | 0.3 wt % | Phenyl-2,4,6-trimethylb enzoylpho sphinate (LAP) | | | | n/a | 1.2 ± 0.3 | | [137] |
| (PEGDA) | | 0.5 wt% | | | | | n/a | 0.5 ± 0.1 | | |
| Poly(ethyle ne glycol) diacrylate (PEGDA) | Methyl methacryl ate (MMA) | 50 %mol | | | | | 32.68 | 2.78 | | [136] |
| | Butyl methacryl ate (BMA) | 15 %mol | | | | | | | | |
| | Methyl methacryl ate (MMA) | 70 %mol | | | | | 260.41 | 10.81 | | |
| | Butyl methacryl ate (BMA) | 22 %mol | | | | | | | | |

Biodegradable scaffolds can actually be fabricated as naturally derived and synthetic. Previous studies indicate that most natural biomaterials, such as collagen, chitosan, hyaluronic acid, elastin and gelatin, are suitable for the development of liver, nerve, bone and heart tissue [6]. Natural polymers can be classified as the first biomaterials to be used clinically in tissue engineering scaffolds. They can be categorized as enzymes (cellulose, amylose) or polynucleotides (DNA, RNA) [31]. Natural polymers of biomaterials have excellent biocompatibility and potential viability. They also establish bioactive characteristics and better cell interactions, which enable them to improve the efficiency of the cells in the biological system. However, one of limitations of natural polymers is their poor mechanical properties, as the natural materials have limited physical and mechanical stability. Therefore, they are not preferred for load-bearing scaffold applications. By using synthetic polymers, the issues associated with natural polymers can be eliminated by using synthetic polymers, as their physical and chemical properties can be changed and repeatedly produced [6].

Synthetic polymers are incredibly beneficial in the biomedical industry due to their properties such as chemical modification, excellent biocompatibility, high versatility, optimal mechanical properties, porosity, degradation time and mechanical characteristics that can be tailored for targeted applications under control [138,139]. Other than that, synthetic polymers cost less than natural polymers for biological scaffolds, which can be manufactured in huge quantities and have a significant shelf life. Synthetic polymers are commonly split into two groups: non-biodegradable and biodegradable. In tissue engineering, the most frequently used biodegradable synthetic polymers for 3D scaffolds are poly(lactic acid) (PLA) and poly(glycolic acid) (PGA), polycaprolactone (PCL), poly(lactic-co-lycolide) (PLGA) copolymers and so on; while an example group of non-biodegradable polymers includes polyvinyl alcohol (PVA), polyhydroxyethyl methacrylate (PHEMA), poly(N-isopropylacrylamide) (PNIPAM) and others.

## 5. Biocompatibility Test

Biocompatibility is a term that covers many aspects of a material, including its physical, mechanical, and chemical properties, as well as its potential in cytotoxicity, therefore there are no significant injuries or toxic effects on the biological function of cells which can possibly inhibit the beneficial properties of a cell ECM scaffold. The term biocompatibility is also defined not only by the lack of the cytotoxicity of a biomaterial but also by the bio-functionality of the material, which enables it to support cell–biomaterial interactions [133,140]. In tissue engineering applications, a scaffold must be non-toxic and biologically compatible so that cells can safely adhere, proliferate, and differentiate within the scaffold [134]. Toxicity and biocompatibility tests are needed to evaluate a scaffold material in facilitating cell proliferation and differentiation, secreting an extracellular matrix and carrying biomolecular signals for cell communication [135]. In measuring biocompatibility, there are several varieties of tests that are currently used to identify whether new materials are biologically acceptable. These tests are classified as the in-vitro, in-vivo, and usage tests.

The most common biocompatibility test for a tissue engineering scaffold is an in-vitro test. The testing is performed outside a living organism, requiring the placement of a material or a component of a material in contact with a cell, enzyme, or some other isolated biological system. In-vitro tests can be roughly subdivided into those that measure cytotoxicity or cell growth. The cytotoxicity tests assess cell death caused by a material by measuring cell number or growth before and after exposure to materials. Membrane-permeability tests are used to measure cytotoxicity by the ease with which a dye can pass through a cell membrane, because membrane permeability is equivalent to or very nearly equivalent to cell death [136]. Some in-vitro tests for biocompatibility use the biosynthetic or enzymatic activity of cells to assess cytotoxic response. A standard method to analyse the biocompatibility properties describes details in the standard testing ISO 10993, which includes a series of guidelines for analyzing biocompatibility and medical devices.

On the other hand, in the field of biocompatibility, some scientists questioned the usefulness of in-vitro and animal tests due to the apparent lack of correlation with usage tests and the clinical history of materials. Furthermore, barriers between the material and tissues may exist in usage tests or clinical use but may not exist in the in-vitro or animal tests. Thus, it is important to remember that each type of test has been designed to measure different aspects of a biological response to a material, and correlation is not always to be expected. Among the biocompatibility tests, the in-vitro tests have several significant advantages over other types of biocompatibility tests (Table 7). In-vitro tests are relatively quick with a lower cost than animal or usage tests, are well-suited to large-scale screening, and can be tightly controlled [136].

**Table 7.** Advantages and disadvantages of biocompatibility tests [136].

| Test | Advantages | Disadvantages |
|---|---|---|
| In-vitro tests | Fast testing Expensive Can be standardized Large-scale screening Good experimental control Excellence for mechanisms of interactions | Relevance to in vivo is questionable |
| In-vivo tests | Allows complex systemic interactions Responds more comprehensively than in-vitro tests More relevant than in-vitrotests | Expensive Time consuming Legal/ethical concerns Difficult to control Difficult to interpret and quantify |
| Usage tests | Relevance to use of the material is assured | Expensive Time consuming Difficult to control Difficult to interpret and quantify |

The concept of correlation between the in-vitro and in-vivo tests has been reported, confirming the advantage of in-vitro tests as a system to select the biomaterials. In cytotoxicity testing, the same type of cells is used. The testing of scaffolds must have cells derived from the tissue origin to ensure a better simulation of the clinical situation [137]. For example, scaffolds derived from orthopedic tissues should be tested on osteoblasts or osteoblast-like cells; cardiovascular-derived scaffolds should be assayed using endothelial cells or cardiomyocytes. For in-vitro cytotoxicity screening, the recommended testing methods include (i) indirect contact assay or the extraction method, and (ii) a direct contact assay. The guidelines on inspecting the biocompatibility of materials for medical applications is set in the International Standard ISO10993 (International Organization for Standardization, 1999), with priority being given to cell-culture-based in-vitro tests using both the direct and indirect contact approaches.

*5.1. Indirect Contact Assay*

The indirect contact assay technique applies cell counting, dye-binding, metabolic impairment, or membrane integrity as endpoints of the cytotoxicity test or assay to assess the short-term cytotoxicity of medical devices [138]. The objectives of the extraction test are to evaluate changes in cell morphology and growth inhibition, and determine whether cells are metabolically active [139].

An ISO guideline (10993- 5:2009) refers to the MTT cytotoxicity assay. MTT is a colorimetric method that measures the reduction of water-soluble yellow 3-(4,5-dimethylthiazol-2-yl)-2,5-diphenyl tetrazolium bromide by mitochondrial succinate dehydrogenase into an insoluble, blue-violet formazan. The number of viable cells correlates to the color intensity determined by photometric measurements after dissolving the formazan. The tested material is considered non-cytotoxic if the percentage of the viable cell is greater than or equal to 70% of the untreated control [124].

*5.2. Direct Contact Assay*

There are many limitations in the current tests used to check the effects of decellularized scaffolds on cells where the results can interfere with the presence of the biomaterial. It is important to understand that unleachable toxic substances that do not pass into the extraction medium can only be proven by direct cell contact [100]. In a direct contact assay, the sample is placed in direct contact with cells by surface culturing. Cells are examined at different time points for signs of toxicity by morphological examination and viability tests [140].

Most of the researchers preferred to test the cytocompatibility of decellularized scaffolds using direct contact assay, as it allows physiological changes made through the interactions of cells with a biomaterial, compared to the MTT assay, which only focuses on toxicity at a cellular level with less consideration of the molecular level [132].

## 6. Summary

This review summarized the application and advantages of the additive manufacturing (AM) technique via photo-polymerization 3D printing as a versatile platform in scaffold fabrication. Even though there are many techniques offered for scaffold fabrication, there are lots of characteristics and requirements that need to be considered in providing a scaffold with good mechanical, internal structure architecture, and biocompatibility properties. Photo-polymerization 3D printing has been reviewed as the most versatile technique and has the capability to produce a high accuracy dimensional architecture of a scaffold with flexibility in design.

Overall, an ideal selected fabrication TE scaffold should be carefully considered and capable of controling the variety in characteristic scaffold parameters needed in order to mimic the natural structure and properties of bone tissue.

**Funding:** This research was funded by Universiti Teknologi Malaysia (UTM) under grant UTM R&D Fund (4J506) and UTM Shine (09G94) as well funded by Universiti Kuala Lumpur under Short Term Research Grant (STRG 18038).

**Institutional Review Board Statement:** Ethical review and approval were waived for this study where this study did not require ethical approval.

**Informed Consent Statement:** Not applicable.

**Data Availability Statement:** No new data were created or analyzed in this study. Data sharing is not applicable to this article.

**Acknowledgments:** The authors wish to thank the Ministry of Higher Education (MOHE), Universiti Teknologi Malaysia (UTM) and Research Management Center, UTM for the financial support to this work under scheme UTM R&D Fund (4J506) and UTMShine (09G94). Many thanks to Universiti Kuala Lumpur for research funding under Short Term Research Grant (STRG 18038).

**Conflicts of Interest:** The authors declare no conflict of interest.

## References

1. Leong, K.F.; Chua, C.K.; Sudarmadji, N.; Yeong, W.Y. Engineering functionally graded tissue engineering scaffolds. *J. Mech. Behav. Biomed. Mater.* **2008**, *1*, 140–152. [CrossRef] [PubMed]
2. Yeong, W.Y.; Chua, C.K.; Leong, K.F.; Chandrasekaran, M. Rapid prototyping in tissue engineering: Challenges and potential. *Trends Biotechnol.* **2004**, *22*, 643–652. [CrossRef] [PubMed]
3. Pluta, K.; Malina, D.; Sobczak-Kupiec, A. Scaffolds for Tissue Engineering. *Tech. Trans. Chem.* **2015**, *1*, 17–27.
4. Karunakaran, R.; Kennedy, J.P. Novel amphiphilic conetworks by synthesis and crosslinking of allyl-telechelic block copolymers. *J. Polym. Sci. Part A Polym. Chem.* **2008**, *46*, 4254–4257. [CrossRef]
5. Chandel, A.K.S.; Shimizu, A.; Hasegawa, K.; Ito, T. Advancement of Biomaterial-Based Postoperative Adhesion Barriers. *Macromol. Biosci.* **2021**, *21*, 2000395. [CrossRef]
6. Li, J.; Wang, K.; Wang, J.; Yuan, Y.; Wu, H. High-tough hydrogels formed via Schiff base reaction between PAMAM dendrimer and Tetra-PEG and their potential as dual-function delivery systems. *Mater. Today Commun.* **2022**, *30*, 103019. [CrossRef]
7. Wang, M. Materials selection and scaffold fabrication for tissue engineering in orthopaedics. In *Advanced Bioimaging Technologies in Assessment of the Quality of Bone and Scaffold Materials*; Springer: Berlin/Heidelberg, Germany, 2007; pp. 259–288.
8. Chandel, A.K.S.; Bera, A.; Nutan, B.; Jewrajka, S.K. Reactive compatibilizer mediated precise synthesis and application of stimuli responsive polysaccharides-polycaprolactone amphiphilic co-network gels. *Polymer* **2016**, *99*, 470–479. [CrossRef]
9. Nutan, B.; Chandel, A.K.S.; Bhalani, D.V.; Jewrajka, S.K. Synthesis and tailoring the degradation of multi-responsive amphiphilic conetwork gels and hydrogels of poly(β-amino ester) and poly(amido amine). *Polymer* **2017**, *111*, 265–274. [CrossRef]
10. Nutan, B.; Chandel, A.K.S.; Jewrajka, S.K. Liquid prepolymer-based in situ formation of degradable poly(ethylene glycol)-linked-poly(caprolactone)-linked-poly(2-dimethylaminoethyl) methacrylate amphiphilic conetwork gels showing polarity driven gelation and bioadhesion. *ACS Appl. Bio Mater.* **2018**, *1*, 1606–1619. [CrossRef]
11. Sachlos, E.; Czernuszka, J.T. Making tissue engineering scaffolds work. Review: The application of solid freeform fabrication technology to the production of tissue engineering scaffolds. *Eur. Cell Mater.* **2003**, *5*, 29–40. [CrossRef]
12. Langer, R.; Vacanti, J.P. Tissue engineering. *Science* **1993**, *260*, 920–926. [CrossRef] [PubMed]
13. Zhu, W.; Ma, X.; Gou, M.; Mei, D.; Zhang, K.; Chen, S. 3D printing of functional biomaterials for tissue engineering. *Curr. Opin. Biotechnol.* **2016**, *40*, 103–112. [CrossRef] [PubMed]
14. Gauvin, R.; Chen, Y.C.; Lee, J.W.; Soman, P.; Zorlutuna, P.; Nichol, J.W.; Bae, H.; Chen, S.; Khademhosseini, A. Microfabrication of complex porous tissue engineering scaffolds using 3D projection stereolithography. *Biomaterials* **2012**, *33*, 3824–3834. [CrossRef]
15. Mondschein, R.J.; Kanitkar, A.; Williams, C.B.; Verbridge, S.S.; Long, T.E. Polymer structure-property requirements for stereolithographic 3D printing of soft tissue engineering scaffolds. *Biomaterials* **2017**, *140*, 170–188. [CrossRef] [PubMed]
16. Osama, A.A.; Darwish, S.M. Fabrication of Tissue Engineering Scaffolds Using Rapid Prototyping Techniques. *Int. J. Ind. Manuf. Eng.* **2011**, *5*, 2317–2325.
17. Ning, Z.; Che, X. Biofabrication of Tissue Scaffolds. In *Advances in Biomaterials Science and Biomedical Applications*; Pignatello, R., Ed.; Books on Demand: Norderstedt, Germany, 2013; pp. 315–328.
18. Subia, B.; Kundu, J. Biomaterial Scaffold Fabrication Techniques for Potential Tissue Engineering Applications. In *Tissue Engineering*; Books on Demand: Norderstedt, Germany, 2010; pp. 141–159.
19. Langer, R.; Tirrell, D.A. Designing Materials for Biology and Medicine. *Nature* **2004**, *428*, 487–492. [CrossRef]
20. Andy, P.; Popis, K.Č.J. Spin to Win: Polymers in Regenerative Medicine. *Mater. Today* **2011**, *14*, 88–95.
21. Moroni, L.; Schrooten, J.; Truckenmüller, R.; Rouwkema, J.; Sohier, J.; Van Blitterswijk, C.A. Tissue Engineering: An Introduction. In *Tissue engineering*, 2nd ed.; Academic Press: Cambridge, MA, USA, 2014; pp. 1–21.
22. Ngadiman, N.H.A.; Noordin, M.Y.; Idris, A.; Kurniawan, D. A review of evolution of electrospun tissue engineering scaffold: From two dimensions to three dimensions. *Proc. Inst. Mech. Eng. Part H J. Eng. Med.* **2017**, *231*, 597–616. [CrossRef]
23. Ikada, Y. Chapter 1—Scope of Tissue Engineering. In *Interface Science and Technology*; Academic Press: Cambridge, MA, USA, 2006; pp. 1–89.
24. Carvalho, J.L.; de Carvalho, P.H.; Gomes, D.A.; de Goes, A.M. Innovative Strategies for Tissue Engineering. In *Advances in Biomaterials Science and Biomedical Applications*; Books on Demand: Norderstedt, Germany, 2013; pp. 295–313.
25. Porter, J.R.; Ruckh, T.T.; Popat, K.C. Bone tissue engineering: A review in bone biomimetics and drug delivery strategies. *Biotechnol. Prog.* **2009**, *25*, 1539–1560. [CrossRef]
26. Hutmacher, D.W.; Schantz, T.; Zein, I.; Ng, K.W.; Teoh, S.H.; Tan, K.C. Mechanical properties and cell cultural response of polycaprolactone scaffolds designed and fabricated via fused deposition modelling. *J. Biomed. Mater. Res.* **2001**, *55*, 203–216. [CrossRef]
27. Ronca, A.; Ambrosio, L.; Grijpma, D.W. Preparation of designed poly(D,L-lactide)/nanosized hydroxyapatite composite structures by stereolithography. *Acta Biomater.* **2013**, *9*, 5989–5996. [CrossRef] [PubMed]
28. Hollister, S.J.; Maddox, R.D.; Taboas, J.M. Optimal design and fabrication of scaffolds to mimic tissue properties and satisfy biological constraints. *Biomaterials* **2002**, *23*, 4095–4103. [CrossRef]
29. Banoriya, D.; Purohit, R.; Dwivedi, R.K. Advanced Application of Polymer based Biomaterials. *Mater. Today Proc.* **2017**, *4*, 3534–3541. [CrossRef]
30. Bracaglia, L.G.; Smith, B.T.; Watson, E.; Arumugasaamy, N.; Mikos, A.G.; Fisher, J.P. 3D printing for the design and fabrication of polymer-based gradient scaffolds. *Acta Biomater.* **2017**, *56*, 3–13. [CrossRef]

31. Dhandayuthapani, B.; Yoshida, Y.; Maekawa, T.; Kumar, D.S. Polymeric scaffolds in tissue engineering application: A review. *Int. J. Polym. Sci.* **2011**, *2011*, 290602. [CrossRef]
32. Cima, L.G.; Vacanti, J.P.; Vacanti, C.; Ingber, D.; Mooney, D.; Langer, R. Tissue Engineering by Cell Transplantation Using Degradable Polymer Substrates. *J. Biomech. Eng.* **1991**, *113*, 143–151. [CrossRef] [PubMed]
33. Chen, W.; Ma, J.; Zhu, L.; Morsi, Y.; Al-Deyab, S.S.; Mo, X. Superelastic, superabsorbent and 3D nanofiber-assembled scaffold for tissue engineering. *Colloids Surf. B Biointerfaces* **2016**, *142*, 165–172. [CrossRef]
34. Nejad, Z.M.; Zamanian, A.; Saeidifar, M.; Vanaei, H.R.; Amoli, M.S. 3D Bioprinting of Polycaprolactone-Based Scaffolds for Pulp-Dentin Regeneration: Investigation of Physicochemical and Biological Behavior. *Polymers* **2021**, *13*, 4442. [CrossRef]
35. Vanaei, S.; Parizi, M.S.; Salemizadehparizi, F.; Vanaei, H.R. An overview on materials and techniques in 3D bioprinting toward biomedical application. *Eng. Regen.* **2021**, *2*, 1–18. [CrossRef]
36. Farahani, R.D.; Dubé, M.; Therriault, D. Three-Dimensional Printing of Multifunctional Nanocomposites: Manufacturing Techniques and Applications. *Adv. Mater.* **2016**, *28*, 5794–5821. [CrossRef]
37. Vlasea, M.; Shanjani, Y.; Basalah, A.; Toyserkani, E. Additive Manufacturing of Scaffolds for Tissue Engineering of Bone and Cartilage. *Int. J. Adv. Manuf. Technol.* **2015**, *13*, 124–141.
38. Sin, D.; Miao, X.; Liu, G.; Wei, F.; Chadwick, G.; Yan, C.; Friis, T. Polyurethane (PU) scaffolds prepared by solvent casting/particulate leaching (SCPL) combined with centrifugation. *Mater. Sci. Eng. C* **2010**, *30*, 78–85. [CrossRef]
39. Prasad, A.; Sankar, M.R.; Katiyar, V. State of Art on Solvent Casting Particulate Leaching Method for Orthopedic Scaffolds Fabrication. *Mater. Today Proc.* **2017**, *4*, 898–907. [CrossRef]
40. Thavornyutikarn, B.; Chantarapanich, N.; Sitthiseripratip, K.; Thouas, G.A.; Chen, Q. Bone tissue engineering scaffolding: Computer-aided scaffolding techniques. *Prog. Biomater.* **2014**, *3*, 61–102. [CrossRef] [PubMed]
41. Sultana, N. Fabrication techniques and properties of scaffolds. In *Springer Briefs in Applied Sciences and Technology*; Springer: Berlin/Heidelberg, Germany, 2013; pp. 19–42.
42. Ma, P.X. Scaffolds for tissue fabrication. *Mater. Today* **2004**, *7*, 30–40. [CrossRef]
43. Sughanthy, A.P.; Ansari, M.N.; Siva, A.P.S. Review on Bone Scaffold Fabrication Methods. *Int. Res. J. Eng. Technol.* **2015**, *2*, 1232–1238.
44. Trachtenberg, J.E.; Kasper, F.K.; Mikos, A.G. Polymer Scaffold Fabrication. In *Principles of Tissue Engineering*, 4th ed.; Academic Press: Cambridge, MA, USA, 2013; pp. 423–440.
45. Chronakis, I.S. *Micro-/Nano-Fibers by Electrospinning Technology: Processing, Properties and Applications*, 1st ed.; Elsevier Ltd.: Amsterdam, The Netherlands, 2010.
46. Persano, L.; Camposeo, A.; Tekmen, C.; Pisignano, D. Industrial upscaling of electrospinning and applications of polymer nanofibers: A review. *Macromol. Mater. Eng.* **2013**, *298*, 504–520. [CrossRef]
47. Mohammadzadehmoghadam, S.; Dong, Y.; Davies, I.J. Recent progress in electrospun nanofibers: Reinforcement effect and mechanical performance. *J. Polym. Sci. Part B Polym. Phys.* **2015**, *53*, 1171–1212. [CrossRef]
48. Agarwal, S.; Wendorff, J.H.; Greiner, A. Progress in the field of electrospinning for tissue engineering applications. *Adv. Mater.* **2009**, *21*, 3343–3351. [CrossRef]
49. Yu, Y.; Hua, S.; Yang, M.; Fu, Z.; Teng, S.; Niu, K.; Zhao, Q.; Yi, C. Fabrication and characterization of electrospinning/3D printing bone tissue engineering scaffold. *RSC Adv.* **2016**, *6*, 110557–110565. [CrossRef]
50. Do, A.V.; Khorsand, B.; Geary, S.M.; Salem, A.K. 3D Printing of Scaffolds for Tissue Regeneration Applications. *Adv. Healthc. Mater.* **2015**, *4*, 1742–1762. [CrossRef]
51. Wu, G.H.; Hsu, S.H. Review: Polymeric-based 3D printing for tissue engineering. *J. Med. Biol. Eng.* **2015**, *35*, 285–292. [CrossRef] [PubMed]
52. Mota, R.C.D.A.G.; da Silva, E.O.; de Lima, F.F.; de Menezes, L.R.; Thiele, A.C.S. 3D Printed Scaffolds as a New Perspective for Bone Tissue Regeneration: Literature Review. *Mater. Sci. Appl.* **2016**, *7*, 430–452.
53. Wang, X.; Ao, Q.; Tian, X.; Fan, J.; Wei, Y.; Hou, W.; Tong, H.; Bai, S. 3D bioprinting technologies for hard tissue and organ engineering. *Materials* **2016**, *9*, 802. [CrossRef] [PubMed]
54. Bártolo, P.J.S. State of the art of solid freeform fabrication for soft and hard tissue engineering. In *Design and Nature III: Comparing Design in Nature with Science and Engineering*; WIT Press: Billerica, MA, USA, 2006; Volume 1, pp. 233–243.
55. Chia, H.N.; Wu, B.M. Recent advances in 3D printing of biomaterials. *J. Biol. Eng.* **2015**, *9*, 4. [CrossRef] [PubMed]
56. An, J.; Teoh, J.E.M.; Suntornnond, R.; Chua, C.K. Design and 3D Printing of Scaffolds and Tissues. *Engineering* **2015**, *1*, 261–268. [CrossRef]
57. Kantaros, A.; Chatzidai, N.; Karalekas, D. 3D printing-assisted design of scaffold structures. *Int. J. Adv. Manuf. Technol.* **2016**, *82*, 559–571. [CrossRef]
58. Liu, W.; Li, Y.; Liu, J.; Niu, X.; Wang, Y.; Li, D. Application and Performance of 3D Printing in Nanobiomaterials. *J. Nanomater.* **2013**, *2013*, 681050. [CrossRef]
59. Bose, S.; Vahabzadeh, S.; Bandyopadhyay, A. Bone tissue engineering using 3D printing. *Mater. Today* **2013**, *16*, 496–504. [CrossRef]
60. Roseti, L.; Parisi, V.; Petretta, M.; Cavallo, C.; Desando, G.; Bartolotti, I.; Grigolo, B. Scaffolds for Bone Tissue Engineering: State of the art and new perspectives. *Mater. Sci. Eng. C* **2017**, *78*, 1246–1262. [CrossRef]
61. Lipowiecki, M.; Ryvolova, M.; Töttösi, Á.; Kolmer, N.; Naher, S.; Brennan, S.A.; Vázquez, M.; Brabazon, D. Permeability of rapid prototyped artificial bone scaffold structures. *J. Biomed. Mater. Res. Part A* **2014**, *102*, 4127–4135. [CrossRef]

62. Bandyopadhyay, A.; Vahabzadeh, S.; Shivaram, A.; Bose, S. Three-dimensional printing of biomaterials and soft materials. *MRS Bull.* **2015**, *40*, 1162–1169. [CrossRef]
63. Manapat, J.Z.; Chen, Q.; Ye, P.; Advincula, R.C. 3D Printing of Polymer Nanocomposites via Stereolithography. *Macromol. Mater. Eng.* **2017**, *302*, 1600553. [CrossRef]
64. Childers, E.P.; Wang, M.O.; Becker, M.L.; Fisher, J.P.; Dean, D. 3D printing of resorbable poly(propylene fumarate) tissue engineering scaffolds. *MRS Bull.* **2015**, *40*, 119–126. [CrossRef]
65. Hokmabad, V.R.; Davaran, S.; Ramazani, A.; Salehi, R. Design and fabrication of porous biodegradable scaffolds: A strategy for tissue engineering. *J. Biomater. Sci. Polym.* **2017**, *28*, 1797–1825. [CrossRef]
66. Seck, T.M.; Melchels, F.P.; Feijen, J.; Grijpma, D.W. Designed biodegradable hydrogel structures prepared by stereolithography using poly(ethylene glycol)/poly(D,L-lactide)-based resins. *J. Control Release* **2010**, *148*, 34–41. [CrossRef]
67. Gu, B.K.; Choi, D.J.; Park, S.J.; Kim, M.S.; Kang, C.M.; Kim, C.H. 3-Dimensional Bioprinting for Tissue Engineering Applications. *Biomater. Res.* **2016**, *20*, 12. [CrossRef]
68. Wang, X.; Jiang, M.; Zhou, Z.; Gou, J.; Hui, D. 3D printing of polymer matrix composites: A review and prospective. *Compos. Part B Eng.* **2017**, *110*, 442–458. [CrossRef]
69. Wurm, M.C.; Möst, T.; Bergauer, B.; Rietzel, D.; Neukam, F.W.; Cifuentes, S.C.; Wilmowsky, C.V. In-vitro evaluation of Polylactic acid (PLA) manufactured by fused deposition modeling. *J. Biol. Eng.* **2017**, *11*, 29. [CrossRef]
70. Velu, R.; Singamneni, S. Evaluation of the influences of process parameters while selective laser sintering PMMA powders. *Proc. Inst. Mech. Eng. Part C J. Mech. Eng. Sci.* **2014**, *229*, 603–613. [CrossRef]
71. Abrego, C.J.G.; Dedroog, L.; Deschaume, O.; Wellens, J.; Vananroye, A.; Lettinga, M.P.; Patterson, J.; Bartic, C. Multiscale Characterization of the Mechanical Properties of Fibrin and Polyethylene Glycol (PEG) Hydrogels for Tissue Engineering Applications. *Macromol. Chem. Phys.* **2022**, *223*, 2100366. [CrossRef]
72. Mota, C.; Puppi, D.; Chiellini, F.; Chiellini, E. Additive manufacturing techniques for the production of tissue engineering constructs. *J. Tissue Eng. Regen. Med.* **2015**, *9*, 174–190. [CrossRef] [PubMed]
73. O'Brien, C.M.; Holmes, B.; Faucett, S.; Zhang, L.G. Three-Dimensional Printing of Nanomaterial Scaffolds for Complex Tissue Regeneration. *Tissue Eng. Part B* **2015**, *21*, 103–114. [CrossRef] [PubMed]
74. Annabi, N.; Nichol, J.W.; Zhong, X.; Ji, C.; Koshy, S.; Khademhosseini, A.; Dehghani, F. Controlling the Porosity and Microarchitecture of Hydrogels for Tissue Engineering. *Tissue Eng. Part B* **2010**, *16*, 371–383. [CrossRef] [PubMed]
75. Stansbury, J.W.; Idacavage, M.J. 3D printing with polymers: Challenges among expanding options and opportunities. *Dent. Mater.* **2016**, *32*, 54–64. [CrossRef]
76. Billiet, T.; Vandenhaute, M.; Schelfhout, J.; Van Vlierberghe, S.; Dubruel, P. A review of trends and limitations in hydrogel-rapid prototyping for tissue engineering. *Biomaterials* **2012**, *33*, 6020–6041. [CrossRef]
77. Ferry, P.W.M.; Feijen, J.; Grijpma, D.W. A review on stereolithography and its applications in biomedical engineering. *Biomaterials* **2010**, *31*, 6121–6130.
78. Hollister, S.; Bergman, T. Biomedical applications of integrated additive/subtractive manufacturing. *Addit. Manuf. Res.* **2004**, *1001*, 55–62.
79. Skoog, S.A.; Goering, P.L.; Narayan, R.J. Stereolithography in tissue engineering. *J. Mater. Sci. Mater. Med.* **2014**, *25*, 845–856. [CrossRef]
80. Zorlutuna, P.; Annabi, N.; Camci-Unal, G.; Nikkhah, M.; Cha, J.M.; Nichol, J.W.; Manbachi, A.; Bae, H.; Chen, S.; Khademhosseini, A. Microfabricated biomaterials for engineering 3D tissues. *Adv. Mater.* **2012**, *24*, 1782–1804. [CrossRef]
81. Pawar, A.A.; Saada, G.; Cooperstein, I.; Larush, L.; Jackman, J.A.; Tabaei, S.R.; Cho, N.J.; Magdassi, S. High-performance 3D printing of hydrogels by water-dispersible photoinitiator nanoparticles. *Sci. Adv.* **2016**, *2*, e1501381. [CrossRef] [PubMed]
82. Anseth, K.S.; Quick, D.J. Polymerizations of multifunctional anhydride monomers to form highly crosslinked degradable networks. *Macromol. Rapid Commun.* **2001**, *22*, 564–572. [CrossRef]
83. Choong, Y.Y.C.; Maleksaeedi, S.; Eng, H.; Su, P.C.; Wei, J. Curing characteristics of shape memory polymers in 3D projection and laser stereolithography. *Virtual Phys. Prototyp.* **2016**, *12*, 77–84. [CrossRef]
84. Gross, B.C.; Erkal, J.L.; Lockwood, S.Y.; Chen, C.; Spence, D.M. Evaluation of 3D printing and its potential impact on biotechnology and the chemical sciences. *Anal. Chem.* **2014**, *86*, 3240–3253. [CrossRef]
85. Low, Z.X.; Chua, Y.T.; Ray, B.M.; Mattia, D.; Metcalfe, I.S.; Patterson, D.A. Perspective on 3D printing of separation membranes and comparison to related unconventional fabrication techniques. *J. Membr. Sci.* **2017**, *523*, 596–613. [CrossRef]
86. Pynaert, R.; Buguet, J.; Croutxé-Barghorn, C.; Moireau, P.; Allonas, X. Effect of reactive oxygen species on the kinetics of free radical photopolymerization. *Polym. Chem.* **2013**, *4*, 2475–2479. [CrossRef]
87. He, Y.; Wu, Y.; Fu, J.Z.; Gao, Q.; Qiu, J.J. Developments of 3D Printing Microfluidics and Applications in Chemistry and Biology: A Review. *Electroanalysis* **2016**, *28*, 1658–1678. [CrossRef]
88. Lutz, J.F. Strategies for Controlling Sequences In Radical Chain-Growth Polymerizations. *Polym. Prepr.* **2011**, *52*, 727.
89. Liska, R.; Schuster, M.; Inführ, R.; Turecek, C.; Fritscher, C.; Seidl, B.; Schmidt, V.; Kuna, L.; Haase, A.; Varga, F.; et al. Photopolymers for rapid prototyping. *J. Coat. Technol. Res.* **2007**, *4*, 505–510. [CrossRef]
90. Pan, Y.; Zhou, C.; Chen, Y. A Fast Mask Projection Stereolithography Process for Fabricating Digital Models in Minutes. *J. Manuf. Sci. Eng.* **2012**, *134*, 051011. [CrossRef]
91. Growth, C.; Graham, J.W.; Redmond, W.R. Three-Dimensional Printing Technology. *J. Clin. Orthod.* **2014**, *48*, 475–485.

92. Melchiorri, A.J.; Hibino, N.; Best, C.A.; Yi, T.; Lee, Y.U.; Kraynak, C.A.; Kimerer, L.K.; Krieger, A.; Kim, P.; Breuer, C.K.; et al. 3D-Printed Biodegradable Polymeric Vascular Grafts. *Adv. Healthc. Mater.* **2016**, *5*, 319–325. [CrossRef]

93. Valentinčič, J.; Peroša, M.; Jerman, M.; Sabotin, I.; Lebar, A. Low cost printer for DLP stereolithography. *J. Mech. Eng.* **2017**, *63*, 559–566. [CrossRef]

94. Türeyen, E.B.; Ali, Z.; Karpat, Y.; ÇaNmaNcı, M. Fabrication of High Aspect Ratio Polymer Structures Using a Digital Micro Mirror Device Based Stereo Lithography Technique. In Proceedings of the 8th International Conference and Exhibiton on Design and Production of Machines Dies/Molds, Aydin, Turkey, 18–21 June 2015; pp. 229–234.

95. Wang, W.L.; Cheah, C.M.; Fuh, J.Y.H.; Lu, L. Influence of process parameters on stereolithography part shrinkage. *Mater. Des.* **1996**, *17*, 205–213. [CrossRef]

96. Clarrisa, Y.Y.; Saeed, M.; Hengky, E. Review of Multi-material Additive Manufacturing. In Proceedings of the 2nd International Conference on Progress in Additive Manufacturing, Singapore, 16–19 May 2016; pp. 294–299.

97. Yanvan, T. Stereolithography Cure Process Modeling. In *Georgia Institute of Technology*; ProQuest Dissertations Publishing: Ann Arbor, MI, USA, 2005.

98. Channasanon, S.; Kaewkong, P.; Uppanan, P.; Chantaweroad, S.; Sitthiseripratip, K.; Tanodekaew, S.; Chantarapanich, N. Acrylic-based Stereolithographic Resins: Effect of Scaffold Architectures on Biological Response. *J. Life Sci. Technol.* **2013**, *1*, 158–162. [CrossRef]

99. Lee, J.W.; Lan, P.X.; Kim, B.; Lim, G.; Cho, D.W. Fabrication and characteristic analysis of a poly(propylene fumarate) scaffold using micro-stereolithography technology. *J. Biomed. Mater. Res. Part B Appl. Biomater.* **2008**, *87*, 1–9. [CrossRef]

100. Loh, Q.L.; Choong, C. Three-Dimensional Scaffolds for Tissue Engineering Applications: Role of Porosity and Pore Size. *Tissue Eng. Part B Rev.* **2013**, *19*, 485–502. [CrossRef]

101. Niesler, F.; Hermatschweiler, M.; Werner, A. Additive manufacturing with NIR lasers forms micro-sized parts. *Laser Focus World* **2014**, *50*, 39–42.

102. Cooke, M.N.; Fisher, J.P.; Dean, D.; Rimnac, C.; Mikos, A.G. Use of stereolithography to manufacture critical-sized 3D biodegradable scaffolds for bone ingrowth. *J. Biomed. Mater. Res.* **2003**, *64*, 65–69. [CrossRef]

103. Song, X.; Zhang, Z.; Chen, Z.; Chen, Y. Porous Structure Fabrication Using a Stereolithography-Based Sugar Foaming Method. *J. Manuf. Sci. Eng.* **2016**, *139*, 15–31. [CrossRef]

104. Cheng, Y.-L.; Lee, M.-L. Development of dynamic masking rapid prototyping system for application in tissue engineering. *Rapid Prototyp. J.* **2009**, *15*, 29–41. [CrossRef]

105. Abdelaal, O.A.M.; Darwish, S.M.H. Review of rapid prototyping techniques for tissue engineering scaffolds fabrication. *Adv. Struct. Mater.* **2013**, *29*, 33–54.

106. Karalekas, D.; Aggelopoulos, A. Study of shrinkage strains in a stereolithography cured acrylic photopolymer resin. *J. Mater. Processing Technol.* **2003**, *136*, 146–150. [CrossRef]

107. Raman, R.; Bashir, R. Chapter 6—Stereolithographic 3D Bioprinting for Biomedical Applications. In *Essentials of 3D Biofabrication and Translation*; Elsevier Inc.: Amsterdam, The Netherlands, 2015; pp. 89–121.

108. Corbel, S.; Dufaud, O.; Roques-Carmes, T. *Stereolithography: Materials, Processes and Applications*; Bartolo, P.J., Ed.; Springer: New York, NY, USA, 2011.

109. Shie, M.Y.; Chang, W.C.; Wei, L.J.; Huang, Y.H.; Chen, C.H.; Shih, C.T.; Chen, Y.W.; Shen, Y.F. 3D printing of cytocompatible water-based light-cured polyurethane with hyaluronic acid for cartilage tissue engineering applications. *Materials* **2017**, *10*, 136. [CrossRef] [PubMed]

110. Jansen, J.; Melchels, F.P.; Grijpma, D.W.; Feijen, J. Fumaric Acid Monoethyl Ester-Functionalized Poly (D,L-lactide)/N-vinyl-2-pyrrolidone Resins for the Preparation of Tissue Engineering Scaffolds by Stereolithography. *Biomacromolecules* **2009**, *10*, 214–220. [CrossRef]

111. Dean, D.; Wallace, J.; Siblani, A.; Wang, M.O.; Kim, K.; Mikos, A.G.; Fisher, J.P. Continuous digital light processing (cDLP): Highly accurate additive manufacturing of tissue engineered bone scaffolds. *J. Virtual Phys. Prototyp.* **2012**, *7*, 13–24. [CrossRef]

112. Lee, K.W.; Wang, S.; Fox, B.C.; Ritman, E.L.; Yaszemski, M.J.; Lu, L. Poly(propylene fumarate) bone tissue engineering scaffold fabrication using stereolithography: Effects of resin formulations and laser parameters. *Biomacromolecules* **2007**, *8*, 1077–1084. [CrossRef]

113. Santos, A.R.C.; Almeida, H.A.; Bártolo, P.J. Additive manufacturing techniques for scaffold-based cartilage tissue engineering. *J. Virtual Phys. Prototyp.* **2013**, *8*, 175–186. [CrossRef]

114. Ronca, A.; Ambrosio, L. Polymer based scaffolds for tissue regeneration by stereolithography. *Adv. Biomater. Devices Med.* **2017**, *4*, 1–15.

115. Melchels, F.P.W.; Feijen, J. A poly(D,L-lactide) resin for the preparation of tissue engineering scaffolds by stereolithography. *Biomaterials* **2009**, *30*, 3801–3809. [CrossRef]

116. Yang, S.; Leong, K.F.; Du, Z.; Chua, C.K. The design of scaffolds for use in tissue engineering. *Tissue Eng.* **2001**, *7*, 679–689. [CrossRef] [PubMed]

117. Acosta, H.L.; Stelnicki, E.J.; Rodriguez, L.; Slingbaum, L.A. Use of absorbable poly (D,L) lactic acid plates in cranial-vault remodeling: Presentation of the first case and lessons learned about its use. *Cleft Palate-Craniofac. J.* **2005**, *42*, 333–339. [CrossRef] [PubMed]

118. Silva, M.M.C.G.; Cyster, L.A.; Barry, J.J.A.; Yang, X.B.; Oreffo, R.O.C.; Grant, D.M.; Scotchford, C.A.; Howdle, S.M.; Shakesheff, K.M.; Rose, F.R.A.J. The effect of anisotropic architecture on cell and tissue infiltration into tissue engineering scaffolds. *Biomaterials* **2006**, *27*, 5909–5917. [CrossRef] [PubMed]
119. Cheung, H.Y.; Lau, K.T.; Lu, T.P.; Hui, D. A critical review on polymer-based bio-engineered materials for scaffold development. *Compos. Part B Eng.* **2007**, *38*, 291–300. [CrossRef]
120. Luo, Y.; Dolder, C.K.; Walker, J.M.; Mishra, R.; Dean, D.; Becker, M.L. Synthesis and Biological Evaluation of Well-Defined Poly(propylene fumarate) Oligomers and Their Use in 3D Printed Scaffolds. *Biomacromolecules* **2016**, *17*, 690–697. [CrossRef] [PubMed]
121. Cimen, Z.; Babadag, S.; Odabas, S.; Altuntas, S.; Demirel, G.; Demirel, G.B. Injectable and Self-Healable pH-Responsive Gelatin–PEG/Laponite Hybrid Hydrogels as Long-Acting Implants for Local Cancer Treatment. *ACS Appl. Polym. Mater.* **2021**, *3*, 3504–3518. [CrossRef]
122. Chandel, A.K.S.; Nutan, B.; Raval, I.H.; Jewrajka, S.K. Self-assembly of partially alkylated dextran-graft-poly [(2-dimethylamino) ethyl methacrylate] copolymer facilitating hydrophobic/hydrophilic drug delivery and improving conetwork hydrogel properties. *Biomacromolecules* **2018**, *19*, 1142–1153. [CrossRef]
123. Melchels, F.P.; Barradas, A.M.; Van Blitterswijk, C.A.; De Boer, J.; Feijen, J.; Grijpma, D.W. Effects of the architecture of tissue engineering scaffolds on cell seeding and culturing. *Acta Biomater.* **2010**, *6*, 4208–4217. [CrossRef]
124. Melchels, F.P.W.; Grijpma, D.W.; Feijen, J. Properties of Porous Structures prepared by Stereolithography using a Polylactide Resin. *J. Control Release* **2009**, *132*, 71–73. [CrossRef]
125. Grijpma, D.W.; Hou, Q.; Feijen, J. Preparation of biodegradable networks by photo crosslinking lactide, ε-caprolactone and trimethylene carbonate-based oligomers functionalized with fumaric acid monoethyl ester. *Biomaterials* **2005**, *26*, 2795–2802. [CrossRef]
126. Lee, J.W.; Ahn, G.; Kim, D.S.; Cho, D.W. Development of nano- and microscale composite 3D scaffolds using PPF/DEF-HA and micro-stereolithography. *Microelectron. Eng.* **2009**, *86*, 1465–1467. [CrossRef]
127. Lee, J.W.; Lan, P.X.; Kim, B.; Lim, G.; Cho, D.W. 3D scaffold fabrication with PPF/DEF using micro-stereolithography. *Microelectron. Eng.* **2007**, *84*, 1702–1705. [CrossRef]
128. Lan, P.X.; Lee, J.W.; Seol, Y.J.; Cho, D.W. Development of 3D PPF/DEF scaffolds using micro-Stereolithography and surface modification. *J. Mater. Sci. Mater. Med.* **2009**, *20*, 271–279. [CrossRef] [PubMed]
129. Palaganas, N.B.; Mangadlao, J.D.; de Leon, A.C.C.; Palaganas, J.O.; Pangilinan, K.D.; Lee, Y.J.; Advincula, R.C. 3D printing of photocurable cellulose nanocrystal composite for fabrication of complex architectures via stereolithography. *ACS Appl. Mater. Interfaces* **2017**, *9*, 34314–34324. [CrossRef]
130. Ronca, A.; Maiullari, F.; Milan, M.; Pace, V.; Gloria, A.; Rizzi, R.; De Santis, R.; Ambrosio, L. Surface functionalization of acrylic based photocrosslinkable resin for 3D printing applications. *Bioact. Mater.* **2017**, *2*, 131–137. [CrossRef]
131. Lipowiecki, M.; Brabazon, D. Design of Bone Scaffolds Structures for Rapid Prototyping with Increased Strength and Osteoconductivity. *Adv. Mater. Res.* **2009**, *83*, 914–922. [CrossRef]
132. Hussein, K.H.; Park, K.M.; Kang, K.S.; Woo, H.M. Biocompatibility evaluation of tissue-engineered decellularized scaffolds for biomedical application. *Mater. Sci. Eng. C* **2016**, *67*, 766–778. [CrossRef]
133. Isabel, C.C.M.P. Polymer Biocompatibility. In *Polymerization*; Books on Demand: Norderstedt, Germany, 2012; pp. 47–62.
134. Rahyussalim, A.J.; Kurniawati, T.; Aprilya, D.; Anggraini, R.; Ramahdita, G.; Whulanza, Y. Toxicity and biocompatibility profile of 3D bone scaffold developed by Universitas Indonesia: A preliminary study. *AIP Conf. Proc.* **2017**, *1817*, 020004-1–020004-6. [CrossRef]
135. Wang, H.; Li, Y.; Zuo, Y.; Li, J.; Ma, S.; Cheng, L. Biocompatibility and osteogenesis of biomimetic nano- hydroxyapatite/polyamide composite scaffolds for bone tissue engineering. *Biomaterials* **2007**, *28*, 3338–3348. [CrossRef]
136. Sakaguchi, R.; Ferracane, J.; Powers, J. Biocompatibility and Tissue Reaction to Biomaterials. In *Craig's Restorative Dental Materials*; Sakaguchi, R.L., Ed.; Elsevier Inc.: Amsterdam, The Netherlands, 2012; pp. 109–133.
137. Harmand, M.F.; Bordenave, L.; Bareille, R.; Naji, A.; Jeandot, R.; Rouais, F.; Ducassou, D. In vitro evaluation of an epoxy resin's cytocompatibility using cell lines and human differentiated cells. *J. Biomater. Sci. Polym.* **1991**, *2*, 67–79. [CrossRef]
138. Murray, P.E.; Godoy, C.G.; Godoy, F.G. How is the biocompatibilty of dental biomaterials evaluated? *Med. Oral Patol. Oral Cirugía Bucal* **2007**, *12*, 258–266.
139. Salgado, A.J.; Coutinho, O.P.; Reis, R.L. Novel Starch-Based Scaffolds for Bone Tissue Engineering: Cytotoxicity, Cell Culture, and Protein Expression. *Tissue Eng.* **2004**, *10*, 465–474. [CrossRef] [PubMed]
140. Gupta, S.K.; Dinda, A.K.; Potdar, P.D.; Mishra, N.C. Fabrication and characterization of scaffold from cadaver goat-lung tissue for skin tissue engineering applications. *Mater. Sci. Eng. C* **2013**, *33*, 4032–4038. [CrossRef] [PubMed]

*Review*

# Functionalized Hemodialysis Polysulfone Membranes with Improved Hemocompatibility

Elena Ruxandra Radu [1,2] and Stefan Ioan Voicu [1,2,*]

[1] Advanced Polymer Materials Group, University Politehnica of Bucharest, 1-7 Gh. Polizu Street, 011061 Bucharest, Romania; radu.elena.ruxandra@gmail.com

[2] Department of Analytical Chemistry and Environmental Engineering, Faculty of Applied Chemistry and Materials Science, University Politehnica of Bucharest, 1-7 Gh. Polizu Street, 011061 Bucharest, Romania

* Correspondence: Correspondence: stefan.voicu@upb.ro

**Abstract:** The field of membrane materials is one of the most dynamic due to the continuously changing requirements regarding the selectivity and the upgradation of the materials developed with the constantly changing needs. Two membrane processes are essential at present, not for development, but for everyday life—desalination and hemodialysis. Hemodialysis has preserved life and increased life expectancy over the past 60–70 years for tens of millions of people with chronic kidney dysfunction. In addition to the challenges related to the efficiency and separative properties of the membranes, the biggest challenge remained and still remains the assurance of hemocompatibility—not affecting the blood during its recirculation outside the body for 4 h once every two days. This review presents the latest research carried out in the field of functionalization of polysulfone membranes (the most used polymer in the preparation of membranes for hemodialysis) with the purpose of increasing the hemocompatibility and efficiency of the separation process itself with a decreasing impact on the body.

**Keywords:** hemodialysis; covalent functionalization; composite membranes; polysulfone

**Citation:** Radu, E.R.; Voicu, S.I. Functionalized Hemodialysis Polysulfone Membranes with Improved Hemocompatibility. *Polymers* **2022**, *14*, 1130. https://doi.org/10.3390/polym14061130

Academic Editor: Carlos A. García-González

Received: 22 February 2022
Accepted: 9 March 2022
Published: 11 March 2022



## 1. Introduction

Since the beginning of their existence, humans have been preoccupied with providing for their primary needs such as food and shelter. As these necessities were resolved, increasing the quality of life and solving health problems became priorities. The advances made in scientific knowledge at the beginning of the last century led to an exponential increase in the knowledge we have and implicitly provided access to solve an increasing number of problems. One of the major health problems that found its solution at the beginning of the last century is chronic kidney dysfunction. The kidneys are those paired organs responsible for cleaning the blood by forming urine and eliminating waste from the body through it, especially the excess water, salts and metabolites resulting from the processing of proteins—urea, uric acid and creatinine. The decrease in the activity of these organs is initially indicated by the onset of acute kidney dysfunction (which can be controlled medicinally) and affects the sufferer's life regimen, in the advanced cases causing chronic kidney dysfunction. In this case, the function of the kidneys is substituted with the help of polymer membranes for hemodialysis, which for four hours once every two days, filter the blood, removing from it what should have been removed by the kidneys in 48 h. It is estimated that currently in the world, between 4.9 and 9.7 million people need dialysis [1]. The main causes that lead to the appearance of chronic kidney dysfunction are diabetes, high blood pressure and pollution. Due to this last reason, it is assumed that the number of patients who will need this therapeutic procedure in the future will increase exponentially in the near future. The field of membranes for hemodialysis is one of the most dynamic fields, in which it is necessary to adapt both the materials used and the procedures for obtaining them. Besides the direct need for membrane materials for the therapeutic process, hemodialysis is a process that generates large quantities of water that in turn must be purified. The dialysis used during the procedure can lead in the case of a single dialysis center to the production of approximately 80 tons per year of water with a high content of urea, uric acid and creatinine, which in turn must be purified by concentrating these substances through other membrane processes, the most used at present being forward osmosis [2,3]. Figure 1 shows the separation scheme used in the hemodialysis process, as well as the scheme of further purification of the waters resulting from the therapeutic process for obtaining pure water that can be reused, as a rule, also for dialysis.

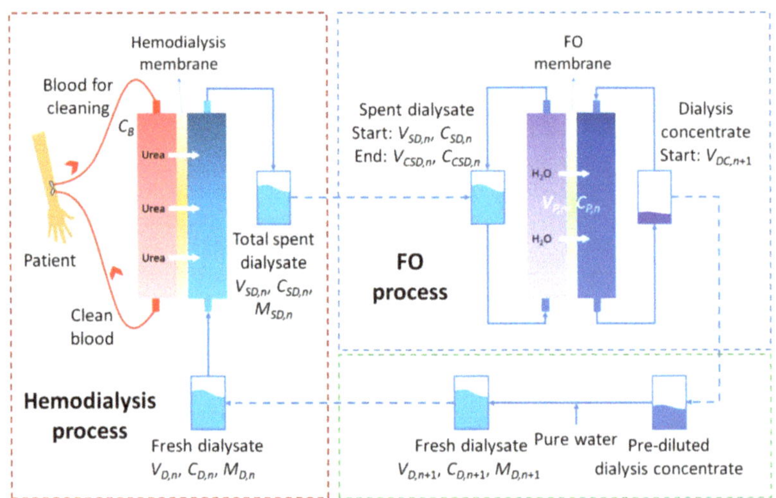


**Figure 1.** Scheme of the combination of the hemodialysis and spent dialysate recovery by osmotic dilution. A partial amount of water in the spent dialysate spontaneously moves toward the dialysis concentrate, as a consequence of the osmotic pressure gradient across the FO membrane. After FO, a certain amount of pure water is added to further dilute the dialysis concentrate. The dialysate without recovered water from the spent dialysate is used in the first hemodialysis session (reproduced with permission after [2]).

Membrane materials, initially widely used for water filtration and drinkable water production [4–6], have found widespread applicability in the last 50 years in other areas, such as filtering in the food industry, the energy industry [7,8], catalysis [9], sensors [10], etc. One of the areas in which the importance of membrane materials has grown exponentially lately, is biomedical sciences [11]. Thus, membranes emerged for filtering and concentrating proteins, both by manipulating porosity [12] and by the modification of the active surface, as well as by the synthesis of composite membranes. Membranes for osseointegration have already found practical use especially in stomatology where they are used to favor the integration of metal implants into the bone [13–15], while other composite membrane materials are used in tissue engineering to obtain various scaffold-type structures [16–19].

Hemodialysis membranes represent the most important class of membranes for biomedical applications and the second largest field for the applicability of membranes produced in the world at present, the first being desalination [20]. The quality conditions that these materials must meet are related both to the efficiency of the separation and to the biocompatibility character of the synthesized final material. Unlike other materials used in bioengineering, the hemocompatibility requirement is a mandatory character for these materials.

Many review papers have been published in the field of hemodialysis over the years due to the high practical interest in this field. The subjects presented are divided into many areas under the same domain. The medical facilities and the management of dialyzed patients are very important concerns, since the procedure requires logistics and consumables for every patient once in two days [21]. This is one of the reasons due to which, in terms of management and logistics there is a large increase in scientific community's interest for developing machines and technical solutions for home care [22–25]. This remains for the moment an impossibility despite the efforts for solving this. Cardiovascular implications of this medical procedure, such as medical conditions and complications associated with kidney failure [26,27] or vascular access for procedure itself [28], were subjects for a large number of studies published as review papers [29]. Environmental contamination as a main source of kidney failure has been also extensively investigated and presented in review papers [30,31]. From the perspective of materials science, the main interest is credited to polymeric membranes used for the procedure. In comparison with other medical applications for polymeric membranes (such as drug delivery systems, for examples), these membranes must have high efficiency for the separation of certain species (urea, uric acid and creatinine), but the interaction with others must be reduced as much as possible (e.g., other proteins or elements from blood or active pharmaceutical compounds used as complementary treatment for kidney failure or for the treatment of other organs affected by this) [32]. The preferentially used polymer for this application is polysulfone, used for more than five decades, due to its remarkable chemical and physical properties [33]. Recent advances in the field of new polymer synthesis investigated the possibility for at least partially replacing polysulfone, without success until now, but generated a large amount of research results and valuable data [34–36]. The control of biofouling [37] represents a key point for membrane characteristics. For this reason, in relation to the engineering of the hemodialysis process, some reviews has been published that present surface phenomena and interactions and also technical solutions to address hydrophilicity and hydrophobicity, repulsion or attraction forces between ions and membrane surface and also explanations from a physical chemistry point of view for hydration layers and friction-reduction properties [38–40]. In terms of hemocompatibility, several methods can be used for improving this property in hemodialysis membranes [41]. A summarization of these methods is presented in Figure 2.

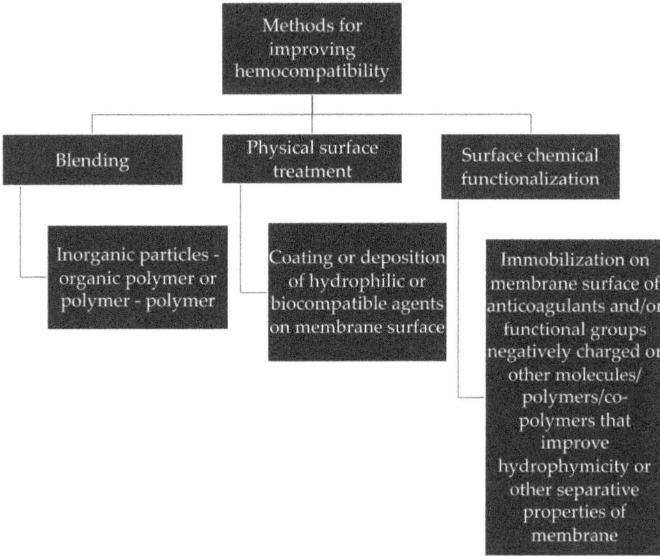


**Figure 2.** Schematic representation of most common methods for improving hemocompatibility.

The most common methods are blending (inorganic particles with polymeric membrane or polymers/co-polymers with polymeric membrane, in which the filler increases the surface hydrophilicity), physical surface treatment (in order to deposit at the surface of the membrane hydrophilic or biocompatible layers under a physical process that turns the hydrophobic surface into a hydrophilic one) and surface chemical functionalization. Chemical functionalization allows the immobilization of various chemical species that improve the separation performances, increase hydrophilicity or improve hemocompatibility of the hemodialysis membranes. The novelty of the present review is given by the new approach in the field of studies related to polysulfone functionalized membranes for hemodialysis from the perspective of functionalization reactions at the surface of the membrane that improve hemocompatibility. Reactions conditions and the influence of immobilized molecules over performances of separation and hemocompatibility are presented and some future trends in the field of functionalized polysulfone membranes for hemodialysis are also discussed at the end of the review.

## 2. Polysulfone Functionalized Membranes for Hemodialysis

Hemocompatibility and anticoagulant properties are the most significant qualities that need to be considered when developing advanced hemodialysis membranes [42]. Biomaterials that come into contact with blood should not initiate the process of blood clotting by adsorption of blood proteins on the surface of the biomaterial, leading to the formation of the thrombus. It was found that the hydrophobic surface character of these biomaterials could lead to the adherence of plasma proteins to the hydrophobic surfaces forming plasma clotting [43]. In the last decade, many materials were studied for their hemocompatibility, such as polyvinyl alcohol (PVA) [43,44], cellulose triacetate (CTA) [43,45], polymethylmethacrylate (PMMA) [38,46], polyacrylonitrile (PAN) [47,48], polysulfone (PSF) [9,20,49], polyethersulfone (PES) [43,50,51] and polyamide (PA) [46,52]. Polysulfone (PSF) is a thermoplastic polymer used in the biomedical field as a dialysis membrane due to physicochemical properties, such as thermal stability, chemical resistance, decent mechanical strength, great processability and biocompatibility [43].

The drawback of PSF is having a hydrophobic nature, which could be conducive to protein adsorption, platelet adhesion and clotting enzymes, which directly leads to thrombosis [38]. The functionalization of PSF surfaces changes the hydrophilicity of the surfaces, leading to an increase in the hemocompatibility and anticoagulant properties of the PSF membrane [41,53]. The functionalization of the PSF could be achieved at every step of the forming phase of the polymers, such as the functionalization of PSF in solution followed by membrane synthesis or formation, and functionalization of the surface of PSF membrane [54]. Further, in functionalization, functional groups could be used, such as small compounds that could be linked directly to the PSF or by using different linker molecules and macromolecules by the grafting of chemical functions onto an aromatic ring followed by immobilization using a linker molecule [49,55]. The PSF does not have any vacant functional groups and in order to perform functionalization, additional functional groups must be added by grafting onto the aromatic rings, such as the introduction of a sulfonate group into the polymer structure through electrophilic substitution via sulfonation [47]. The addition of the sulfonic group increases the membrane's permeability, hydrophilicity, hemocompatibility, anti-fouling behavior and antimicrobial properties [56,57]. The sulfonation treatment is capable of influencing the morphology of PSF membranes, such that neat PSF membranes have a smoother surface and sulfonated PSF presents a significant increase in the roughness of the surfaces [34]. The sulfonated group will have a negative charge [1].

Aydemir Sezer et al. [58] presented the sulfonation treatment of PSF that could replace the heparin-based functionalization of PSF via the introduction of the sulfonic acid group, $SO_3H$, into the structure of a molecule using a chlorosulfonic agent as a sulfonating agent. A core—shell electrospun structure based on PSF core and sulfonated PSF shell was investigated. The sulfonation treatment of PSF leads to an increase of the hydrophilicity of PSF indicating the improvement of blood compatibility due to hydrophilic groups on the surface of sulfonated PSF. The mechanical properties, hemocompatibility and biocompatibility properties of sulfonated PSF are improved in comparison with those of neat PSF [58]. In this study, Mahlicli et Altinkaya [59] modified PSF with alpha-lipoid acid (ALA), which is an antioxidant obtained from the reduced form of dihydrolipoic acid (DHLA), in order to obtain a hemodialysis membrane. Before preparing the support membrane, the PSF was modified via sulfonation in order to induce negatively charged groups ($SO_3^-$). ALA was immobilized onto the hemodialysis membranes used in order to prevent hemodialysis-induced oxidative stress. The described immobilization method was based on site-specific binding of a carboxylic group of ALA to an ammonium group of the anchoring polyelectrolyte layer—PEI—through electrostatic interactions. It was reported that the best antioxidant effect was seen in the case of ALA placed between two layers of PEI. In addition, reduced stress oxidation, protein adsorption and the platelet activation of the membranes were also observed [59]. Chien et al. [60] described polysulfone modification through sulfonation with sulfuric acid, decreasing the contact angle from 91.3° to 87.2° and the deposition of polyelectrolytes. Moreover, the obtained composites of PSF and the deposition of poly(acrylic acid) (PAA)/poly(allylamine hydrochloride) (PAH) multilayer films were studied for hemocompatibility through the determination of the number and morphology of adhered platelets. The results showed that the addition of the PAA and PAH layers increased the extension of platelet spreading [41]. Aydemir Sezer et al. [58] presented the sulfonation of PSF with chlorosulfonic acid resulting in a decrease of the contact angle in comparison with pristine PSF, from 133° ± 13° for pristine PSF to 125° ± 12° for sulfonated PSF. Xie et al. [61] reported the influence of sulfonated PSF with different degrees of sulfonation (20%, 30% and 50%), which was used as an additive for polyvinyl chloride (PVC). The PVC/SPSF membrane composites present an improvement in the permeability and antifouling properties of the PVC. It was shown that the degree of sulfonation of PSF influences the topology and the roughness of the obtained membranes, directly proportional with the degree of sulfonation. The increased roughness is a consequence of increasing pore size due to the larger size of the polymer lean region during the liquid–liquid phase separation leading to the larger pore size with the increased sulfonation degree. The reported hydrophilicity of sulfonated PSF via air bubble contact angle measurements was increased due to the surface segregation of sulfonate groups resulting an antifouling surface for organic foulants. BSA was used in order to study protein adsorption on the membrane's surface. This has resulted in a decreased adsorption due to the increased hydrophilicity of the composite membranes [61].

Functionalization of PSF with sulfonated hydroxypropyl chitosan leads to an increase of the membrane's biocompatibility and antimicrobial properties due to the presence of chitosan through a Schiff base reaction carried out between the attached acetaldehyde and ortho-aminophenol (OAP) or meta-aminophenol (MAP) of PSF [62]. Firstly, chitosan was treated with propylene/isopropanol oxide for hydroxypropyl grafting, followed by sulfonation with formamide/chlorosulfonic acid in order to obtain hydroxypropyl chitosan sulfonate. After the functionalization, the hydrophilicity was improved, resulting in better anticoagulant properties and antimicrobial properties, which recommend it as membrane for hemodialysis [63]. Tu et al. [64] modified PSF with covalently grafted acrylic acid and sulfonated hydroxypropyl chitosan for better hemocompatibility and anticoagulant properties (Figure 3). The acrylic acid was first grafted on PSF by adding the –COOH functional group, which will react with sulfonated hydroxypropyl chitosan. In comparison with the Schiff base reaction previously used in the functionalization of PSF with sulfonated hydroxypropyl chitosan, this time, covalently grafted acrylic acid and sulfonated hydroxypropyl chitosan were used. The results showed a decrease of the contact angle from 86° for neat PSF to 22° for the modified membrane with sulfonated hydroxypropyl chitosan and acrylic acid, leading to an improved hydrophilicity of the functionalized membrane, which results in an improved performance against protein contamination. The bovine serum albumin (BSA) adsorption was lower after the grafting with both acrylic acid and sulfonated chitosan due to the increased hydrophilicity. Ganj et al. [65] functionalized PSF by grafting the acrylic acid via free graft polymerization. After the functionalization, the contact angle of the composite PSF—acrylic acid membrane was reduced by 30%. Further, the flux recovery ratio was increased by 32.2%, which could indicate that the modified membrane has antifouling properties [65].

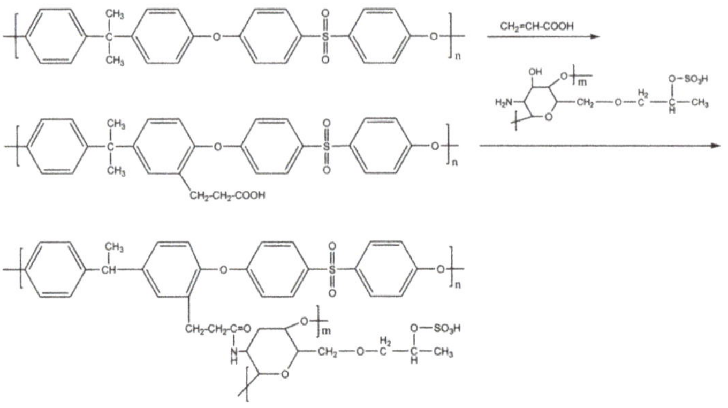

**Figure 3.** The steps of the modification of the PSF membrane (reproduced with permission after [64]).

Yan et al. [66] functionalized PSF with sulfonated hydroxypropyl chitosan and 4-(chloromethyl)benzoic acid in order to improve hemocompatibility. The 4-(chloromethyl) benzoic acid was grafted on PSF via the Friedel—Crafts alkylation reaction, providing available carboxyl groups, followed by the grafting of sulfonated hydroxypropyl chitosan via esterification (Figure 4). The functionalized membrane presented increased hydrophilicity and remarkable hemocompatibility, and the hemolysis rate decreased after functionalization. The contact angle decreased after grafting, from 86° for neat PSF to 59° after functionalization through the addition of carboxyl groups into sulfone after adding 4-(chloromethyl)benzoic acid. The reaction time of PSF functionalized membrane with 4-(chloromethyl) benzoic acid and sulfonated hydroxypropyl chitosan had a decreasing effect on the contact angle, such that, after 12 h of reaction time, the contact angle decreased up to 47° and, after 24 h of reaction time, the contact angle decreased up to 32°. These results showed an exceptional increase of the hydrophilicity, which could lead to an improved hemocompatibility.

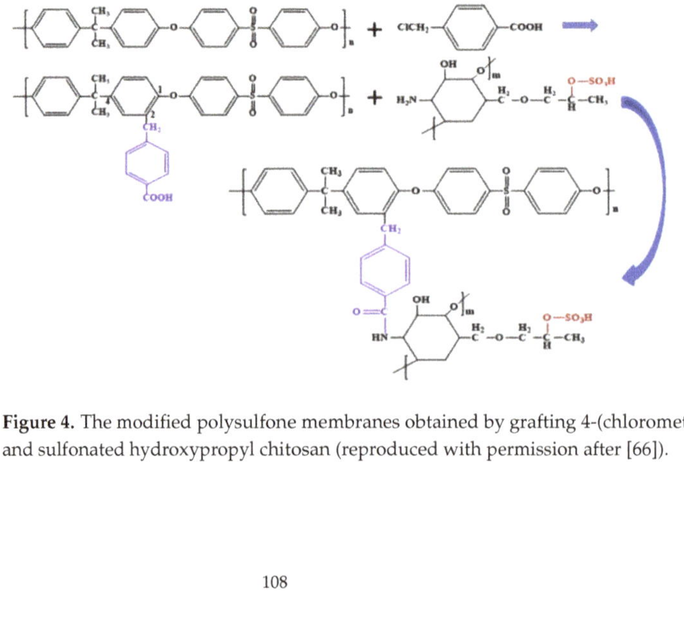

**Figure 4.** The modified polysulfone membranes obtained by grafting 4-(chloromethyl)benzoic acid and sulfonated hydroxypropyl chitosan (reproduced with permission after [66]).

In addition to sulfonation, another modification method of the PSF is chloromethylation, which will induce the appearance of numerous functional groups on the PSF in order to increase hydrophilicity, such as hydroxyl group ($-OH$), azide group ($-N_3$), amino ($-NH_2$), carboxyl ($-COOH$) and sulfo ($-SO_3H$) groups [54,67]. Through chloromethylation of the PSF, a surface benzyl chloride group is introduced as an active ATRP initiator [68]. The biocompatibility of PSF is improved by the grafting of negative carboxyl and sulfonic groups. Dong et al. [69] grafted poly(poly(ethylene glycol) methyl ether methacrylate) and poly(glycidyl methacrylate) on chloromethylated polysulfone in order to increase the antifouling properties (Figure 5). After grafting, the BSA absorption decreased from ~50 to ~5 µg/cm$^2$. Yue et al. [70] described PSF functionalization with zwitterionic poly (sulfobetaine methacrylate) (PSBMA) via SI-ATRP. Initially, the PSF was chloromethylated, and afterward, the sulfobetaine methacrylate (SBMA) monomer was grafted on polysulfone (PSF-g-PSBMA). The roughness of the functionalized membranes increased and contact angle values decreased after functionalization, thus the chloromethylated PSF contact angle slightly decreased up to 66°. A significant decrease of the contact angle was observed in the case of PSF-g-PSBMA, with a value up to 25°, due to sulfobetaine groups, which could form a hydration layer via electrostatic interaction in addition to the hydrogen bond (Figure 6) [70]. Liu et al. [71] present the chloromethylation of PSF in order to prepare the anion-exchange membranes (AEMs), which are polymer electrolytes that are able to conduct anions such as $SO_4^{2-}$, $OH^-$ and $Cl^-$. The AEM structure, properties and morphologies are controlled by chloromethylation conditions.

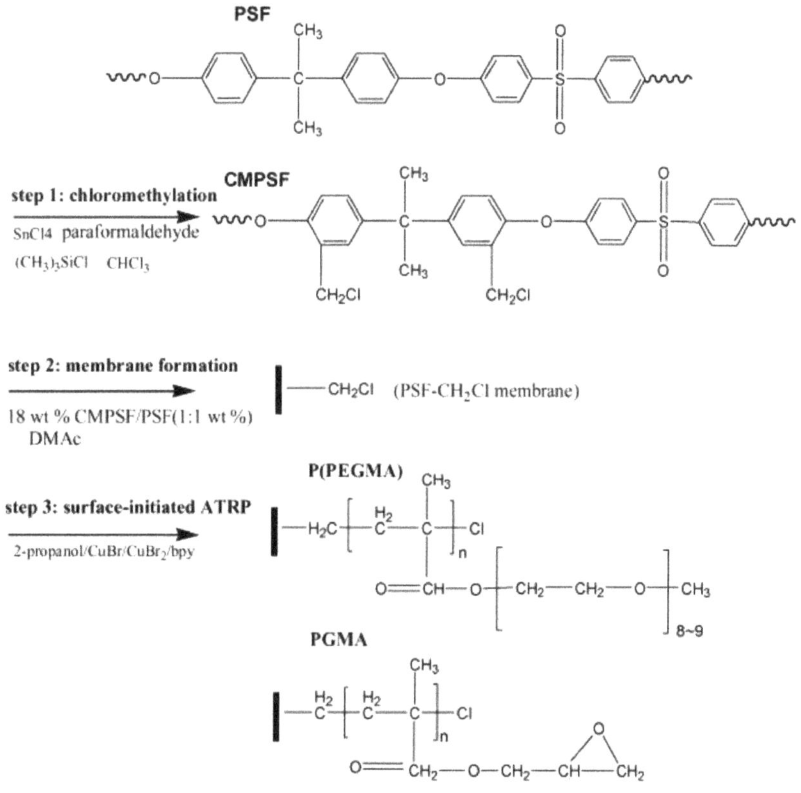

**Figure 5.** Process of surface-initiated ATRP from the polysulfone membrane (reproduced with permission after [69]).

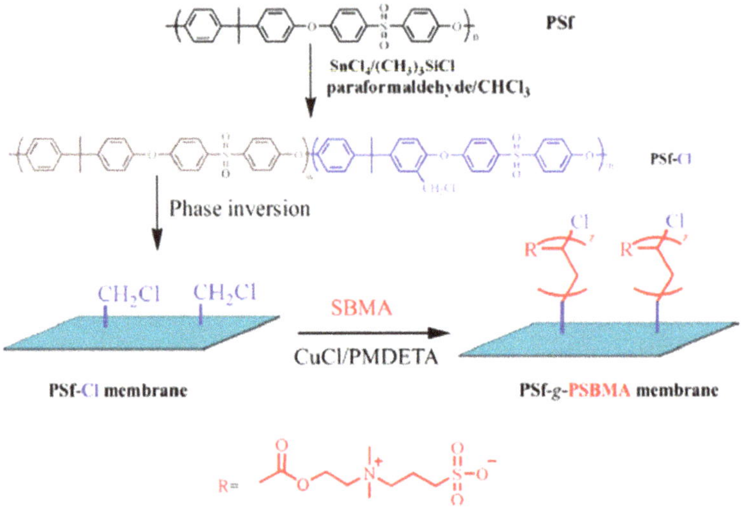

**Figure 6.** Scheme illustration of the preparation of PSF-g-PSBMA membrane (reproduced with permission after [70]).

The zwitterionic polymers, such as poly(sulfobetaine methacrylate) (PSBMA), were grafted from a chloromethylated PSF membrane through surface-initiated atom transfer radical polymerization (SI-ATRP), in order to reduce polymer adsorption and overcome platelet adhesion, but lightly prolonged clotting [72]. The zwitterionic polymers present both positive and negative charge on the same side chain maintaining neutrality, which makes them an excellent inhibitor for the adhesion on the membrane surface of the plasma protein [72]. The chlorine groups (–Cl) were attached to the PSF via the chloromethylation reaction and used as initiators for the ATRP reaction using as reagents paraformaldehyde, chlorotrimethylsilane ($(CH_3)_3SiCl$) and tin (IV) chloride ($SnCl_4$), followed by the synthetization of block zwitterionic polymer via SI-ATRP (Figure 7). The modified PSF with grafted zwitterionic showed better antifouling properties and hemocompatibility in comparison with neat PSF [73]. Maggay et al. [74] proposed a PSF membrane modified with zwitterionic polymer, a copolymer made of styrene and 4-vinylpyridine units via a dual-bath procedure, and studied the antifouling properties of the modified membrane. The results showed that the biofouling was reduced by 87% in comparison with the pristine PSF membrane after incubating the membranes with *E. coli* and with a 90% reduction of biofouling in the case of whole blood.

In addition, the combination of zwitterionic polymers and the negatively charge given by the sulfonic and carboxyl groups, increases the hydrophilicity of the PSF, resulting in a good antifouling property and hemocompatibility. Xiang et al. [49] also proposed zwitterionic polymers of poly(sulfobetaine methacrylate) (PSBMA) and negatively charged polymers of poly(sodium p-styrene sulfonate) (PNaSS) and/or poly(sodium methacrylate) (PNaMAA) to functionalize PSF membranes via click chemistry in one step (Figure 8). The resulting polymers were obtained via ATRP and linked at the surface of the azido-functionalized PSF membrane via click chemistry. The obtained functionalized membranes presented increased hydrophilicity, due to the decrease of contact angle values, good resistance to protein adsorption and platelet and bacterial adhesion. Further, the addition of negative charge improved the anticoagulant properties due to the synergistic effect of the sulfonic and carboxyl group.

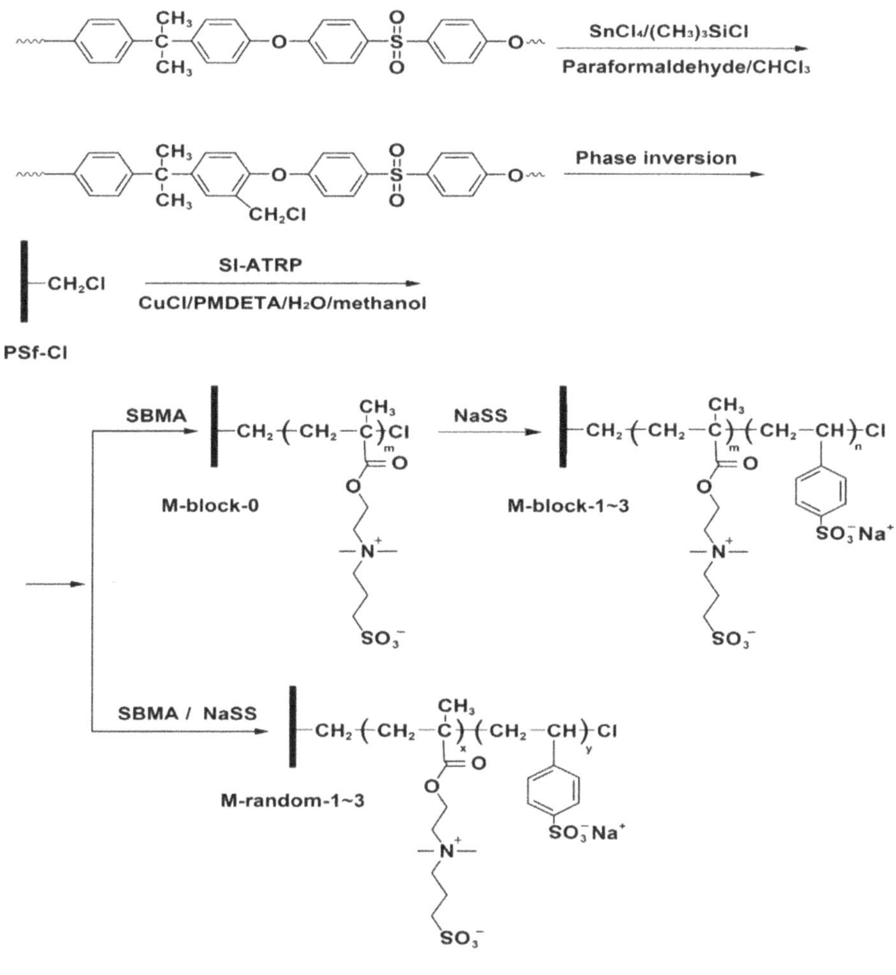

**Figure 7.** The preparation of PSF—Cl membrane, and P(SBMA-b-NaSS) and P(SBMA-co-NaSS) grafted membranes (reproduced with permission after [73]).

A surface modification of PSF via on plasma treatment was reported [75]. The surface hydrophilicity was improved via plasma treatment, such that a decrease of the contact angle was observed on increasing the oxygen plasma treatment time. Through plasma treatment, oxygen-containing polar groups were introduced to the membrane surfaces resulting in an increase in hydrophilicity behavior on the PSF membrane [76]. The low-temperature plasma treatment (LTPT) was extensively used in membrane surface modification. Zheng et al. [77] reported a successful surface modification of the PSF membrane via LTPT and grafting acrylic acid (AA) and PEG, 2-methacryloyloxyethyl phosphorylcholine (MPC), heparin and collagen (Figure 9). Zhang et al. [78] described a surface modification of the PSF via ammonia–oxygen ($NH_3$–$O_2$) plasma treatment and found that the hydrophilicity of membrane was improved in comparison with pristine PSF membrane. Furthermore, the antifouling properties of the modified PSF membrane surface via low-temperature water vapor plasma was studied. After plasma modification, oxidation occurred, leading to improving wettability of the PSF membrane [79].

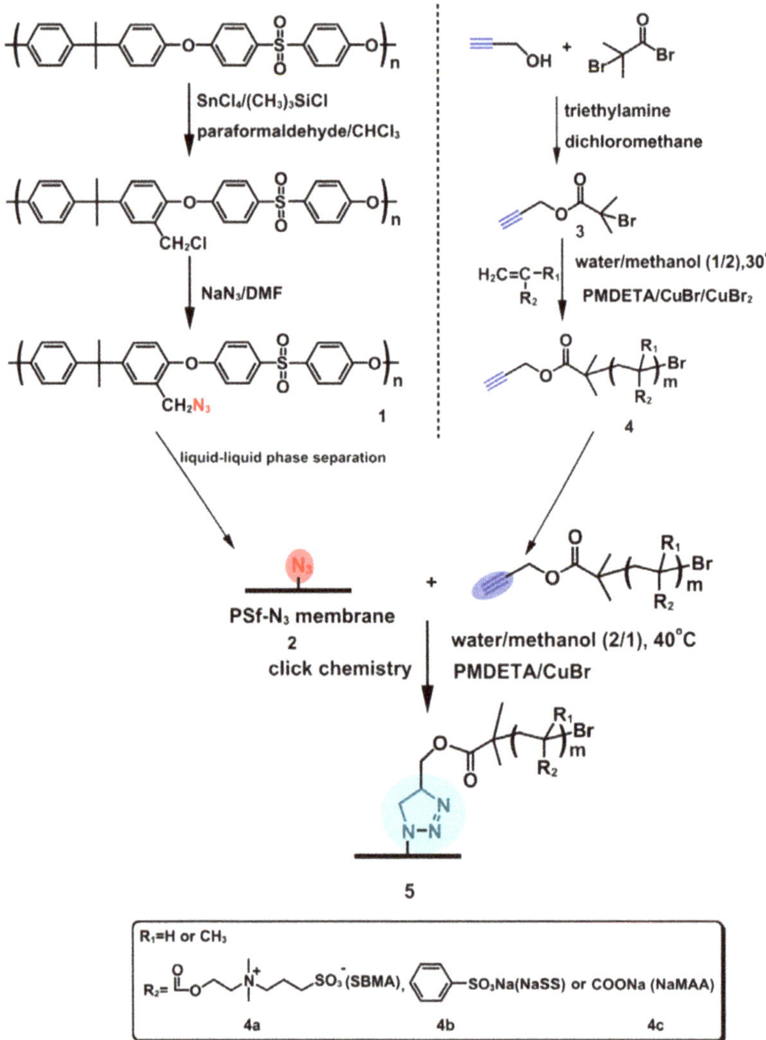

**Figure 8.** Synthesis of functionalize azido-polysulfone, alkynyl-functionalized polymers and grafted polysulfone membrane via click chemistry (reproduced with permission after [49]).

Anticoagulation properties are crucial when speaking about the hemocompatibility of hemodialysis membranes. The functionalization of PSF membrane with different anticoagulant compounds was studied in order to improve hemocompatibility. Compounds such as heparin [80], polyethylene glycol (PEG) [81], polyvinyl pyrrolidone (PVP) [82] and sulfobetaine metacrylate (SBMA) [49] have been studied for PSF functionalization [66]. Moreover, the heparin-modified PSF membrane was studied as a solution for upsurging the hemocompatibility and anticoagulant properties of the PSF membrane [64,83,84]. Heparin is a linear, sulfated glycosaminoglycan produced by mast cells, having repeating monomeric disaccharides of uronic acid and glucosamine in a 1,4-linkage, and with a three-dimensional structure in a helical form [85,86]. Heparin is used as an anticoagulant agent, administrated to prevent dialyzer and circuit clotting [87–90]. Heparin is used worldwide and is approved by the Federal Drug Administration (FDA) for the treatment of deep vein thrombosis and pulmonary embolism [86,89].

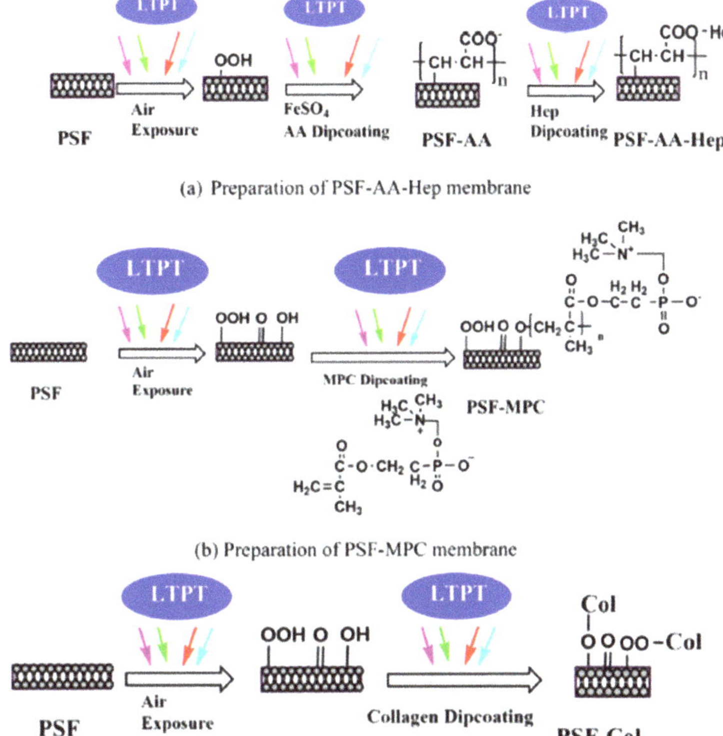



**Figure 9.** Modification process of surface grafting (reproduced with permission after [77]).

The heparin immobilization on PSF membrane is called a heparization treatment, and it is obtained via chemical amination, electrostatic self-assembled layers and plasma treatment [91]. Mostly, heparin immobilization is obtained through chemical amination by a chemically activated PSF membrane of the amino group, followed by heparization treatment. Li et al. [91] studied the covalent immobilization of heparin onto PSF using atmospheric pressure glow discharge (APGD) in order to chemically bind heparin molecules on the PSF surface. Huang et al. [84] described covalent immobilization of heparin on the PSF membrane for the particular adsorption of low-density lipoprotein (LDL). For PSF functionalization, the activation with consecutive chloromethyl ether and ethylenediamine treatments were needed in order to obtain available functional groups linked to heparin (Figure 10). After heparin immobilization, the hydrophilicity of the PSF membrane was improved. Furthermore, compared to the pristine PSF membrane, the heparin functionalized PSF membrane greatly enhanced the adsorption of LDL.

The drawback of the use of heparin is the fact that the only source of heparin is from animal tissues, leading to a possible risk of virus contamination and adverse effects, thrombocytopenia for long-term treatments and hemorrhages for patients with a large administrated dose of heparin [85]. It was also shown that heparin only inhibits plasma-free thrombin but does not inhibit clot-bound thrombin [47].

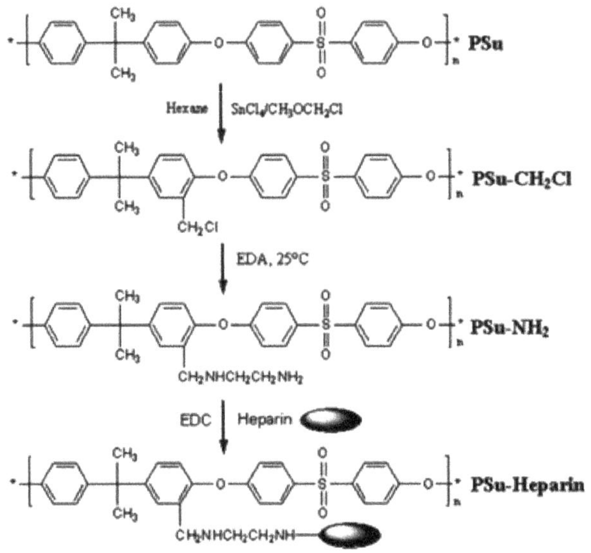

**Figure 10.** Scheme of the chemical activation of PSF film (reproduced with permission after [84]).

Lately, heparin-mimic compounds with a better control over structure, sulfation and purity have been developed. Lin et al. [92] developed PSF functionalized with sulfonated citric chitosan, which has a structure and groups similar to heparin to improve the hemocompatibility of PSF. Sulfonated citric chitosan was obtained via *N*-acylation and sulfonation to take the negative carboxyl and sulfonic groups and attach them to chloroacylated PSF, which previously was treated by introducing –COCH$_2$Cl groups into the PSF, followed by the preparation of chloroacylated membranes via the phase separation method. The obtained membranes present outstanding hydration capacity, which highly diminishes the amount of protein adsorption. Wang et al. [93] proposed an increase of hydrophilicity and anticoagulation properties of the PSF membrane by grafting onto the PSF surface a heparin-like polymer, sulfonated dihydroxypropyl chitosan (SDHPCS), via alkalization of chitosan, etherification and sulfonation (Figure 11). The reactivity of PSF was achieved by using chloroacetyl chloride. The results showed that the structure of functionalized PSF is not destroyed by the functionalization, but in comparison with pristine PSF, functionalized PSF has increased hydrophilicity, lower BSA adsorption and enhanced blood compatibility than before.

Liu et al. were the first to propose vorapaxar-modified polysulfone (VMPSF) for increased hemocompatibility. Vorapaxar is a protease-activated receptor 1 (PAR1) that is able to stop the cascade of reactions required for clotting. The composite membrane was obtained via the liquid–liquid phase inversion method. The functionalized membrane with vorapaxar presents a decrease of the contact angle values, leading to an increase in hydrophilicity [47]. Qi et al. [94] suggested a functionalization of PSF with resveratrol as an antioxidant compound for an improved hemodialysis-induced oxidative stress. Resveratrol is a plant-based extracted antioxidant used as an inhibitor of the oxidation reaction caused by free radicals. The functionalization is obtained via an immersion precipitation phase inversion method. The functionalized PSF membrane with resveratrol has increased hydrophilicity; improved water flux of the membrane; excellent and strong free radical scavenging ability; improved resistance of protein adhesion and excellent urea (90.33%), creatinine (89.50%) and lysozyme retention (74.60%). Another natural antioxidant used in PSF functionalization is silibilin. The functionalization was obtained via the immersion precipitation phase inversion method. The introduction of silibilin in the PSF matrix improves the hydrophilicity, antioxidant properties and hemocompatibility [95].

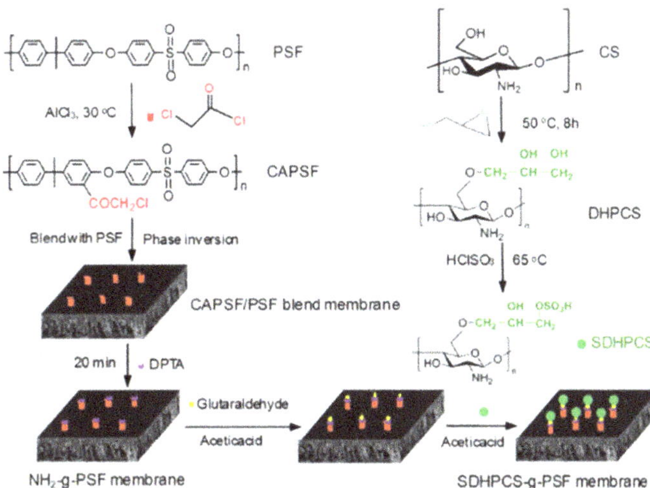



**Figure 11.** The preparation process of SDHPCS-g-PSF membrane (reproduced with permission after [93], Scheme 1).

Nanofunctionalization is a novel direction in obtaining PSF with greater properties developed by the addition of different nanoparticles, such as $TiO_2$ [96], $SiO_2$ [97], $Al_2O_3$ [98], iron oxide [99] and carbon-based nanoparticles [100–102]. PSF functionalized with carbon nanotubes has been previously reported. Firstly, the PSF was chloromethylated, followed by functionalization with different types of nanotubes, single walled (SWNT) and double walled (DWNT). The functionalized membrane presents a large pore size and relatively small compact structure, which could be used in blood filtration as it favors the flow [100]. Mahat et al. [101] reported polysulfone functionalization with carbon quantum dots in order to improve the hydrophilicity of pristine PSF and antimicrobial properties. Zheng et al. [103] proposed an increase of antifouling properties of polysulfone via carbon nanoparticles obtained from agricultural waste of corn stalks. The modified membranes improved hydrophilicity by increasing the pore size and modified roughness leading to improved membrane permeability and antifouling properties.

Polysulfone functionalization with clay nanoparticles was studied. Recently, Ouradi et al. [104] reported a nanofunctionalization of PSF with poly(acrylonitrileco-sodium methallyl sulfonate) copolymer (AN69) (which is a hydrophilic material) and montmorillonite clay (MMT) for hemodialysis applications. It was reported that the nanocomposite membrane presents hydrophilic properties, good thermal sterilization resistance, better water permeability and a good capacity of protein adsorption. AN69 induces a negative charge on the PSF surface, and MMT reduces the free volume between PSF chains leading to increased permeability due to the rise of hydrophilic character. The addition of MMT also increases the thermal stability of the membrane, promoting a great thermal sterilization resistance [104–106]. Likewise, clay-polysulfone nanocomposites were developed by Ma et al. [107] via the phase inversion method. *N,N*-Dimethyl acetamide (DMAc), deionized water and PEG 400 were used as solvent, coagulant and pore-forming agent, respectively. The nanocomposite membrane displayed an asymmetric structure, with increased hydrophilicity due to the addition of MMT.

Mansur et al. [108] functionalized polysulfone membrane with silica nanoparticles ($SiO_2$) in order to remove the protein uremic toxin by adsorption and alpha mangostin (α-mangostin), which is an antioxidant bioactive compound extracted from mangosteen pericarp. After functionalization, the obtained composite membranes were more hydrophilic in comparison with pristine polysulfone membrane, having an increased permeability and biocompatibility. Song et al. [109] modified polysulfone with poly(1-vinylpyrrolidone)-grafted silica nanoparticles (PVP-g-silica). The contact angle values of the composite membrane decreased with the increase of PVP-g-silica content, inducing a hydrophilic behavior. The antifouling properties were reported as being improved in comparison with those of the polysulfone/nanosilica composite membrane.

In the past decade, the functionalization with iron oxide has attracted attention due to its favorable properties for biomedical applications, such as biocompatibility, chemical stability and nontoxicity [110,111]. The usage of iron oxide improves the separation performance. Said et al. [112] reported the obtaining of polysulfone membrane functionalized with iron oxide for better biocompatibility and being able to remove the middle molecule uremic toxin effectively. It was shown that even though a high concentration of iron oxide was added, it displayed a reduction of protein adsorption and platelet adhesion while maintaining normal blood coagulation time and admissible complement activation [112]. Unfortunately, iron oxide nanoparticles have a tendency to form aggregates. For better dispersion, it was reported that the addition of citric acid could improve iron oxide dispersion in the polymer matrix [113–115].

$TiO_2$ nanofunctionalization is used due to the super hydrophilicity behavior of $TiO_2$ nanoparticles, UV resistance, antimicrobial activity and biocompatibility [116]. Nevertheless, many applications of polysulfone functionalized with $TiO_2$ are in the wastewater purification field, due to its electrochemical properties, chemical stability, non-toxicity, powerful anti-oxidizing power [117–119].

Lately, the reduction of uremic toxin retention through functionalized PSF was studied [56,108,120–122]. The uremic toxins are organic or inorganic substances that are accumulated in the body fluids of patients with acute or chronic kidney disease and impaired kidney function [123]. The accumulation of these uremic toxins has a dangerous effect on the physiological function, leading to the appearance of intoxication, resulting in the deterioration of the clinical conditions [43,124]. The uremic toxins, such as urea and creatinine, are water-soluble compounds with a small molecular weight (<0.5 kDa), which are conventionally removed using dialysis [43,124–126]. Activated carbon, hydrated zirconium-oxide, hydrated zirconium-phosphate and activated aluminum silicate were used as the absorbent for removing uremic toxins [52,121]. Abidin et al. [56] proposed a dual-layer hollow-fiber membrane based on PSF/amino-silanized poly(methyl methacrylate) for uremic toxin separation. These sandwich-structured membranes were formed by using two layers of PSF and polyvinylpyrrolidone (PVP), which is used for pore formation, and an outer layer from silanized PMMA with 3-(aminopropyl) triethoxysilane (APTES), for the introduction of adsorption properties to the composite membrane. After silanization, the adsorption capacity was improved, with the composite membrane being able to remove urea and also to filter larger lysozyme molecules. Further, Abidin et al. [121] reported upgraded composite membranes based on PSF/activated carbon (AC) as an inner layer and PSF/PMMA as an outer layer, developed for urea and creatinine removal. The addition of AC increases the creatinine adoption.

Chen et al. [127] obtained a PSF-block-polyethylene glycol (PEG) membrane via nonsolvent-induced phase separation (NIPS). The PEG blocks are covalently bound to the PSF blocks, resulting in a composite membrane with an increased permeability (the pure water permeability (PWP) was reported being 225 L m$^{-2}$ h$^{-1}$ bar$^{-1}$), compared with PSF functionalized with $TiO_2$ nanoparticles and chitosan (PWP being 31 L m$^{-2}$ h$^{-1}$ bar$^{-1}$) [128] and PSF functionalized with $TiO_2$ nanoparticles and HEMA (PWP being 115 L m$^{-2}$ h$^{-1}$ bar$^{-1}$) [129], but presenting similar BSA retention (almost 90%) compared with the aforementioned composite membrane. Said et al. [112] reported that the $PSF/Fe_2O_3$ composite membranes obtained a great BSA and urea retention (99.9% and 82%), good lysozyme retention (46,7%), but a lower a permeability (PWP being 110.47 L m$^{-2}$ h$^{-1}$ bar$^{-1}$). Mansur et al. [108] reported that the addition of $SiO_2$/alpha-mangostin nanoparticles increased urea and creatinine clearance (92.48% and 87.71%), compared with neat PSF membrane.

The removal of the uremic toxins with a larger molecular weight is still a challenge. Kohlová et al. [122] proposed removal of the middle-size uremic toxins (molecular weight 0.5–15 kDa) via PSF functionalized with hydrophilic additives, such as polyethylene glycol (PEG) and polyvinylpyrrolidone (PVP). The functionalized PSF membrane allowed a selective molecular separation, attributed to the dense top skin layer of the membrane, with a partial removal of middle-size molecules, and bovine serum albumin (BSA) rejection close to 100%. Table 1 summarizes the previously presented functionalized polysulfone membranes and their main properties.

**Table 1.** Functionalized membrane and induced properties.

| | Membrane | Properties | Ref. |
|---|---|---|---|
| 1 | Sulfonated polysulfone | Hemocompatibility<br>Biocompatibility | [58] |
| 2 | Acrylic acid and sulfonated hydroxypropyl chitosan functionalized polysulfone | Anticoagulant properties<br>Antifouling properties<br>Hemocompatibility<br>Hydrophilicity | [64] |
| 3 | Sulfonated polysulfone/PVC | Permeability<br>Antifouling properties<br>Hydrophilicity | [61] |
| 4 | Alpha-lipoic acid (ALA) functionalized polysulfone | Antioxidant activity<br>Antifouling properties | [59] |
| 5 | 4-(chloromethyl)benzoic acid and sulfonated hydroxypropyl chitosan functionalized polysulfone | Hemocompatibility<br>Biocompatibility<br>Antifouling property | [66] |
| 6 | Chloromethylated polysulfone functionalized with poly(ethylene glycol)monomethacrylate (PEGMA) and 2-hydroxyethyl methacrylate (HEMA) | Antifouling properties | [68] |
| 7 | Zwitterionic copolymers of P(SBMA-b-NaSS) and P(SBMA-co-NaSS) functionalized polysulfone | Antifouling property<br>Hemocompatibility<br>Resistance to platelet adhesion<br>Anticoagulant property | [73] |
| 8 | Zwitterionic polymer of poly(sulfobetaine methacrylate) (PSBMA) functionalized polysulfone | Antifouling property<br>Hemocompatibility<br>Cytocompatibility | [70] |
| 9 | Zwitterionic polymer of poly(sulfobetaine methacrylate) (PSBMA), negatively charged polymers of poly(sodium methacrylate) (PNaMAA) and/or poly(sodium p-styrene sulfonate) (PNaSS) functionalized polysulfone | Hydrophilicity<br>Antifouling property<br>Good resistance to protein adsorption, platelet adhesion and bacterial adhesion<br>Anticoagulant property | [49] |
| 10 | Acrylic acid (AA) with heparin, 2-methacryloyloxyethyl phosphorylcholine (MPC), and collagen functionalized polysulfone | Hemocompatibility | [77] |
| 11 | Ammonia–oxygen ($NH_3$–$O_2$) plasma-treated polysulfone | Hydrophilicity<br>Antifouling property | [78] |
| 12 | Plasma functionalized polysulfone | Wettability | [79] |
| 13 | Chlorodimethyl ether and ethylenediamine functionalized polysulfone | Hydrophilicity<br>Selectivity | [84] |
| 14 | Sulfonated citric chitosan functionalized polysulfone | Hemocompatibility<br>Anticoagulation properties | [92] |

**Table 1.** *Cont.*

|  | Membrane | Properties | Ref. |
|---|---|---|---|
| 15 | Heparin functionalized polysulfone | Hydrophilicity<br>Hemocompatibility | [91] |
| 16 | AN69/MMT functionalized polysulfone | Hemocompatibility<br>Hydrophilicity<br>Good capacity of protein adsorption<br>Thermal stability<br>Permeability | [105] |
| 17 | Resveratrol functionalized polysulfone | Hydrophilicity<br>Free radical scavenging properties<br>Resistance to protein adhesion<br>Hemocompatibility<br>Antioxidant properties | [94] |
| 18 | Silibilin functionalized polysulfone | Antioxidant properties<br>Hydrophilicity<br>Hemocompatibility | [95] |
| 19 | Polyamide/SiO$_2$ functionalized polysulfone | Excellent stability<br>Hydrophilicity | [97] |
| 20 | Carbon quantum dot functionalized polysulfone | Hydrophilicity | [101] |
| 21 | AN69/clay composite functionalized polysulfone | Permeability<br>Hydrophilicity<br>Thermal stability | [105] |
| 22 | Montmorillonite functionalized polysulfone | Hydrophilicity<br>Thermal stability<br>Tensile properties | [106] |
| 23 | Iron oxide nanoparticle functionalized polysulfone | Hemocompatibility<br>Biocompatibility | [112] |
| 24 | TiO$_2$—graphene oxide functionalized polysulfone | Antifouling<br>Antibacterial | [119] |
| 25 | PSF-activated carbon and PSF/PMMA | Good uremic toxin adsorption<br>Hydrophilicity | [121] |
| 26 | Polysulfone-block-poly (ethyleneglycol) | Hydrophilicity<br>Great permeability<br>Antifouling property | [127] |

### 3. Conclusions and Future Perspectives

Some of the recent approaches in the field of polysulfone functionalized membranes for hemodialysis from the perspective of functionalization reactions at the surface of the membrane that improve hemocompatibility have been presented in this review. Reactions conditions and the influence of immobilized molecules over performances of separation and hemocompatibility were presented with an accent on the surface chemistry conditions and improved properties that have been modified for blood filtration applications.

Future perspectives in relation to membranes for hemodialysis can be divided into several categories.

1. The nature of the polymer for the manufacture of membranes. At the beginning of the research to obtain dialysis membranes, the most used polymers were cellulose derivatives, especially nitrocellulose (which was found in large quantities due to its use as an explosive powder). With the discovery of polysulfone, its physical and chemical properties established it as the main polymer for obtaining membranes for hemodialysis. However, in the current context of the circular economy and the use of as many materials of natural origin as possible, we could see a return of green polymers and an intensification of the research on the use of these polymers, such

as cellulose derivatives, chitosan, alginate or even starch. The current methods of synthesis of membrane materials and the possibilities of unconventional chemical modification, such as reactions in plasma or laser surface changes, can solve in this time many of the technological challenges that seemed impossible to solve in the 1950s–1960s.

2. Modification of the membrane surface to improve the hemotoxic character. The wide range of molecules that are continuously developed in the pharmaceutical industry, as well as the increasingly easy access to different methods of membrane functionalization on a scale industrial, could ensure in a reasonable time horizon, the use of more and more advanced membranes. The current problem is not the existence of technological solutions but the cost price that would be reached if these solutions were applied. Quantitatively, hemodialysis membranes are the second application for membranes produced worldwide, and an uncontrolled cost would have cascading effects from influencing capacities of production, up to the costs of the health system. Innovations in technology and chemical engineering will be able to solve such problems, allowing us to obtain functionalized membranes with other molecules, which will make synthesis easier.

3. Development of a niche field—one-day hemodialysis. In addition to treating chronic kidney failure, hemodialysis finds wider and wider applications in the purification of waste and the removal from the body of elements or substances whose concentration in the body increases following accidental poisoning. This category includes heavy metals, an organic substance used as intermediates in various syntheses or other substances from niche industries, which can reach the body following an accident. The challenges related to this type of application are far from solved. In the case of heavy metal retention, the problem is not the synthesis of a membrane that retains that heavy metal, but the fact that the membrane should retain the metal alone without removing from the blood other cations of biological interest (such as $Na^+$, $K^+$).

4. Coupling of hemodialysis with other biomedical processes. Patients who suffer from chronic kidney dysfunction usually have other associated conditions. A great challenge lies in obtaining membranes for hemodialysis that also have the ability to release pharmaceutically active substances, so that, during the medical procedure of blood filtration to be carried out, the treatment of associated diseases can be managed simultaneously, thus increasing the quality of life of these patients. This would be especially beneficial for those who suffer from a form of cancer for which chemotherapy is required. In addition, the release of analgesics would make the procedure more bearable, especially by alleviating the inflammatory effects that occur and are maintained for several hours after.

**Author Contributions:** Conceptualization, S.I.V. and E.R.R.; resources, S.I.V.; data curation, S.I.V. and E.R.R.; writing—original draft preparation, S.I.V. and E.R.R.; writing—review and editing, S.I.V. and E.R.R.; supervision, S.I.V.; project administration, S.I.V.; funding acquisition, S.I.V. All authors have read and agreed to the published version of the manuscript.

**Funding:** This work was supported by a grant of the Ministry of Research, Innovation and Digitization, CNCS/CCCDI—UEFISCDI, project number PN-III-P4-ID-PCE-2020-1154, Hemodialysis combined with stimuli responsive drug delivery—a new generation of polymeric membranes for advanced biomedical applications within PNCDI III.

**Institutional Review Board Statement:** Not applicable.

**Informed Consent Statement:** Not applicable.

**Data Availability Statement:** Not applicable.

**Conflicts of Interest:** The authors declare no conflict of interest.

# References


1. Liyanage, T.; Ninomiya, T.; Jha, V.; Neal, B.; Patrice, H.M.; Okpechi, I.; Zhao, M.H.; Lv, J.; Garg, A.X.; Knight, J.; et al. Worldwide access to treatment for end-stage kidney disease: A systematic review. *Lancet* **2015**, *385*, 1975–1982. [CrossRef]
2. Dou, P.; Donato, D.; Guo, H.; Zhao, S.; He, T. Recycling water from spent dialysate by osmotic dilution: Impact of urea rejection of forward osmosis membrane on hemodialysis duration. *Desalination* **2020**, *496*, 114605. [CrossRef]
3. Cath, T.Y.; Childress, A.E.; Elimelech, M. Forward osmosis: Principles, applications, and recent developments. *J. Membr. Sci.* **2006**, *281*, 70–87. [CrossRef]
4. Rana, A.K.; Gupta, V.K.; Saini, A.K.; Voicu, S.I.; Abdellattifaand, M.H.; Thakur, V.K. Water desalination using nanocelluloses/cellulose derivatives based membranes for sustainable future. *Desalination* **2021**, *520*, 115359. [CrossRef]
5. Voicu, S.I.; Thakur, V.K. Graphene-based composite membranes for nanofiltration: Performances and future perspectives. *Emergent Mater.* **2021**, 1–13. [CrossRef]
6. Pandele, A.M.; Iovu, H.; Orbeci, C.; Tuncel, C.; Miculescu, F.; Nicolescu, A.; Deleanu, C.; Voicu, S.I. Surface modified cellulose acetate membranes for the reactive retention of tetracycline. *Sep. Purif. Technol.* **2020**, *249*, 117145. [CrossRef]
7. Serbanescu, O.S.; Pandele, A.M.; Miculescu, F.; Voicu, S.I. Synthesis and Characterization of Cellulose Acetate Membranes with Self-Indicating Properties by Changing the Membrane Surface Color for Separation of Gd(III). *Coatings* **2020**, *10*, 468. [CrossRef]
8. Serbanescu, O.S.; Pandele, A.M.; Oprea, M.; Semenescu, A.; Thakur, V.K.; Voicu, S.I. Crown Ether-Immobilized Cellulose Acetate Membranes for the Retention of Gd (III). *Polymers* **2021**, *13*, 3978. [CrossRef]
9. Muhulet, A.; Tuncel, C.; Miculescu, F.; Pandele, A.M.; Bobirica, C.; Orbeci, C.; Bobirica, L.; Palla-Papavlu, A.; Voicu, S.I. Synthesis and characterization of polysulfone-TiO$_2$ decorated MWCNT composite membranes by sonochemical method. *Appl. Phys. A* **2020**, *126*, 233. [CrossRef]
10. Voicu, Ş.I.; Pandele, A.; Tanasă, E.; Rughinis, R.; Crica, L.; Pilan, L.; Ionita, M. The impact of sonication time through polysulfone-graphene oxide composite films properties. *Dig. J. Nanomater. Biostructures* **2013**, *8*, 1389–1394.
11. Chiulan, I.; Heggset, E.B.; Voicu, Ş.I.; Chinga-Carrasco, G. Photopolymerization of Bio-Based Polymers in a Biomedical Engineering Perspective. *Biomacromolecules* **2021**, *22*, 1795–1814. [CrossRef] [PubMed]
12. Voicu, S.; Dobrica, A.; Sava, S.; Ivan, A.; Naftanaila, L.J.J.o.O.; Materials, A. Cationic surfactants-controlled geometry and dimensions of polymeric membrane pores. *J. Optoelectron. Adv. Mater.* **2012**, *14*, 923–928.
13. Pandele, A.M.; Constantinescu, A.; Radu, I.C.; Miculescu, F.; Ioan Voicu, S.; Ciocan, L.T. Synthesis and Characterization of PLA-Micro-structured Hydroxyapatite Composite Films. *Materials* **2020**, *13*, 274. [CrossRef] [PubMed]
14. Oprea, M.; Voicu, S.I. Recent Advances in Applications of Cellulose Derivatives-Based Composite Membranes with Hydroxyapatite. *Materials* **2020**, *13*, 2481. [CrossRef] [PubMed]
15. Voicu, S.I.; Thakur, V.K. Aminopropyltriethoxysilane as a linker for cellulose-based functional materials: New horizons and future challenges. *Curr. Opin. Green Sustain. Chem.* **2021**, *30*, 100480. [CrossRef]
16. Oprea, M.; Voicu, S.I. Cellulose Composites with Graphene for Tissue Engineering Applications. *Materials* **2020**, *13*, 5347. [CrossRef]
17. Erdal, N.B.; Adolfsson, K.H.; Pettersson, T.; Hakkarainen, M. Green Strategy to Reduced Nanographene Oxide through Microwave Assisted Transformation of Cellulose. *ACS Sustain. Chem. Eng.* **2018**, *6*, 1246–1255. [CrossRef]
18. Erdal, N.B.; Hakkarainen, M. Construction of Bioactive and Reinforced Bioresorbable Nanocomposites by Reduced Nano-Graphene Oxide Carbon Dots. *Biomacromolecules* **2018**, *19*, 1074–1081. [CrossRef]
19. Erdal, N.B.; Yao, J.G.; Hakkarainen, M. Cellulose-Derived Nanographene Oxide Surface-Functionalized Three-Dimensional Scaffolds with Drug Delivery Capability. *Biomacromolecules* **2019**, *20*, 738–749. [CrossRef]
20. Pandele, A.M.; Oprea, M.; Dutu, A.A.; Miculescu, F.; Voicu, S.I. A Novel Generation of Polysulfone/Crown Ether-Functionalized Reduced Graphene Oxide Membranes with Potential Applications in Hemodialysis. *Polymers* **2022**, *14*, 148. [CrossRef]
21. Malavade, T.S.; Dey, A.; Chan, C.T. Nocturnal Hemodialysis: Why Aren't More People Doing It? *Adv. Chronic Kidney Dis.* **2021**, *28*, 184–189. [CrossRef] [PubMed]
22. Lavoie-Cardinal, M.; Nadeau-Fredette, A.-C. Physical Infrastructure and Integrated Governance Structure for Home Hemodialysis. *Adv. Chronic Kidney Dis.* **2021**, *28*, 149–156. [CrossRef] [PubMed]
23. Miller, B.W. Reconciling the Current Status of Home Hemodialysis With the 2019 Executive Order—Realistic or Obtainable? *Adv. Chronic Kidney Dis.* **2021**, *28*, 124–128. [CrossRef]
24. Gupta, N. Strategic Planning for Starting or Expanding a Home Hemodialysis Program. *Adv. Chronic Kidney Dis.* **2021**, *28*, 143–148. [CrossRef] [PubMed]
25. Bieber, S.D.; Young, B.A. Home Hemodialysis: Core Curriculum 2021. *Am. J. Kidney Dis.* **2021**, *78*, 876–885. [CrossRef] [PubMed]
26. Han, X.; Zhang, S.; Chen, Z.; Adhikari, B.K.; Zhang, Y.; Zhang, J.; Sun, J.; Wang, Y. Cardiac biomarkers of heart failure in chronic kidney disease. *Clin. Chim. Acta* **2020**, *510*, 298–310. [CrossRef]
27. Stopper, H.; Bankoglu, E.E.; Marcos, R.; Pastor, S. Micronucleus frequency in chronic kidney disease patients: A review. *Mutat. Res./Rev. Mutat. Res.* **2020**, *786*, 108340. [CrossRef]
28. Vachharajani, T.J.; Taliercio, J.J.; Anvari, E. New Devices and Technologies for Hemodialysis Vascular Access: A Review. *Am. J. Kidney Dis.* **2021**, *78*, 116–124. [CrossRef]
29. Tarrass, F.; Benjelloun, O.; Benjelloun, M. Towards zero liquid discharge in hemodialysis. Possible issues. *Nefrología* **2021**, *41*, 620–624. [CrossRef]

30. Ramanathan, S.; Gopinath, S.C.B.; Arshad, M.K.M.; Poopalan, P. Nanostructured aluminosilicate from fly ash: Potential approach in waste utilization for industrial and medical applications. *J. Clean. Prod.* **2020**, *253*, 119923. [CrossRef]

31. Massey, I.Y.; Yang, F.; Ding, Z.; Yang, S.; Guo, J.; Tezi, C.; Al-Osman, M.; Kamegni, R.B.; Zeng, W. Exposure routes and health effects of microcystins on animals and humans: A mini-review. *Toxicon* **2018**, *151*, 156–162. [CrossRef] [PubMed]

32. Stamatialis, D.F.; Papenburg, B.J.; Gironés, M.; Saiful, S.; Bettahalli, S.N.M.; Schmitmeier, S.; Wessling, M. Medical applications of membranes: Drug delivery, artificial organs and tissue engineering. *J. Membr. Sci.* **2008**, *308*, 1–34. [CrossRef]

33. Zydney, A.L. New developments in membranes for bioprocessing—A review. *J. Membr. Sci.* **2021**, *620*, 118804. [CrossRef]

34. Nunes, S.P.; Culfaz-Emecen, P.Z.; Ramon, G.Z.; Visser, T.; Koops, G.H.; Jin, W.; Ulbricht, M. Thinking the future of membranes: Perspectives for advanced and new membrane materials and manufacturing processes. *J. Membr. Sci.* **2020**, *598*, 117761. [CrossRef]

35. Galiano, F.; Briceño, K.; Marino, T.; Molino, A.; Christensen, K.V.; Figoli, A. Advances in biopolymer-based membrane preparation and applications. *J. Membr. Sci.* **2018**, *564*, 562–586. [CrossRef]

36. Wang, Z.-G.; Wan, L.-S.; Xu, Z.-K. Surface engineerings of polyacrylonitrile-based asymmetric membranes towards biomedical applications: An overview. *J. Membr. Sci.* **2007**, *304*, 8–23. [CrossRef]

37. Jhaveri, J.H.; Murthy, Z.V.P. A comprehensive review on anti-fouling nanocomposite membranes for pressure driven membrane separation processes. *Desalination* **2016**, *379*, 137–154. [CrossRef]

38. Schlenoff, J.B. Zwitteration: Coating Surfaces with Zwitterionic Functionality to Reduce Nonspecific Adsorption. *Langmuir* **2014**, *30*, 9625–9636. [CrossRef]

39. Lin, W.; Klein, J. Control of surface forces through hydrated boundary layers. *Curr. Opin. Colloid Interface Sci.* **2019**, *44*, 94–106. [CrossRef]

40. Erfani, A.; Seaberg, J.; Aichele, C.P.; Ramsey, J.D. Interactions between Biomolecules and Zwitterionic Moieties: A Review. *Biomacromolecules* **2020**, *21*, 2557–2573. [CrossRef]

41. Salimi, E.; Ghaee, A.; Ismail, A.F.; Othman, M.H.D.; Sean, G.P. Current Approaches in Improving Hemocompatibility of Polymeric Membranes for Biomedical Application. *Macromol. Mater. Eng.* **2016**, *301*, 771–800. [CrossRef]

42. Weber, M.; Steinle, H.; Golombek, S.; Hann, L.; Schlensak, C.; Wendel, H.P.; Avci-Adali, M. Blood-Contacting Biomaterials: In Vitro Evaluation of the Hemocompatibility. *Front. Bioeng. Biotechnol.* **2018**, *6*, 99. [CrossRef] [PubMed]

43. Vanholder, R.; Van Laecke, S.; Glorieux, G. What is new in uremic toxicity? *Pediatr. Nephrol.* **2008**, *23*, 1211–1221. [CrossRef] [PubMed]

44. Azhar, O.; Jahan, Z.; Sher, F.; Niazi, M.B.K.; Kakar, S.J.; Shahid, M. Cellulose acetate-polyvinyl alcohol blend hemodialysis membranes integrated with dialysis performance and high biocompatibility. *Mater. Sci. Eng. C* **2021**, *126*, 112127. [CrossRef] [PubMed]

45. Eduok, U.; Abdelrasoul, A.; Shoker, A.; Doan, H. Recent developments, current challenges and future perspectives on cellulosic hemodialysis membranes for highly efficient clearance of uremic toxins. *Mater. Today Commun.* **2021**, *27*, 102183. [CrossRef]

46. Koh, E.; Lee, Y.T. Development of an embossed nanofiber hemodialysis membrane for improving capacity and efficiency via 3D printing and electrospinning technology. *Sep. Purif. Technol.* **2020**, *241*, 116657. [CrossRef]

47. Liu, W.; Fu, X.; Liu, Y.-F.; Su, T.; Peng, J. Vorapaxar-modified polysulfone membrane with high hemocompatibility inhibits thrombosis. *Mater. Sci. Eng. C* **2021**, *118*, 111508. [CrossRef]

48. Zhang, W.; Yue, P.; Zhang, H.; Yang, N.; Li, C.; Li, J.H.; Meng, J.; Zhang, Q. Surface modification of AO-PAN@OHec nanofiber membranes with amino acid for antifouling and hemocompatible properties. *Appl. Surf. Sci.* **2019**, *475*, 934–941. [CrossRef]

49. Xiang, T.; Lu, T.; Xie, Y.; Zhao, W.-F.; Sun, S.-D.; Zhao, C.-S. Zwitterionic polymer functionalization of polysulfone membrane with improved antifouling property and blood compatibility by combination of ATRP and click chemistry. *Acta Biomater.* **2016**, *40*, 162–171. [CrossRef]

50. Cisse, I.; Oakes, S.; Sachdev, S.; Toro, M.; Lutondo, S.; Shedden, D.; Atkinson, K.M.; Shertok, J.; Mehan, M.; Gupta, S.K.; et al. Surface Modification of Polyethersulfone (PES) with UV Photo-Oxidation. *Technologies* **2021**, *9*, 36. [CrossRef]

51. Abdelrasoul, A.; Shoker, A. Induced hemocompatibility of polyethersulfone (PES) hemodialysis membrane using polyvinylpyrrolidone: Investigation on human serum fibrinogen adsorption and inflammatory biomarkers released. *Chem. Eng. Res. Des.* **2022**, *177*, 615–624. [CrossRef]

52. Wester, M.; Simonis, F.; Lachkar, N.; Wodzig, W.K.; Meuwissen, F.J.; Kooman, J.P.; Boer, W.H.; Joles, J.A.; Gerritsen, K.G. Removal of urea in a wearable dialysis device: A reappraisal of electro-oxidation. *Artif. Organs* **2014**, *38*, 998–1006. [CrossRef] [PubMed]

53. Song, X.; Ji, H.; Zhao, W.; Sun, S.; Zhao, C. Hemocompatibility enhancement of polyethersulfone membranes: Strategies and challenges. *Adv. Membr.* **2021**, *1*, 100013. [CrossRef]

54. Serbanescu, O.S.; Voicu, S.I.; Thakur, V.K. Polysulfone functionalized membranes: Properties and challenges. *Mater. Today Chem.* **2020**, *17*, 100302. [CrossRef]

55. Mollahosseini, A.; Abdelrasoul, A.; Shoker, A. A critical review of recent advances in hemodialysis membranes hemocompatibility and guidelines for future development. *Mater. Chem. Phys.* **2020**, *248*, 122911. [CrossRef]

56. Abidin, M.N.Z.; Goh, P.S.; Said, N.; Ismail, A.F.; Othman, M.H.D.; Abdullah, M.S.; Ng, B.C.; Hasbullah, H.; Sheikh Abdul Kadir, S.H.; Kamal, F.; et al. Polysulfone/amino-silanized poly(methyl methacrylate) dual layer hollow fiber membrane for uremic toxin separation. *Sep. Purif. Technol.* **2020**, *236*, 116216. [CrossRef]

57. Ashour, G.R.; Hussein, M.A.; Sobahi, T.R.; Alamry, K.A.; Alqarni, S.A.; Rafatullah, M. Modification of Sulfonated Polyethersulfone Membrane as a Selective Adsorbent for Co(II) Ions. *Polymers* **2021**, *13*, 3569. [CrossRef]

58. Aydemir Sezer, U.; Ozturk, K.; Aru, B.; Yanıkkaya Demirel, G.; Sezer, S. A design achieved by coaxial electrospinning of polysulfone and sulfonated polysulfone as a core-shell structure to optimize mechanical strength and hemocompatibility. *Surf. Interfaces* **2018**, *10*, 176–187. [CrossRef]

59. Mahlicli, F.Y.; Altinkaya, S.A. Immobilization of alpha lipoic acid onto polysulfone membranes to suppress hemodialysis induced oxidative stress. *J. Membr. Sci.* **2014**, *449*, 27–37. [CrossRef]

60. Chien, H.-W.; Wu, S.-P.; Kuo, W.-H.; Wang, M.-J.; Lee, C.; Lai, J.-Y.; Tsai, W.-B. Modulation of hemocompatibility of polysulfone by polyelectrolyte multilayer films. *Colloids Surf. B Biointerfaces* **2010**, *77*, 270–278. [CrossRef]

61. Xie, Y.X.; Wang, K.K.; Yu, W.H.; Cui, M.B.; Shen, Y.J.; Wang, X.Y.; Fang, L.F.; Zhu, B.K. Improved permeability and antifouling properties of polyvinyl chloride ultrafiltration membrane via blending sulfonated polysulfone. *J. Colloid Interface Sci.* **2020**, *579*, 562–572. [CrossRef] [PubMed]

62. Zhang, D.; Gao, B.; Cui, K. Photoluminescence property of polymer—Rare earth complexes containing acetaldehyde/aminophenol type bidentate Schiff base ligand. *J. Coord. Chem.* **2017**, *70*, 3275–3292. [CrossRef]

63. Liu, T.M.; Xu, J.J.; Qiu, Y.R. A novel kind of polysulfone material with excellent biocompatibility modified by the sulfonated hydroxypropyl chitosan. *Mater. Sci. Engineering. C Mater. Biol. Appl.* **2017**, *79*, 570–580. [CrossRef]

64. Tu, M.-M.; Xu, J.-J.; Qiu, Y.-R. Surface hemocompatible modification of polysulfone membrane via covalently grafting acrylic acid and sulfonated hydroxypropyl chitosan. *RSC Adv.* **2019**, *9*, 6254–6266. [CrossRef]

65. Ganj, M.; Asadollahi, M.; Mousavi, S.A.; Bastani, D.; Aghaeifard, F. Surface modification of polysulfone ultrafiltration membranes by free radical graft polymerization of acrylic acid using response surface methodology. *J. Polym. Res.* **2019**, *26*, 231. [CrossRef]

66. Yan, S.; Tu, M.-M.; Qiu, Y.-R. The hemocompatibility of the modified polysulfone membrane with 4-(chloromethyl)benzoic acid and sulfonated hydroxypropyl chitosan. *Colloids Surf. B Biointerfaces* **2020**, *188*, 110769. [CrossRef] [PubMed]

67. Xiang, T.; Xie, Y.; Wang, R.; Wu, M.; Sun, S.; Zhao, C. Facile chemical modification of polysulfone membrane with improved hydrophilicity and blood compatibility. *Mater. Lett.* **2014**, *137*, 192–195. [CrossRef]

68. Li, L.; Yan, G.; Wu, J. Modification of polysulfone membranes via surface-initiated atom transfer radical polymerization and their antifouling properties. *J. Appl. Polym. Sci.* **2009**, *111*, 1942–1946. [CrossRef]

69. Dong, H.-B.; Xu, Y.-Y.; Yi, Z.; Shi, J.-L. Modification of polysulfone membranes via surface-initiated atom transfer radical polymerization. *Appl. Surf. Sci.* **2009**, *255*, 8860–8866. [CrossRef]

70. Yue, W.-W.; Li, H.-J.; Xiang, T.; Qin, H.; Sun, S.-D.; Zhao, C.-S. Grafting of zwitterion from polysulfone membrane via surface-initiated ATRP with enhanced antifouling property and biocompatibility. *J. Membr. Sci.* **2013**, *446*, 79–91. [CrossRef]

71. Liu, Y.; Wang, J. Preparation of anion exchange membrane by efficient functionalization of polysulfone for electrodialysis. *J. Membr. Sci.* **2020**, *596*, 117591. [CrossRef]

72. Chen, S.-H.; Chang, Y.; Lee, K.-R.; Wei, T.-C.; Higuchi, A.; Ho, F.-M.; Tsou, C.-C.; Ho, H.-T.; Lai, J.-Y. Hemocompatible Control of Sulfobetaine-Grafted Polypropylene Fibrous Membranes in Human Whole Blood via Plasma-Induced Surface Zwitterionization. *Langmuir* **2012**, *28*, 17733–17742. [CrossRef] [PubMed]

73. Xiang, T.; Zhang, L.S.; Wang, R.; Xia, Y.; Su, B.H.; Zhao, C.S. Blood compatibility comparison for polysulfone membranes modified by grafting block and random zwitterionic copolymers via surface-initiated ATRP. *J. Colloid Interface Sci.* **2014**, *432*, 47–56. [CrossRef] [PubMed]

74. Maggay, I.V.B.; Aini, H.N.; Lagman, M.M.G.; Tang, S.-H.; Aquino, R.R.; Chang, Y.; Venault, A. A Biofouling Resistant Zwitterionic Polysulfone Membrane Prepared by a Dual-Bath Procedure. *Membranes* **2022**, *12*, 69. [CrossRef]

75. Kheirieh, S.; Asghari, M.; Afsari, M. Application and modification of polysulfone membranes. *Rev. Chem. Eng.* **2018**, *34*, 657–693. [CrossRef]

76. Kim, K.; Lee, K.; Cho, K.; Park, C. Surface modification of polysulfone ultrafiltration by oxygen plasma treatment. *J. Membr. Sci.* **2002**, *199*, 135–145. [CrossRef]

77. Zheng, Z.; Wang, W.; Huang, X.; Fan, W.; Li, L. Surface modification of polysulfone hollow fiber membrane for extracorporeal membrane oxygenator using low-temperature plasma treatment. *Plasma Process. Polym.* **2018**, *15*, 1700122. [CrossRef]

78. Zhang, X.; Zhang, S.; Wang, Y.; Zheng, Y.; Han, Y.; Lu, Y. Polysulfone membrane treated with $NH_3$–$O_2$ plasma and its property. *High Perform. Polym.* **2017**, *30*, 1139–1144. [CrossRef]

79. Pegalajar-Jurado, A.; Mann, M.N.; Maynard, M.R.; Fisher, E.R. Hydrophilic Modification of Polysulfone Ultrafiltration Membranes by Low Temperature Water Vapor Plasma Treatment to Enhance Performance. *Plasma Process. Polym.* **2016**, *13*, 598–610. [CrossRef]

80. Wang, L.; Fang, F.; Liu, Y.; Li, J.; Huang, X. Facile preparation of heparinized polysulfone membrane assisted by polydopamine/polyethyleneimine co-deposition for simultaneous LDL selectivity and biocompatibility. *Appl. Surf. Sci.* **2016**, *385*, 308–317. [CrossRef]

81. Dizman, C.; Demirkol, D.O.; Ates, S.; Torun, L.; Sakarya, S.; Timur, S.; Yagci, Y. Photochemically prepared polysulfone/poly(ethylene glycol) amphiphilic networks and their biomolecule adsorption properties. *Colloids Surf. B Biointerfaces* **2011**, *88*, 265–270. [CrossRef] [PubMed]

82. Causserand, C.; Pellegrin, B.; Rouch, J.-C. Effects of sodium hypochlorite exposure mode on PES/PVP ultrafiltration membrane degradation. *Water Res.* **2015**, *85*, 316–326. [CrossRef] [PubMed]

83. Huang, X.J.; Guduru, D.; Xu, Z.K.; Vienken, J.; Groth, T. Blood compatibility and permeability of heparin-modified polysulfone as potential membrane for simultaneous hemodialysis and LDL removal. *Macromol. Biosci.* **2011**, *11*, 131–140. [CrossRef] [PubMed]

84. Huang, X.J.; Guduru, D.; Xu, Z.K.; Vienken, J.; Groth, T. Immobilization of heparin on polysulfone surface for selective adsorption of low-density lipoprotein (LDL). *Acta Biomater.* **2010**, *6*, 1099–1106. [CrossRef]
85. Sasisekharan, R.; Venkataraman, G. Heparin and heparan sulfate: Biosynthesis, structure and function. *Curr. Opin. Chem. Biol.* **2000**, *4*, 626–631. [CrossRef]
86. Paluck, S.J.; Nguyen, T.H.; Maynard, H.D. Heparin-Mimicking Polymers: Synthesis and Biological Applications. *Biomacromolecules* **2016**, *17*, 3417–3440. [CrossRef]
87. Fischer, K.-G. Essentials of anticoagulation in hemodialysis. *Hemodial. Int.* **2007**, *11*, 178–189. [CrossRef]
88. Wang, H.; Li, J.; Liu, F.; Li, T.; Zhong, Y.; Lin, H.; He, J. Enhanced hemocompatibility of flat and hollow fiber membranes via a heparin free surface crosslinking strategy. *React. Funct. Polym.* **2018**, *124*, 104–114. [CrossRef]
89. Banik, N.; Yang, S.-B.; Kang, T.-B.; Lim, J.-H.; Park, J. Heparin and Its Derivatives: Challenges and Advances in Therapeutic Biomolecules. *Int. J. Mol. Sci.* **2021**, *22*, 524. [CrossRef]
90. Hao, C.; Sun, M.; Wang, H.; Zhang, L.; Wang, W. Chapter Two—Low molecular weight heparins and their clinical applications. In *Progress in Molecular Biology and Translational Science*; Zhang, L., Ed.; Academic Press: Cambridge, MA, USA, 2019; Volume 163, pp. 21–39.
91. Li, J.; Huang, X.J.; Ji, J.; Lan, P.; Vienken, J.; Groth, T.; Xu, Z.K. Covalent heparin modification of a polysulfone flat sheet membrane for selective removal of low-density lipoproteins: A simple and versatile method. *Macromol. Biosci.* **2011**, *11*, 1218–1226. [CrossRef]
92. Lin, B.; Liu, K.; Qiu, Y. Preparation of modified polysulfone material decorated by sulfonated citric chitosan for haemodialysis and its haemocompatibility. *R. Soc. Open Sci.* **2021**, *8*, 210462. [CrossRef] [PubMed]
93. Wang, C.; Lin, B.; Qiu, Y. Enhanced hydrophilicity and anticoagulation of polysulfone materials modified via dihydroxypropyl, sulfonic groups and chitosan. *Colloids Surf. B Biointerfaces* **2022**, *210*, 112243. [CrossRef] [PubMed]
94. Qi, X.; Yang, N.; Luo, Y.; Jia, X.; Zhao, J.; Feng, X.; Chen, L.; Zhao, Y. Resveratrol as a plant type antioxidant modifier for polysulfone membranes to improve hemodialysis-induced oxidative stress. *Mater. Sci. Eng. C* **2021**, *123*, 111953. [CrossRef] [PubMed]
95. Yang, N.; Jia, X.; Wang, D.; Wei, C.; He, Y.; Chen, L.; Zhao, Y. Silibinin as a natural antioxidant for modifying polysulfone membranes to suppress hemodialysis-induced oxidative stress. *J. Membr. Sci.* **2018**, *574*, 86–89. [CrossRef]
96. Fang, Y.; Duranceau, S.J. Study of the Effect of Nanoparticles and Surface Morphology on Reverse Osmosis and Nanofiltration Membrane Productivity. *Membranes* **2013**, *3*, 196. [CrossRef]
97. Liu, Q.; Wu, X.; Zhang, K. Polysulfone/Polyamide-$SiO_2$ Composite Membrane with High Permeance for Organic Solvent Nanofiltration. *Membranes* **2018**, *8*, 89. [CrossRef]
98. Sherugar, P.; Naik, N.S.; Padaki, M.; Nayak, V.; Gangadharan, A.; Nadig, A.R.; Déon, S. Fabrication of zinc doped aluminium oxide/polysulfone mixed matrix membranes for enhanced antifouling property and heavy metal removal. *Chemosphere* **2021**, *275*, 130024. [CrossRef]
99. Chai, P.V.; Mahmoudi, E.; Teow, Y.H.; Mohammad, A.W. Preparation of novel polysulfone-$Fe_3O_4$/GO mixed-matrix membrane for humic acid rejection. *J. Water Process. Eng.* **2017**, *15*, 83–88. [CrossRef]
100. Nechifor, G.; Voicu, S.I.; Nechifor, A.C.; Garea, S. Nanostructured hybrid membrane polysulfone-carbon nanotubes for hemodialysis. *Desalination* **2009**, *241*, 342–348. [CrossRef]
101. Mahat, N.A.; Shamsudin, S.A.; Jullok, N.; M'Radzi, A.H. Carbon quantum dots embedded polysulfone membranes for antibacterial performance in the process of forward osmosis. *Desalination* **2020**, *493*, 114618. [CrossRef]
102. Kang, Y.; Obaid, M.; Jang, J.; Ham, M.-H.; Kim, I.S. Novel sulfonated graphene oxide incorporated polysulfone nanocomposite membranes for enhanced-performance in ultrafiltration process. *Chemosphere* **2018**, *207*, 581–589. [CrossRef] [PubMed]
103. Zheng, Z.; Chen, J.; Wu, J.; Feng, M.; Xu, L.; Yan, N.; Xie, H. Incorporation of Biomass-Based Carbon Nanoparticles into Polysulfone Ultrafiltration Membranes for Enhanced Separation and Anti-Fouling Performance. *Nanomaterials* **2021**, *11*, 2303. [CrossRef] [PubMed]
104. Ouradi, A.; Nguyen, Q.T.; Benaboura, A. Polysulfone—AN69 blend membranes and its surface modification by polyelectrolyte-layer deposit—Preparation and characterization. *J. Membr. Sci.* **2014**, *454*, 20–35. [CrossRef]
105. Ouradi, A.; Cherifi, N.; Nguyen, Q.T.; Benaboura, A. Preliminary study of the prepared polysulfone/AN69/clay composite membranes intended for the hemodialysis application. *Chem. Pap.* **2020**, *74*, 2133–2144. [CrossRef]
106. Anadão, P.; Sato, L.F.; Wiebeck, H.; Valenzuela-Díaz, F.R. Montmorillonite as a component of polysulfone nanocomposite membranes. *Appl. Clay Sci.* **2010**, *48*, 127–132. [CrossRef]
107. Ma, Y.; Shi, F.; Wang, Z.; Wu, M.; Ma, J.; Gao, C. Preparation and characterization of PSf/clay nanocomposite membranes with PEG 400 as a pore forming additive. *Desalination* **2012**, *286*, 131–137. [CrossRef]
108. Mansur, S.; Othman, M.H.D.; Abidin, M.N.Z.; Ismail, A.F.; Abdul Kadir, S.H.S.; Goh, P.S.; Hasbullah, H.; Ng, B.C.; Abdullah, M.S.; Mustafar, R. Enhanced adsorption and biocompatibility of polysulfone hollow fibre membrane via the addition of silica/alpha-mangostin hybrid nanoparticle for uremic toxins removal. *J. Environ. Chem. Eng.* **2021**, *9*, 106141. [CrossRef]
109. Song, H.J.; Kim, C.K. Fabrication and properties of ultrafiltration membranes composed of polysulfone and poly(1-vinylpyrrolidone) grafted silica nanoparticles. *J. Membr. Sci.* **2013**, *444*, 318–326. [CrossRef]
110. Soenen, S.J.; Himmelreich, U.; Nuytten, N.; De Cuyper, M. Cytotoxic effects of iron oxide nanoparticles and implications for safety in cell labelling. *Biomaterials* **2011**, *32*, 195–205. [CrossRef]

111. Wu, W.; Wu, Z.; Yu, T.; Jiang, C.; Kim, W.-S. Recent progress on magnetic iron oxide nanoparticles: Synthesis, surface functional strategies and biomedical applications. *Sci. Technol. Adv. Mater.* **2015**, *16*, 023501. [CrossRef]
112. Said, N.; Abidin, M.N.Z.; Hasbullah, H.; Ismail, A.F.; Goh, P.S.; Othman, M.H.D.; Abdullah, M.S.; Ng, B.C.; Kadir, S.H.S.A.; Kamal, F. Iron oxide nanoparticles improved biocompatibility and removal of middle molecule uremic toxin of polysulfone hollow fiber membranes. *Malays. J. Fundam. Appl. Sci.* **2019**, *136*, 48234. [CrossRef]
113. Said, N.; Zainol Abidin, M.N.; Hasbullah, H.; Ismail, A.; Goh, P.; Hafiz, M.; Othman, M.H.; Abdullah, S.; Sheikh Abdul Kadir, S.H.; Kamal, F. Polysulfone hemodialysis membrane incorporated with Fe$_2$O$_3$ for enhanced removal of middle molecular weight uremic toxin. *Malays. J. Fundam. Appl. Sci.* **2020**, *16*, 1–5. [CrossRef]
114. Said, N.; Hasbullah, H.; Abidin, M.N.Z.; Ismail, A.F.; Goh, P.S.; Othman, M.H.D.; Kadir, S.H.S.A.; Kamal, F.; Abdullah, M.S.; Ng, B.C. Facile modification of polysulfone hollow-fiber membranes via the incorporation of well-dispersed iron oxide nanoparticles for protein purification. *J. Appl. Polym. Sci.* **2019**, *136*, 47502. [CrossRef]
115. Li, L.; Mak, K.Y.; Leung, C.W.; Chan, K.Y.; Chan, W.K.; Zhong, W.; Pong, P.W.T. Effect of synthesis conditions on the properties of citric-acid coated iron oxide nanoparticles. *Microelectron. Eng.* **2013**, *110*, 329–334. [CrossRef]
116. Yang, Y.; Wang, P.; Zheng, Q. Preparation and properties of polysulfone/TiO$_2$ composite ultrafiltration membranes. *J. Polym. Sci. Part B Polym. Phys.* **2006**, *44*, 879–887. [CrossRef]
117. Zhang, S.; Wang, Q.; Dai, F.; Gu, Y.; Qian, G.; Chen, C.; Yu, Y. Novel TiO$_2$ Nanoparticles/Polysulfone Composite Hollow Microspheres for Photocatalytic Degradation. *Polymers* **2021**, *13*, 336. [CrossRef] [PubMed]
118. Yang, Y.; Zhang, H.; Wang, P.; Zheng, Q.; Li, J. The influence of nano-sized TiO$_2$ fillers on the morphologies and properties of PSF UF membrane. *J. Membr. Sci.* **2007**, *288*, 231–238. [CrossRef]
119. Zhu, L.; Wu, M.; Van der Bruggen, B.; Lei, L.; Zhu, L. Effect of TiO$_2$ content on the properties of polysulfone nanofiltration membranes modified with a layer of TiO$_2$—Graphene oxide. *Sep. Purif. Technol.* **2020**, *242*, 116770. [CrossRef]
120. Zailani, M.Z.; Ismail, A.F.; Goh, P.S.; Abdul Kadir, S.H.S.; Othman, M.H.D.; Hasbullah, H.; Abdullah, M.S.; Ng, B.C.; Kamal, F.; Mustafar, R. Immobilizing chitosan nanoparticles in polysulfone ultrafiltration hollow fibre membranes for improving uremic toxins removal. *J. Environ. Chem. Eng.* **2021**, *9*, 106878. [CrossRef]
121. Zainol Abidin, M.N.; Goh, P.S.; Said, N.; Ismail, A.F.; Othman, M.H.D.; Hasbullah, H.; Abdullah, M.S.; Ng, B.C.; Sheikh Abdul Kadir, S.H.; Kamal, F.; et al. Co-Adsorptive Removal of Creatinine and Urea by a Three-Component Dual-Layer Hollow Fiber Membrane. *ACS Appl. Mater. Interfaces* **2020**, *12*, 33276–33287. [CrossRef]
122. Kohlová, M.; Amorim, C.G.; da Nova Araújo, A.; Santos-Silva, A.; Solich, P.; Montenegro, M.C.B.S.M. In vitro assessment of polyethylene glycol and polyvinylpyrrolidone as hydrophilic additives on bioseparation by polysulfone membranes. *J. Mater. Sci.* **2020**, *55*, 1292–1307. [CrossRef]
123. Glassock, R.J.; Massry, S.G. Uremic toxins: An integrated overview of classification and pathobiology. In *Nutritional Management of Renal Disease*, 4th ed.; Kopple, J.D., Massry, S.G., Kalantar-Zadeh, K., Fouque, D., Eds.; Academic Press: Cambridge, MA, USA, 2022; Chapter 6; pp. 77–89.
124. Vanholder, R.; Glorieux, G.; De Smet, R.; Lameire, N. New insights in uremic toxins. *Kidney Int.* **2003**, *63*, S6–S10. [CrossRef] [PubMed]
125. Magnani, S.; Atti, M. Uremic Toxins and Blood Purification: A Review of Current Evidence and Future Perspectives. *Toxins* **2021**, *13*, 246. [CrossRef] [PubMed]
126. Clark, W.R.; Dehghani, N.L.; Narsimhan, V.; Ronco, C. Uremic Toxins and their Relation to Dialysis Efficacy. *Blood Purif.* **2019**, *48*, 299–314. [CrossRef]
127. Chen, W.; Wei, M.; Wang, Y. Advanced ultrafiltration membranes by leveraging microphase separation in macrophase separation of amphiphilic polysulfone block copolymers. *J. Membr. Sci.* **2017**, *525*, 342–348. [CrossRef]
128. Kumar, R.; Isloor, A.; Ismail, A.; Matsuura, T. Synthesis and characterization of novel water soluble derivative of Chitosan as an additive for polysulfone ultrafiltration membrane. *J. Membr. Sci.* **2013**, *440*, 140–147. [CrossRef]
129. Zhang, G.; Lu, S.; Zhang, L.; Meng, Q.; Shen, C.; Zhang, J. Novel polysulfone hybrid ultrafiltration membrane prepared with TiO$_2$-g-HEMA and its antifouling characteristics. *J. Membr. Sci.* **2013**, *436*, 163–173. [CrossRef]

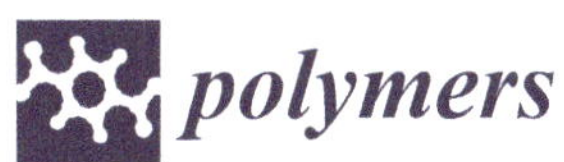 *polymers*

*Article*

# Evolution of Spinal Cord Transection of Rhesus Monkey Implanted with Polymer Synthesized by Plasma Evaluated by Diffusion Tensor Imaging

Axayacatl Morales-Guadarrama [1,2,3], Hermelinda Salgado-Ceballos [4,5], Israel Grijalva [4,5], Juan Morales-Corona [6], Braulio Hernández-Godínez [7], Alejandra Ibáñez-Contreras [7], Camilo Ríos [8], Araceli Diaz-Ruiz [8], Guillermo Jesus Cruz [3], María Guadalupe Olayo [3], Stephanie Sánchez-Torres [4,5], Rodrigo Mondragón-Lozano [5,9], Laura Alvarez-Mejia [4,8], Omar Fabela-Sánchez [2,10] and Roberto Olayo [6,*]

[1] Centro Nacional de Investigación en Imagenología e Instrumentación Médica, Universidad Autónoma Metropolitana Iztapalapa, CDMX, Mexico City 09340, Mexico; amorales@ci3m.mx
[2] Departamento de Ingeniería Eléctrica, Universidad Autónoma Metropolitana Iztapalapa, CDMX, Mexico City 09340, Mexico; ibqfabela@hotmail.com
[3] Departamento de Física, Instituto Nacional de Investigaciones Nucleares, Axapusco 52750, Mexico; gogol1000@hotmail.com (G.J.C.); guadalupe.olayo@hotmail.com (M.G.O.)
[4] Instituto Mexicano del Seguro Social, Unidad de Investigación Médica en Enfermedades Neurológicas, Hospital de Especialidades Centro Médico Nacional Siglo XXI, CDMX, Mexico City 06720, Mexico; melisalce@yahoo.com (H.S.-C.); igrijalvao@yahoo.com (I.G.); phanie85@yahoo.com.mx (S.S.-T.); lau_alvarezmejia@yahoo.com.mx (L.A.-M.)
[5] Centro de Investigación del Proyecto CAMINA A.C., CDMX, Mexico City 14050, Mexico; ruy.lozano@gmail.com
[6] Departamento de Física, Universidad Autónoma Metropolitana Iztapalapa, CDMX, Mexico City 09340, Mexico; jmor@xanum.uam.mx
[7] Investigación Biomédica Aplicada S.A.S. de C.V., CDMX, Mexico City 14240, Mexico; rhpithecus@yahoo.com.mx (B.H.-G.); ibanez.alejandra@hotmail.com (A.I.-C.)
[8] Departamento de Neuroquímica, Instituto Nacional de Neurología y Neurocirugía Manuel Velasco Suárez S.S.A., CDMX, Mexico City 14269, Mexico; crios@correo.xoc.uam.mx (C.R.); aradiazruiz@hotmail.com (A.D.-R.)
[9] Catedrático CONACyT-Instituto Mexicano del Seguro Social, Unidad de Investigación Médica en enfermedades Neurológicas, Hospital de Especialidades, Centro Médico Nacional Siglo XXI, CDMX, Mexico City 06720, Mexico
[10] Departamento de Química Macromoléculas y Nanomateriales, Centro de Investigación en Química Aplicada, Saltillo 25294, Mexico
* Correspondence: oagr@xanum.uam.mx

**Citation:** Morales-Guadarrama, A.; Salgado-Ceballos, H.; Grijalva, I.; Morales-Corona, J.; Hernández-Godínez, B.; Ibáñez-Contreras, A.; Ríos, C.; Diaz-Ruiz, A.; Cruz, G.J.; Olayo, M.G.; et al. Evolution of Spinal Cord Transection of Rhesus Monkey Implanted with Polymer Synthesized by Plasma Evaluated by Diffusion Tensor Imaging. *Polymers* **2022**, *14*, 962. https://doi.org/10.3390/polym14050962

Academic Editors: Andrada Serafim and Stefan Ioan Voicu

Received: 13 January 2022
Accepted: 24 February 2022
Published: 28 February 2022

**Publisher's Note:** MDPI stays neutral with regard to jurisdictional claims in published maps and institutional affiliations.

**Abstract:** In spinal cord injury (SCI) there is damage to the nervous tissue, due to the initial damage and pathophysiological processes that are triggered subsequently. There is no effective therapeutic strategy for motor functional recovery derived from the injury. Several studies have demonstrated neurons growth in cell cultures on polymers synthesized by plasma derived from pyrrole, and the increased recovery of motor function in rats by implanting the polymer in acute states of the SCI in contusion and transection models. In the process of transferring these advances towards humans it is recommended to test in mayor species, such as nonhuman primates, prioritizing the use of non-invasive techniques to evaluate the injury progression with the applied treatments. This work shows the ability of diffusion tensor imaging (DTI) to evaluate the evolution of the SCI in nonhuman primates through the fraction of anisotropy (FA) analysis and the diffusion tensor tractography (DTT) calculus. The injury progression was analysed up to 3 months after the injury day by FA and DTT. The FA recovery and the DTT re-stabilization were observed in the experimental implanted subject with the polymer, in contrast with the non-implanted subject. The parameters derived from DTI are concordant with the histology and the motor functional behaviour.

**Keywords:** plasma polymerization; spinal cord injury; diffusion tensor imaging; rhesus monkey

## 1. Introduction

SCI has a great medical and socioeconomic impact, due to a severe neurological disability [1–3] and, so far it doesn't exist an effective treatment to recovery after the injury. The pathophysiological mechanisms triggered after SCI are complex [2], small animal studies have contributed greatly to a better understanding of these mechanisms. However, it has not been possible to translate those findings effectively to improve treatments for human SCI. Therefore, to facilitate the translation of advances made in the laboratory to the clinic it is recommended the use of mayor species [4–6], including methods which allow the information collected per animal to be maximized in order to reduce the use of animals [7], prioritizing the use of non-invasive techniques to evaluate the injury progression with the applied treatments, according to organizations such as "The National Centre for the Replacement, Refinement & Reduction of Animals in research" (NC3Rs) in the UK's that practice the principles of replacement, reduction and refinement (3Rs).

Polymers such as collagen/silk fibroin, polyethylene glycol, poly-β-hydroxybutyrate, chitosan tube, poly(ε-caprolactone), poly(lactic-co-glycolic acid), and polymers synthesized by plasma derived from pyrrole (PPPyI) have been proposed with encouraging results in the treatment of SCI [8–16], it has been described that polymers implanted after a SCI have beneficial effects such as: promoting functional recovery, axonal remyelination, preservation of nervous tissue adjacent to the epicentre of the injury, decrease in the number of reactive astrocytes, among others; although commonly the aforementioned effects do not occur together, in addition to not being attributed only to the polymer, but have been associated with the combination with drugs, cells or other agents [8–13]. While PPPyI has been reported to have neuroprotective and neuroregenerative effects *per se*. In addition, it was shown that animals with SCI transection (SCIT) implanted with PPPY presented a lower inflammatory response, better integration with the nervous tissue, and greater functional recovery compared to animals administered PPy synthesized by chemical or electrochemical methods. This can be attributed to the plasma synthesis of the polymer, with which polymers with physical-chemical characteristics different are obtained [17]. Additionally, it has been shown that combining rehabilitation with the PPPyI implant promotes the expression of βIII-tubulin (molecule related to nerve plasticity), reduces glial scar formation, favours the preservation of nerve tissue, nerve fibres cross the injured site and recovery of motor function is observed [16]. PPPyI implanted in adult rats with SCI by both, transection and contusion models, favour protection of nervous tissue adjacent to the lesion and the functional recovery of animals, significantly [14–16].

Magnetic Resonance Imaging (MRI) has been used as a non-invasive tool in order to study the SCI in vivo. With conventional MRI, for instance T2-weighted (T2W), it is possible to obtain morphometric information of the injured area, the affected tissue and the conformation of cysts and scar areas [18–20]. However, to evaluate the severity of SCI, more sensitive methods are needed to reveal changes in the neurological structure, as DTI that has been developed and used as an accurate, non-invasive evaluation approach in SCI [19,21,22]. DTI is an MRI technique with the possibility to determine the direction of water diffusion in biological tissues. In white matter, water diffusion is highly directional through axons [23], DTI can be used to assess the microstructure of white matter by FA, which reflects the anisotropy of the diffusion [24]. DTT refers to the estimation of axonal connectivity according to local diffusion properties. In its simplest form, DTT follows the main axis of the nerve fibres and their propagation from one area to another anatomically connected [25–27].

In this work experimental T9 SCIT was performed, and the evolution through time by DTI of SCIT in nonhuman primates (NHP) at the epicentre and around the injury was studied.

## 2. Materials and Methods

### 2.1. Synthesis of Polypyrrole Iodine by Plasma

The monomer used in the polymerizations was pyrrole (99%, Sigma-Aldrich, St. Louis, MO, USA), the dopant was iodine (99.8%, Sigma-Aldrich, St. Louis, MO, USA). Polypyrrole Iodine was synthesized by the plasma polymerization method. The film was kept in the reactor for 24 h. In an iodine atmosphere to neutralize the last free radicals and increase the amount of iodine in the material. The pyrrole and iodine used in the synthesis and the solvents applied to remove the polymers were used without further purification. Once the PPPyI films were synthesized, acetone was applied to separate the films from the substrates and dissolve the remaining oligomers. FT-IR, XPS, TGA and morphological analyses were performed (its characterization has already been reported in previous studies [14,28,29]). When removing the polymer from the reactor, flakes of around 1mm thickness are obtained, films were pulverized in an agate mortar to obtain a fine brown powder, which was compressed at 9 tons for 10 min to form a thin tablet. Electrical conductivity was calculated from resistance measured directly from the tablet [14,28].

### 2.2. Polymer Characterization

The analysis FT-IR of the polymer was obtained with a Perkin-Elmer 2000 Infrared Spectrophotometer (Perkin Elmer-DuraSampIIR II, Waltham, MA, USA) collected directly from the films through 32 scans. PPPyI films were analyzed by X-ray photoelectron spectroscopy (XPS) using an X-ray monochromator from an Al k$\alpha$ (1486.6 eV) source (Thermo hermo Scientific, Waltham, MA, USA).

Resistance of the tablet was measured perpendicular to the polymer surface using a two-probe device, the sample placed between two copper electrodes in a capacitive array. The resistance of the polymer was measured with a multimeter.

### 2.3. Animal Grouping

This study was approved by the National Commission of Scientific Investigation of Mexican Social Security Institute, The Committee for the Care and Use of Laboratory Animals of Proyecto CAMINA A.C. research centre, and Bioethics Committee of National Centre of Investigation in Medical Instrumentation and Imaging. The NHP were treated humanely according with "National Institutes of Health Guide for the Care and Use of Laboratory Animals".

Two female NHP (Macaca Mulatta) from CAMINA A.C. were used. They were selected according to their general healthy aspect, mobility, vital sings, and blood chemistry. The NHPs underwent the SCI, one of them was implanted with the polymer (RHI) and the other one was only injured (RHC). A month before starting the experimental procedures, NHP were moved from their troop to a cage. They were maintained *ad libitum* with *pellets* Purina Monkey Diet 5045® (PMI Nutrition International, St. Louis, MO, USA) and water. To safely handle NHPs, *Pole-and-Collar* systems (Primate Products Inc., Immokalee, FL, USA, EE.UU.) were used.

### 2.4. Experimental Treatment

NHPs were induced with tiletamine zolazepam (Zoletil, Virbac S.A., Carros, France) intramuscularly (4 mg/Kg). Isoflurane anaesthesia (Rhodia Organique Fine Ltd., CDMX, México) was kept at 1.5% providing oxygen mixed with environmental air though an endotracheal tube (approximately at 25 mL/s). Physiological parameters were monitored: rectal temperature, oxygen saturation and ECG, during the surgery. The ventilator was maintained at 15 respirations per minute. For the SCI, the isoflurane concentration increased to 2.5%. Afterwards the NHPs were subjected to a sagittal incision on the skin and to a dissection in the paravertebral muscles to carefully remove one lamina. In order to observe the laminar process of this vertebra, the ninth spinous process was extirpated. Then a laminectomy was carried out, being extended until the facet process. So far, the meninges were kept intact. Once the laminectomy was finished a longitudinal incision was carried out in the meninges. After that, a complete transversal cut in the spinal cord was carried out and the PPPyI implant was introduced into the lesion site in the RHI subject. Afterwards the meninges were sutured with stitches and the surgical incision was sutured in two planes. Immediately after the injury, MRI studies were performed. Ciprofloxacin Lactate (Bayer de México S.A. de C.V., CDMX, México) was administrated intravenously every 12 h the first day after the injury and then a daily intramuscular dose of 15 mg/kg for 6 days, as prophylactic to prevent infections. For Analgesia, a daily dose for two weeks of 100 mg/kg of Acetaminophen (Cilag, México) was administrated orally.

### 2.5. MRI and DTI Scan

MRI studies were performed with a 3.0 T whole-body MRI clinical scanner (Achieva, Philips Medical Systems, Eindhoven, Netherlands) with 4-channel SENSE coil.

T1-Weigthed was obtained: PROSET-CLEAR sequence; TE/TR 6.9/10.9 ms; FOV 196 × 132; matrix 300 × 200; slice thickness 1mm.

T2-Weigthed: VISTA-CLEAR sequence; TE/TR = 115/2500 ms; FOV 196 × 132; matrix 300 × 200; slice thickness 1mm.

DTI was acquired with the following parameters: DTI-high_iso sequence; TE/TR 70/5934 ms; b = 800; 32 directions; FOV 128 × 128; matrix 256 × 256, slice thickness 2 mm.

The studies were carried out: before SCI, the day of the SCI and the 1st, 2nd, and 3rd month after SCI. The data were then further analysed using DSI Studio software (7 January 2021 build, Fang-Cheng, Pittsburgh, PA, USA), The diffusion tensor was calculated. A deterministic fibre tracking algorithm was used [30], the regions of interest were placed at the epicentre (ECn), the rostral direction (Rn), and the caudal direction (Cn) of the injury. The seeding region was placed around the epicentre of injury. The anisotropy threshold was 0.225. The angular threshold was 30 degrees.

### 2.6. Obtaining Tissue

Three months after the SCIT, subjects were anesthetized followed by an intraperitoneal administration of 0.8 mL of heparin, a wide thoracotomy was performed, the ascending aorta was cannulated, and 1000 mL of cool physiological saline solution followed by 2000 mL of 4% paraformaldehyde in phosphate buffer were perfused through the heart. The spinal cord was obtained. The samples were embedded in paraffin. Serial longitudinal sections of 10 μm thickness was cut and stained for histological analysis.

## 3. Results

### 3.1. FT-IR Spectcroscopy

The infrared spectra of PPPyI are shown in Figure 1. Three broad absorption bands are primarily identified and emphasized. Absorption region located in the interval 400–800 cm$^{-1}$, shows the substitutions in the pyrrole rings, associated with the crosslinking between the chains, the partial branching, and the growth of the polymer. The C–I groups originating from iodine doping during polymerization can be identified in the peak centred at 604 cm$^{-1}$. Wide absorption located in the region between 1600 and 1800 cm$^{-1}$, is due to the C–N, C=C and C=0 groups, the peak at 1630 belongs to the amine group. The vibration at 2218 cm$^{-1}$ corresponds to nitrile groups, C≡N, this vibration suggests high dehydrogenation and breakage of some monomeric rings. The aliphatic C–H groups can be assigned to absorption centred at 2935 cm$^{-1}$ and suggest the ring fragmentation, as result of plasma synthesis high-energy collisions. In the region between 3000 and 3800 cm$^{-1}$ may be associated to N–H and O–H groups, the most significant absorption is centred at 3354 cm$^{-1}$, which corresponds to the pyrrole bonds N–H.

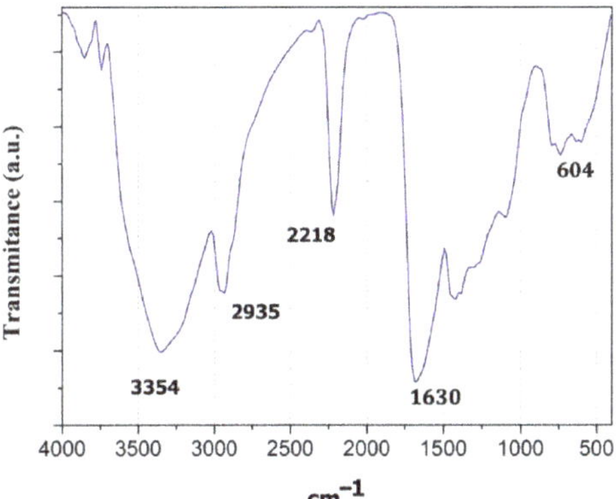

**Figure 1.** Infrared spectrum (FT-IR) of iodine-doped polymer of pyrrole synthesized by plasma.

### 3.2. Elemental Analysis by XPS

Atomic percentage analysis of PPY/I indicated C 77.04%, O 4.86%, N 17.5% and I 0.6%; as is showed in Figure 2. Carbon and Nitrogen are part of the pyrrole structure, while the Iodine has a low participation due to its integration as a dopant during the synthesis. On the other hand, Oxygen can be a consequence of the neutralization of the last free radicals when the reactor is opened and exposed to the atmospheric interaction.

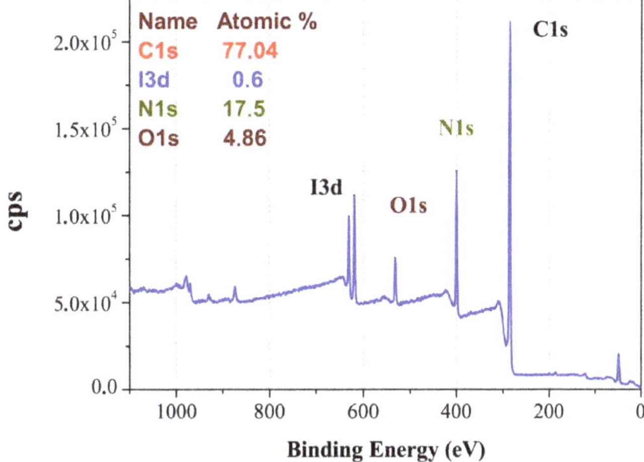

**Figure 2.** XPS survey spectrum for PPPy/I. The peaks correspond to configurations of carbon (C1s), nitrogen (N1s), oxygen (O1s), and iodine (I3d).

### 3.3. Electric Conductivity

The electrical resistance of the PPy/I tablet was 1.4 MOhm at 30% relative humidity and the associated conductivity was $2.2 \times 10^{-10}$ S/cm, these results are consistent with those reported [14]. Some PPPYI, follow an ohmic behaviour, which increases several orders of magnitude when the relative humidity is >60% [28]. This behaviour is desired since the material is implanted in a medium surrounded by body fluids.

### 3.4. Implant Evolution

T2W show the transection site and its changes around the injury (Figure 3). In the rhesus injured (RHC), cysts can be observed in rostral direction from the injury epicentre, increasing size over time. The rhesus injured and implanted (RHI) T2W images show a hyper intense region around the implant, which does not extend.

**Injured**

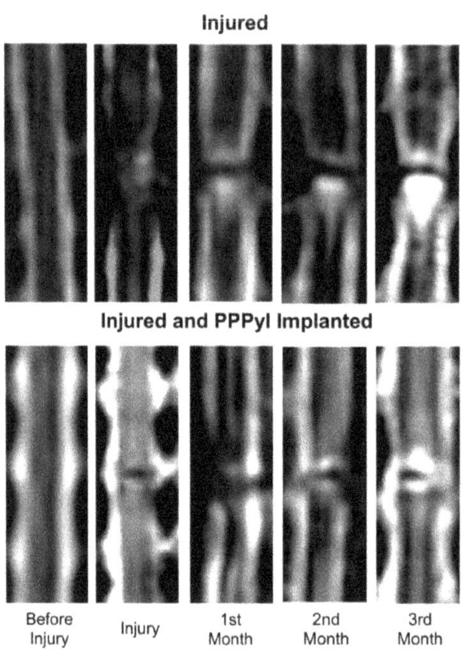



**Figure 3.** T2-weighted magnetic resonance imaging study of Rhesus injured and injured + implanted. T2W sagittal reconstruction images show transection site from before injured up to 3 months after injury.

FA was measured in different areas around the injury site (Figure 4). The values of FA in the different regions measured before the lesion are larger than 0.8 mm for both, RHC and RHI. This value shows a unidirectional diffusion with high FA. After the injury the evolution in time of RHC, and RHI are different. While in RHC a tendency to FA decreasing in the injury area was observed from the 2nd month, in RHI there is an increasing FA. FA for the RHC before the injury was $0.83 \pm 0.02$. The injury day, FA decreased to $0.42 \pm 0.02$ at the injury epicentre, and as the injury was progressing FA drops. One month after the injury FA was $0.32 \pm 0.02$, two months after the injury was $0.24 \pm 0.02$ and 3 months after the injury was $0.14 \pm 0.02$. In RHI, FA was $0.86 \pm 0.02$ before the injury, the day of injury and implant FA decreased to $0.40 \pm 0.02$ at the injury epicentre, one month after that, FA was $0.41 \pm 0.02$, after two months it was $0.64 \pm 0.02$, and three months post injury was $0.74 \pm 0.02$. At the end of the study, the FA in RHC decreased 86% while for RHI it decreased 15%.

DTT of the SCI region allows visualizing changes in the white matter through time (Figure 5). The before injury column shows a well-organized fibre-tracking in both, RHC and RHI. For the first image after the injury, discontinuity in the projection of the tracks is observed in both subjects. In the RHC, fibre-tracking integrity decreases over time (from 1 to 3 months) in rostral and caudal direction, extending to the ends of the analysed region. Restructuring of the fibre-tracing was observed in RHI, from the first month after the injury, showing irregularity and discontinuity in the fibre-tracing mainly in the epicentre of the injury, at the second month homogeneous lateralized fibres were observed at the second month the formation of homogeneous lateral fibres were observed, with tendency to integration, for the third month after the injury, the fibre-tracing is homogeneous in the lateral portion shown in the second month and the tendency to restructure on the opposite side is observed (see Supplementary Video).

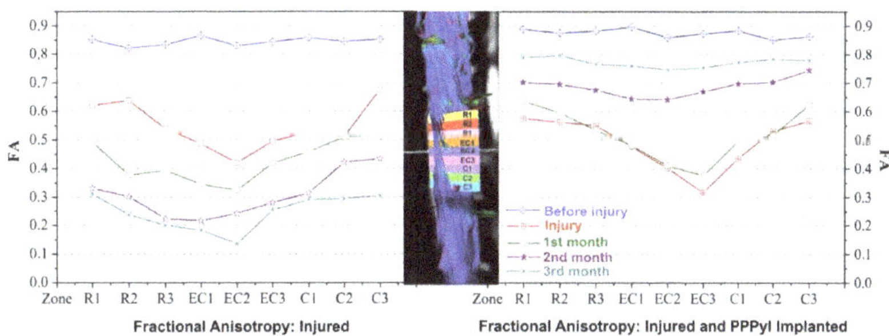

**Figure 4.** Fractional anisotropy of injured area. The FA was measured in 9 different regions of the transection area, shown in the middle of the figure, where ECn corresponds to the epicentre of injury, Rn is the rostral, and Cn is the caudal region. The time evolution graphs of the values of FA of the RHC (**left**) and RHI (**right**) are presented.

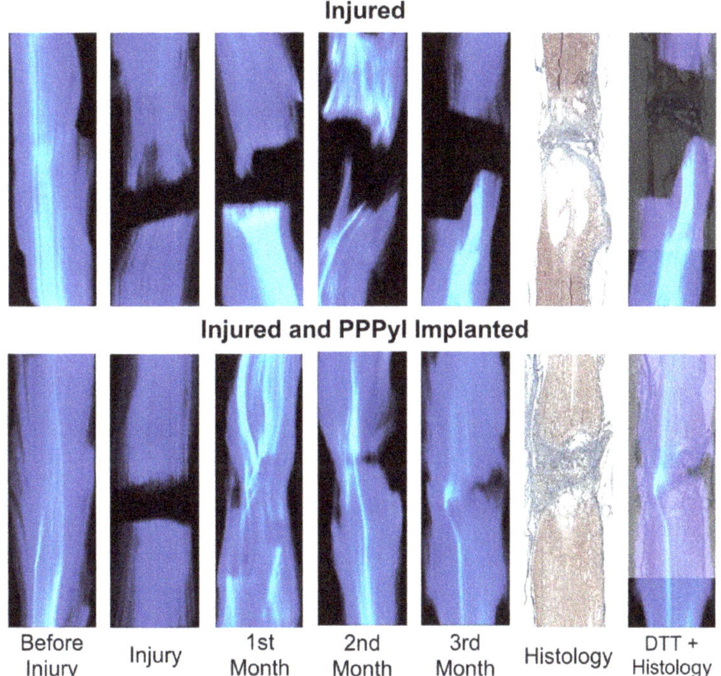


**Figure 5.** Diffusion Tensor Tractography of Spinal Cord Injury through time. The evolution shown is from before the injury to 3 months later. In the injured Rhesus, shown above, the anisotropy decreases with time, showing an increasing discontinuity in the fibre- tracking to both sides of the lesion. In the injured and implanted Rhesus, shown below, tracts are identified through the injured area, showing a gradual decrease in discontinuity and a progressive recovery of the tractography over time.

Histology (Figure 5) show scar formation at the epicentre of the injury, as well as cyst formation in both the rostral and caudal directions. For RHC there is coalescence of cysts forming a large cyst, thus contributing to the loss of histoarchitecture, while for RHI there is no extension of the injury area, and integration of scar, PPPyI, and tissue is observed at the epicentre of the injury.

DTT and Histology overlap (Figure 5) also shows a composition of the DTT and histology, this combination has not total synchronization but allows to compare the morphology predicted by the two techniques, and they show congruency giving a clear idea of the difference in of the evolution of the injury in the subjects.

The recovery of sensitivity and movement in the lower extremities, was observed only in RHI since RHC did not show any motor recovery and presented lacerations, sores, muscle atrophy, and therefore no movement is discussed; evidence is omitted due to the shocking and crude nature of the injuries. In the 2nd month, RHI subject had slight movements of the lower joints, while in the third month, complete flexures were observed, particularly of the relation Hip-Knee-Ankle (see Supplementary Video), Figure 6 shows the movement trajectory of a lower extremity evaluated 3 months after injury, and their kinetics representation.

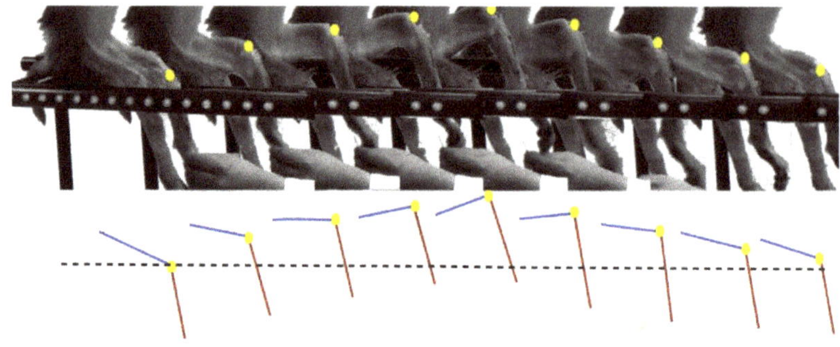

**Figure 6.** Movement recovery in implanted NHP. The one limb movement trajectory of the RHI is observed 3 months after the polymer implantation. Below, the movement kinetics representation of the relation Hip-Knee-Ankle, the dotted line shows the knee initial position.

## 4. Discussion

In this work a SCIT was carried out in non-human primates, one with implanted of PPPyI (RHI) and other just with the injury (RHC), the effect of the PPPyI in RHI is in accordance with the previous work in murines, in which, the use of PPPy in different models of SCI has promoted mechanisms of neuroprotection, contributing in the stimulation of motor plasticity mechanisms and in the histoarchitecture conservation, increasing the motor functional recovery of the experimental subjects [14–16]; DTI showed capacity to evaluate the evolution in vivo of the SCIT in NHP emphasizing the difference between the RHI and RHC.

FA is a sensitive marker for SCI and is strongly related to the severity of the lesion [21]. FA can be an indicator of axonal structure damage with the possibility to quantify the severity and extension of SCI. The decrease in FA reflects axonal loss and nervous degeneration. FA decreased the injury day for the subjects in the epicentre of the injury. Subsequently, FA in RHC kept decreasing in the epicentre, and in all the evaluated regions showing a larger damage, FA value was $0.14 \pm 0.02$ in the epicentre three months after the lesion, agreeing with the tendency in the decrease of FA reported in murine [21,31] and canine [32] with SCI. In RHI a gradual recovery of FA after the second month is observed and the fraction increased up to $0.74 \pm 0.02$ at the epicentre and showed a tendency to recover the value of both, caudal and cephalic, suggesting a re-stabilization of the microenvironment in the lesion zone [31,33].

DTT can differentiate the interrupted nerve fibres from intact regions and can be used as a qualitative indicator of SCI to represent nerve fibres and to observe the spinal cord evolution after an injury, the fibres tracking is directly related to the change in FA, since as tracts are damaged the anisotropy decreases [21,31–34]. In both subjects, the DTT projection of the initial spinal cord interruption on the day of the injury was observed. The decrease in anisotropy can be observed in the RHC, where gradually there was an increase in the separation between each extreme of the spinal cord, for the 3rd month after the injury, the space is extensive which is consistent with the space generated by the cysts, according to what is shown in the histology, the DTT + Histology overlap, demonstrates how it is possible to visualize the extent of the lesion in vivo in RHC by means of DTT. In RHI, DTT shows a tendency to recover the track continuity in the injury area because of the FA recovery, showing also local changes in the directionality through time, for the first month, the DTT projection shows a large number of incomplete tracts in the epicentre of the injury, this is attributed to the change in the anisotropy of the region due to the presence of the PPPyI and its interaction with the tissue. Subsequently, greater homogeneity is observed in the area, and the existence of calculated tracts due to the increase in FA in the area, this being coincident with the reappearance in the 2nd month of slight movements in the subject of the lower joints, for the 3rd month, the DTT projection shows a greater number of tracts calculated in the epicentre of the injury as well as changes in the local directionality, analogously in the histology, interaction between tissue and scar is observed in the epicentre of the injury, the DTT + Histology overlap suggests the coincidence of this interaction and the DTT projection, as well as the recovery at the 3rd month after the injury of complete flexures and sensitivity.

The analysis of DTI/DTT has the ability to monitor in vivo the state of the tissue surrounding the epicentre of injury, as well as quantifying the anisotropy of the same region in non-human primates after SCI. In humans, the MRI studies commonly used in SCI are qualitative, limited only to determining the morphology of the Injury, therefore the use of DTI/DTT in the clinic would be an important tool to evaluate and follow up SCI in humans.

In conclusion, diffusion tensor imaging is sensitive to evaluate spinal cord injury, allows monitoring of in vivo injury in nonhuman primates, serving as a tool to evaluate the progression of the injury through time.

**Supplementary Materials:** The following supporting information can be downloaded at: https://www.mdpi.com/article/10.3390/polym14050962/s1.

**Author Contributions:** All authors participated in the planning, conceived and designed the whole study; A.M.-G., R.O. designed and performed the MRI experiments and analysis; H.S.-C., I.G. performed the spinal cord total sections and material implantation, J.M.-C., O.F.-S., G.J.C., M.G.O. collaborated in the synthesis and characterization of the material; B.H.-G., A.I.-C., trained and attended the subjects; C.R., A.D.-R. contributed in material characterization and animal evaluations; A.M.-G., S.S.-T., R.M.-L., L.A.-M. participated in the surgeries, animals evaluations and material characterization. All authors have read and agreed to the published version of the manuscript.

**Funding:** The work was partially supported by the projects: FIS/IMSS/PROT/G15/1444, CONACyT-155239, ICy-T-DF-PIUTE 10-63, 276/2010, CONACyT-FC2015-1 Proyecto 152, CONACyT-CF2019-263993 and CONACyT-CF2019-1311312. Axayacatl Morales-Guadarrama received a scholarship from CONACyT.

**Institutional Review Board Statement:** The animal study protocol was approved by the National Commission of Scientific Investigation of Mexican Social Security Institute (protocol code R-2013-785-078, 26 November 2013) and The Committee for the Care and Use of Laboratory Animals of Proyecto *CAMINA A.C.* research centre (protocol name: "Efecto de un implante polimérico sobre la recuperación funcional de monos Rhesus con lesión por sección completa de la médula espinal", 17 June 2013).

**Informed Consent Statement:** Not applicable.

**Data Availability Statement:** The data presented in this study are available on request from the corresponding author.

# References

1. Strauss, D.; DeVivo, M.; Shavelle, R.; Brooks, J.; Paculdo, D. Economic factors and longevity in spinal cord injury: A reappraisal. *Arch. Phys. Med. Rehabil.* **2008**, *89*, 532. [CrossRef]
2. Ahuja, C.S.; Wilson, J.R.; Nori, S.; Kotter, M.R.N.; Druschel, C.; Curt, A.; Fehlings, M.G. Traumatic spinal cord injury. *Nat. Rev. Dis. Primers* **2017**, *3*, 1–21. [CrossRef]
3. National Spinal Cord Injury Statistical Center. *2020 Annual Statistical Report for the Spinal Cord Injury Model Systems*; University of Alabama at Birmingham: Birmingham, AL, USA, 2021; pp. 1–162.
4. Guízar-Sahagún, G.; Grijalva, I.; Hernández-Godínez, B.; Franco-Bourland, R.E.; Cruz-Antonio, L.; Martínez-Cruz, A.; Ibañez-Contreras, A. New approach for graded compression spinal cord injuries in Rhesus macaque: Method feasibility and preliminary observations. *J. Med. Primatol.* **2011**, *40*, 401–413. [CrossRef]
5. Jannesar, S.; Salegio, E.A.; Beattie, M.S.; Bresnahan, J.C.; Sparrey, C.J. Correlating Tissue Mechanics and Spinal Cord Injury: Patient-Specific Finite Element Models of Unilateral Cervical Contusion Spinal Cord Injury in Non-Human Primates. *J. Neurotrauma* **2021**, *38*, 698–717. [CrossRef]
6. Courtine, G.; Bunge, M.B.; Fawcett, J.W.; Hodgson, J.; McKay, H.; Yang, H.; Zhong, H.; Tuszynski, M.H.; Edgerton, V.R. Can experiments in nonhuman primates expedite the translation of treatments for spinal cord injury in humans? *Nat. Med.* **2007**, *13*, 561. [CrossRef]
7. Festing, S.; Wilkinson, R. The ethics of animal research. *EMBO Rep.* **2007**, *8*, 1–5. [CrossRef]
8. Li, X.-H.; Zhu, X.; Liu, X.-Y.; Xu, H.-H.; Jiang, W.; Wang, J.-J.; Chen, F.; Zhang, S.; Li, R.-X.; Chen, X.-Y.; et al. The corticospinal tract structure of collagen/silk fibroin scaffold implants using 3D printing promotes functional recovery after complete spinal cord transection in rats. *J. Mater. Sci. Mater. Med.* **2021**, *32*, 31. [CrossRef]
9. Estrada, V.; Brazda, N.; Schmitz, C.; Heller, S.; Blazyca, H.; Martini, R.; Müller, H.W. Long-lasting significant functional improvement in chronic severe spinal cord injury following scar resection and polyethylene glycol implantation. *Neurobiol. Dis.* **2014**, *67*, 165–179. [CrossRef]
10. Novikov, L.N.; Novikova, L.N.; Mosahebi, A.; Wiberg, M.; Terenghi, G.; Kellerth, J.-O. A novel biodegradable implant for neuronal rescue and regeneration after spinal cord injury. *Biomaterials* **2002**, *23*, 3369–3376. [CrossRef]
11. Li, X.; Yang, Z.; Zhang, A.; Wang, T.; Chen, W. Repair of thoracic spinal cord injury by chitosan tube implantation in adult rats. *Biomaterials* **2009**, *30*, 1121–1132. [CrossRef]
12. Wong, D.Y.; Leveque, J.-C.; Brumblay, H.; Krebsbach, P.H.; LaMarca, F. Macro-Architectures in Spinal Cord Scaffold Implants Influence Regeneration. *J. Neurotrauma* **2008**, *25*, 1027–1037. [CrossRef]
13. Gelain, F.; Panseri, S.; Antonini, S.; Cunha, C.; Donega, M.; Lowery, J.; Taraballi, F.; Cerri, G.; Montagna, M.; Baldissera, F.; et al. Transplantation of Nanostructured Composite Scaffolds Results in the Regeneration of Chronically Injured Spinal Cords. *ACS NANO* **2010**, *5*, 227–236. [CrossRef]
14. Olayo, R.; Ríos, C.; Salgado-Ceballos, H.; Cruz, G.J.; Morales, J.; Olayo, M.-G.; Alcaraz-Zubeldia, M.; Alvarez, A.L.; Mondragon, R.; Morales, A.; et al. Tissue spinal cord response in rats after implants of polypyrrole and polyethylene glycol obtained by plasma. *J. Mater. Sci. Mater. Med.* **2007**, *19*, 817–826. [CrossRef]
15. Alvarez-Mejía, L.; Morales, J.; Cruz, G.J.; Olayo, M.-G.; Olayo, R.; Diaz-Ruiz, A.; Ríos, C.; Mondragón-Lozano, R.; Sánchez-Torres, S.; Morales-Guadarrama, A.; et al. Functional recovery in spinal cord injured rats using polypyrrole/iodine implants and treadmill training. *J. Mater. Sci. Mater. Med.* **2015**, *26*, 209. [CrossRef]
16. Sánchez-Torres, S.; Diaz-Ruiz, A.; Ríos, C.; Olayo, M.G.; Cruz, G.J.; Olayo, R.; Morales, J.; Mondragón-Lozano, R.; Fabela-Sánchez, O.; Orozco-Barrios, C.; et al. Recovery of motor function after traumatic spinal cord injury by using plasma-synthesized polypyrrole/iodine application in combination with a mixed rehabilitation scheme. *J. Mater. Sci. Mater. Med.* **2020**, *31*, 58. [CrossRef]
17. Álvarez-Mejía, L.; Salgado-Ceballos, H.; Olayo, R.; Cruz, G.J.; Olayo, M.G.; Díaz-Ruiz, A.; Ríos, C.; Mondragón-Lozano, R.; Morales-Guadarrama, A.; Sánchez-Torres, S.; et al. Effect of Pyrrole Implants Synthesized by Different Methods on Spinal Cord Injuries of Rats. *Rev. Mex. Ing. Biomédica* **2015**, *36*, 7–21.
18. Kozlowski, P.; Raj, D.; Liu, J.; Lam, C.; Yung, C.A.; Tetzlaff, W. Characterizing white matter damage in rat spinal cord with quantitative MRI and histology. *J. Neurotrauma* **2008**, *25*, 653–676. [CrossRef]
19. Freund, P.; Seif, M.; Weiskopf, N.; Friston, K.; Fehlings, M.G.; Thompson, A.J.; Curt, A. MRI in traumatic spinal cord injury: From clinical assessment to neuroimaging biomarkers. *Lancet Neurol.* **2019**, *18*, 1123–1135. [CrossRef]
20. Yoon, J.; Lee, N.; Seo, J. Morphometric Magnetic Resonance Imaging Evaluation of Cervical Spinal Canal and Cord in Normal Small-Breed Dogs. *Front. Vet. Sci.* **2021**, *8*, 732953.
21. Li, X.-H.; Li, J.-B.; He, X.-J.; Wang, F.; Huang, S.-L.; Bai, Z.-L. Timing of diffusion tensor imaging in the acute spinal cord injury of rats. *Sci. Rep.* **2015**, *5*, 12639. [CrossRef]
22. Budde, M.D.; Skinner, N.P. Diffusion MRI in acute nervous system injury. *J. Magn. Reson.* **2018**, *292*, 137–148. [CrossRef]
23. Beaulieu, C. The basis of anisotropic water diffusion in the nervous system a technical review. *NMR Biomed. Int. J. Devoted Dev. Appl. Magn. Reson. In Vivo* **2002**, *15*, 435–455. [CrossRef]
24. Basser, J.P.; Mattiello, J.; Lebihan, D. Estimation of the effective selfdiffusion tensor from the NMR spin echo. *J. Magn. Reson. Ser. B* **1994**, *103*, 247–254. [CrossRef]

25. Basser, P.J.; Pajevic, S.; Pierpaoli, C.; Duda, J.; Aldroubi, A. In vivo fiber tractography using Dt-MRI Data. *Magn. Reson. Med.* **2000**, 625–632. [CrossRef]
26. Shi, Y.; Short, S.J.; Knickmeyer, R.C.; Wang, J.; Coe, C.L.; Niethammer, M.; Gilmore, J.H.; Zhu, H.; Styner, M.A. Diffusion Tensor Imaging-Based Characterization of Brain Neurodevelopment in Primates. *Cereb. Cortex* **2012**, *23*, 36–48. [CrossRef]
27. Manzanera Esteve, I.V.; Farinas, A.F.; Pollins, A.C.; Nussenbaum, M.E.; Cardwell, N.L.; Kahn, H.; Does, M.D.; Dortch, R.D.; Thayer, W.P. Noninvasive diffusion MRI to determine the severity of peripheral nerve injury. *Magn. Reson. Imaging* **2021**, *83*, 96–106. [CrossRef]
28. Cruz, G.J.; Morales, J.; Olayo, R. Films obtained by plasma polymerization of pyrrole. *Thin Solid Film.* **1999**, *342*, 119–126. [CrossRef]
29. Osorio-Londoño, D.M.; Godínez-Fernández, J.R.; Acosta-García, M.C.; Morales-Corona, J.; Olayo Gonzalez, R.; Morales-Guadarrama, A. Pyrrole Plasma Polymer-Coated Electrospun Scaffolds for Neural Tissue Engineering. *Polymers* **2021**, *13*, 3876. [CrossRef]
30. Yeh, F.-C.; Verstynen, T.D.; Wang, Y.; Fernández-Miranda, J.C.; Tseng, W.-Y.I. Deterministic Diffusion Fiber Tracking Improved by Quantitative Anisotropy. *PLoS ONE* **2013**, *8*, e80713. [CrossRef]
31. Kelley, B.J.; Harel, N.Y.; Kim, C.-Y.; Papademetris, X.; Coman, D.; Wang, X.; Hasan, O.; Kaufman, A.; Globinsky, R.; Staib, L.H.; et al. Diffusion Tensor Imaging as a Predictor of Locomotor Function after Experimental Spinal Cord Injury and Recovery. *J. Neurotrauma* **2014**, *31*, 1362–1373. [CrossRef]
32. Zhang, Z.; Yao, S.; Xie, S.; Wang, X.; Chang, F.; Luo, J.; Wang, J.; Fu, J. Effect of hierarchically aligned fibrin hydrogel in regeneration of spinal cord injury demonstrated by tractography: A pilot study. *Sci. Rep.* **2016**, *7*, 40017. [CrossRef]
33. Sasiadek, M.J.; Szewczyk, P.; Bladowska, J. Application of diffusion tensor imaging (DTI) in pathological changes of the spinal cord. *Med. Sci. Monit. Int. Med. J. Exp. Clin. Res.* **2012**, *18*, 73–79. [CrossRef]
34. Vedantam, A.; Jirjis, M.; Eckhardt, G.; Sharma, A.; Schmit, B.D.; Wang, M.C.; Ulmer, J.L.; Kurpad, S. Diffusion tensor imaging of the spinal cord: A review. *Coluna/Columna* **2013**, *12*, 64–69. [CrossRef]

 *polymers*

*Article*

# Adult Human Vascular Smooth Muscle Cells on 3D Silk Fibroin Nonwovens Release Exosomes Enriched in Angiogenic and Growth-Promoting Factors

Peng Hu [1,2,†], Anna Chiarini [1,*,†], Jun Wu [3], Zairong Wei [2], Ubaldo Armato [1,3] and Ilaria Dal Prà [1,3,*,†]

1   Human Histology & Embryology Section, Department of Surgery, Dentistry, Paediatrics & Gynaecology, University of Verona Medical School, 37134 Verona, Italy; peng.hu@univr.it (P.H.); uarmato@gmail.com (U.A.)
2   Department of Burns & Plastic Surgery, The Affiliated Hospital of Zunyi Medical University, Zunyi 563000, China; zairongwei@sina.com
3   Department of Burns and Plastic Surgery, Second People's Hospital, University of Shenzhen, Shenzhen 518000, China; junwupro@126.com
*   Correspondence: anna.chiarini@univr.it (A.C.); ilaria.dalpra@univr.it (I.D.P.)
†   These authors contributed equally to this work.

**Abstract:** Background. Our earlier works showed the quick vascularization of mouse skin grafted *Bombyx mori* 3D silk fibroin nonwoven scaffolds (3D-SFnws) and the release of exosomes enriched in angiogenic/growth factors (AGFs) from in vitro 3D-SFnws-stuck human dermal fibroblasts (HDFs). Here, we explored whether coronary artery adult human smooth muscle cells (AHSMCs) also release AGFs-enriched exosomes when cultured on 3D-SFnws in vitro. Methods. Media with exosome-depleted FBS served for AHSMCs and human endothelial cells (HECs) cultures on 3D-SFnws or polystyrene. Biochemical methods and double-antibody arrays assessed cell growth, metabolism, and intracellular TGF-β and NF-κB signalling pathways activation. AGFs conveyed by CD9⁺/CD81⁺ exosomes released from AHSMCs were double-antibody array analysed and their angiogenic power evaluated on HECs in vitro. Results. AHSMCs grew and consumed *D*-glucose more intensely and showed a stronger phosphorylation/activation of TAK-1, SMAD-1/-2/-4/-5, ATF-2, c-JUN, ATM, CREB, and an IκBα phosphorylation/inactivation on SFnws vs. polystyrene, consistent overall with a proliferative/secretory phenotype. SFnws-stuck AHSMCs also released exosomes richer in IL-1α/-2/-4/-6/-8; bFGF; GM-CSF; and GRO-α/-β/-γ, which strongly stimulated HECs' growth, migration, and tubes/nodes assembly in vitro. Conclusions. Altogether, the intensified AGFs exosomal release from 3D-SFnws-attached AHSMCs and HDFs could advance grafts' colonization, vascularization, and take in vivo—noteworthy assets for prospective clinical applications.

**Keywords:** silk fibroin; nonwovens; human; smooth muscle cells; vascular endothelial cells; exosomes; cytokines; chemokines; proliferation; mobilization; angiogenesis

**Citation:** Hu, P.; Chiarini, A.; Wu, J.; Wei, Z.; Armato, U.; Dal Prà, I. Adult Human Vascular Smooth Muscle Cells on 3D Silk Fibroin Nonwovens Release Exosomes Enriched in Angiogenic and Growth-Promoting Factors. *Polymers* **2022**, *14*, 697. https://doi.org/10.3390/polym14040697

Academic Editors: Andrada Serafim and Stefan Ioan Voicu

Received: 18 January 2022
Accepted: 8 February 2022
Published: 11 February 2022



## 1. Introduction

Various insect and arthropod species synthesize complex structural proteins, generically named fibroins [1]. Repeated sequences of three amino acids, Gly-Ser-Gly and Ala-Gly-Ala, denote the biochemical structure of purified silk fibroin (SF) from domesticated *Bombyx mori* silkworm [2]. Macromolecular SF occurs in soluble α-helix or random coil and in insoluble β-sheet forms [3]. Degummed (i.e., sericin-deprived) SF microfibers from silkworms' cocoons are suitable for producing textiles, surgical sutures, and biomaterial scaffolds. Conversely, random coil or α-helix SF forms are not apt for mechanically adequate scaffolds [3]. Notably, SF's intrinsic plasticity and many available SF processing methods have allowed the design of versatile scaffolds proper for biomedical tissue engineering/regeneration applications [4]. In fact, native SF microfibers in β-sheet form enjoy good biomechanical properties, biocompatibility, and biodegradability, while lacking significant cytotoxicity and immunogenicity. In addition, various other β-sheet SF forms, e.g., films, sheets, electrospun mats, hydrogels, and nanofibers, were considered as prospective therapeutic tools to engineer, for instance, skin, cartilage, and corneal tissues [5–8]. Remarkably, the success of biomaterial implants crucially depends, among other factors, on the host's reaction in terms of neovascularization, regenerating tissue organization, and immune/foreign body responses [9]. In relation to translational medicine, another hidden advantage proper of three-dimensional (3D) SF scaffolds is that humans have well over 50 proteins carrying significant stretches of conserved amino acid sequences also present in *Bombyx mori*'s SF. This evolutionary relationship is at the root of the remarkable biocompatibility and lack of immunogenicity of the SF scaffolds [10–12]. Due to these promising features, various SF-based scaffolds/devices have undergone preclinical testing in vitro and in animal models in vivo, some becoming objects of clinical trials, and even entering the market in rare cases [13].

SF scaffolds structured as 3D nonwovens (SFnws) [14] have been our research focus in the skin engineering/regeneration field of endeavour [10,11,15]. Initially, we made two types of SFnws, either by gluing their randomly oriented microfibers with formic acid (FA) or by tangling them via textile carding/needling (C/N) technology [10,11,15]. Once grafted into the subcutaneous tissue of C57/BL6 mice, both FA- and C/N-SFnws guided the successful engineering of a reticular connective tissue integrating the SF microfibers in three-to-six months' time lags [10,11]. Remarkably, by one month after grafting, abundant proliferating capillaries already grew, first along the SF microfibers and next into the intervening voids in close association with fibroblasts and a few macrophages, multinucleated giant cells, and leukocytes. The upshot was a vascularized tissue that lacked any sign of inflammation, foreign body response, fibrosis, or peripheral encapsulation. The biological mechanism(s) underlying the in vivo intense neovascularization of the FA- and C/N-3D-SFnws remained undetermined. However, both 3D-SFnws showed biomechanical shortcomings such as stiffness and fragility [16]. To address the latter, we combined the carding (C) and hydroentangling (HE) textile technologies to produce a third biomechanically more satisfactory SFnws. On such C/HE SFnws we cultured human dermal fibroblasts (HDFs) in vitro to assess whether they would release exosomes carrying any amounts of angiogenic/growth factors (AGFs) [16]. Reports existed that exosomes released from mesenchymal stem cells (MSCs) and endothelial cells (ECs) promoted vascular endothelial cells (ECs) regeneration and angiogenesis [17,18].

Exosomes are nanoscale (30–120 nm in diameter) membranous extracellular vesicles originating in cells' multivesicular bodies [19–21]. They neatly differ from apoptotic bodies [22]. Once released extracellularly, exosomes transport variable combinations of proteins, lipids, DNA, and RNAs [23,24] they shelter from any environmental breakdown mechanism. Thus, exosomes do travel through the extracellular matrix (ECM) and body fluids (blood, cerebrospinal fluid, urine, and saliva) to reach nearby or far away target cells to which they hand over their complex cargoes via interactions with plasma membranes surface receptors or after endocytosis [19]. The various exosome-conveyed agents affect intracellular signalling pathways; homeostatic mechanisms; antigen presentations; inflammatory processes; blood clotting; cell growth, migration, and death; and angiogenesis/vascularization [16,25–27]. Indeed, the exosomes released from the C/HE SFnws-grown HDFs carried heightened amounts of twelve AGFs and potently induced cultured human ECs (HECs) to quickly form abundant endothelial tubes in vitro [16]. One of the queries the latter results raised was whether, besides HDFs, other cell types adhering to C/HE-3D-SFnws-based scaffolds might also release exosomes carrying enriched sets of AGFs.

The smooth muscle cells (SMCs) are relevant to this query as they abound in the layers (*tunicae*) of hollow viscera, vessels included. SMCs embryological origins are multiple, i.e., in the mesoderm's lateral plate (viscera); in its proepicardium derivative (coronary arteries); and in the neural crest (ascending aorta, aortic arch, and pulmonary trunk) [28–32]. SMCs also derive from ECs' transdifferentiation [33]. Typically, SMCs phenotypic plasticity is striking, as it ranges with intermediate graduations from the quiescent/contractile to the proliferating/secretory/migratory one [34,35]. This highly modulable phenotypic diversity is crucially relevant to pathological neointima formation and vascular remodelling [36]. Interestingly, mesoderm lateral plate-derived vascular SMCs supported the survival of complex ECs networks in 3D Matrigel cocultures in vitro [37].

Variously structured SF scaffolds modelled as arterial vessels have supplied promising preclinical results [38,39]. Hence, it seemed worth investigating the interactions between human coronary artery SMCs and C/HE-3D-SFnws in view of prospective applications as artificial bypass grafts or devices for cardiac revascularization [32]. However, as the proepicardium, from which coronary artery SMCs stem, derives from the lateral plate mesoderm [32], we hypothesized that our study might throw light on the behaviour of SMCs inhabiting hollow viscera walls (e.g., airways, intestine, or bladder), which too stem from the mesodermal lateral plate [40,41]. Moreover, SMCs produce and release compounds such as soluble enzymes, growth factors, cytokines, and chemokines, which in their turn affect cell growth and/or differentiation and/or apoptosis, innate immunity, inflammation, angiogenesis, and cancer onset and metastasis [42–47]. Vascular SMCs, too, release exosomes, carrying variable loads of proteins; those so far identified were related to focal adhesion and ECM constituents [48].

Therefore, in this work, we investigated both the activation of growth-relevant intracellular signalling pathways in nontumorigenic adult human SMCs (AHSMCs) cultured on C/HE-3D-SFnws in vitro and their concurrent release of AGFs by way of exosomes, using as comparison terms AHSMCs grown on polystyrene. We report here that various constituents of TGF-β and NF-κB intracellular signalling pathways were phosphorylated and hence activated more intensely in the C/HE-3D-SFnws-stuck AHSMCs than in polystyrene-adhering ones. We also show that the former cells discharged exosomes carrying 43 different AGFs, eight of which in significantly richer amounts than the latter. Moreover, we prove that the AGFs conveyed by exosomes released from C/HE-SFnws-attached AHSMCs powerfully stimulate human microvascular endothelial cells (HECs) to proliferate, migrate, and form tubes and nodes in vitro.

## 2. Materials and Methods

### 2.1. C/HE-3D-SFnws

Comber waste-derived sericin-deprived (via standard degumming) spun silk in staple form (average fibres length, $50 \pm 7$ mm) was used to produce the C/HE-3D-SFnws. A cotton type flat carding machine (width, 100 cm) first processed this SF material. Carding arranged the fibres in bundles preferentially oriented lengthwise, i.e., aligned in the longitudinal direction of the nonwoven web. Next, a mechanical hydroentanglement created numerous bonding points on both surfaces of the carded web (Figure 1a,b) [16].

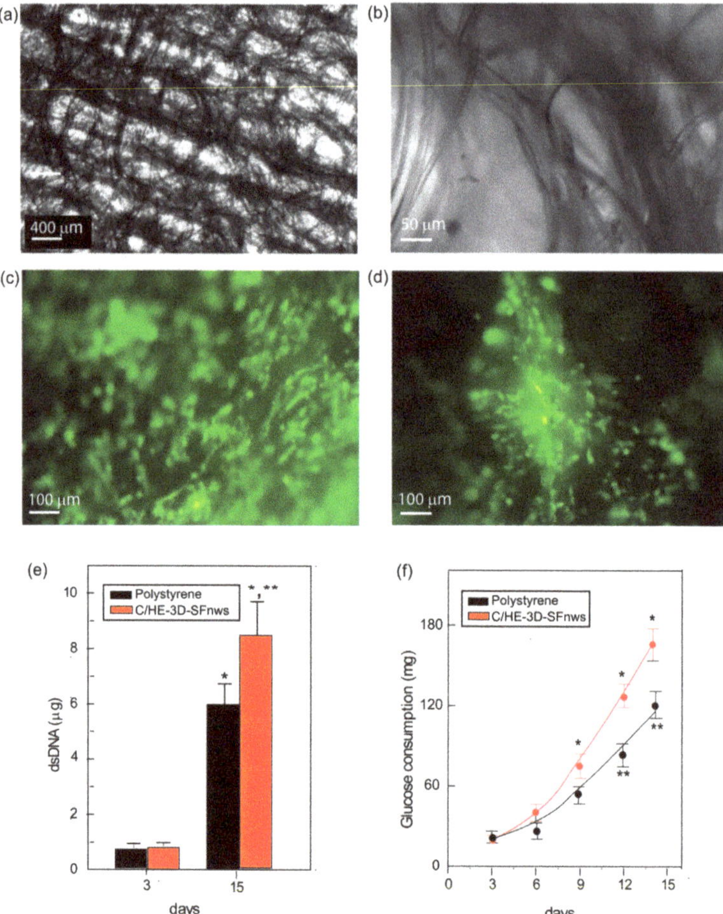

**Figure 1.** Three-dimensional SF nonwovens (C/HE-3D-SFnws) produced via the carding/ hydroentanglement technology are scaffolds supporting the growth of AHSMCs. (**a,b**) In these cell-free C/HE-3D-SFnws fabrics the *SF* microfibers form thick longitudinal bundles along the carding machine direction joined by transversal bridges created by the hydroentanglement technology. Light microscope; at low (50×; **a**), and high (500×; **b**) magnifications. (**c,d**) Five days after seeding intravitally prestained AHSMCs have adhered along an SF microfiber bundle and have also started colonizing the nearby intercalated voids. (**e**) AHSMCs attached to 3D-SFnws intensely grew between 3 and 15 days of staying in vitro as shown by their 10.9-fold increase in number, while the polystyrene-stuck AHSMCs rose by 7.8-fold. Double strand (ds)DNA amounts were assayed as detailed in the Section 2.5 Bars are mean values ± standard deviations (SDs) from three distinct duplicate determinations at each time point. *, $p < 0.05$ vs. day 3; **, $p < 0.05$ between the two groups at day 15. (**f**) The cumulative *D*-glucose consumption of AHSMCs. Between day 3 and 15, its uptake from the growth medium on the part of C/HE-3D-SFnws-bound AHSMCs increased by 8.4-fold vs. the 6.1-fold of polystyrene-attached AHSMCs. *D*-glucose levels of the AHSMC-conditioned media at each time point were assayed as detailed in the Section 2.6. Each dot stands for the mean value ± SDs from three distinct duplicate determinations. *, $p < 0.05$, vs. day 3 values; **, $p < 0.05$ between the time-corresponding values of the two groups.

### 2.2. C/HE-3D-SFnws Samples for In Vitro Cell Cultures

After thorough soaking in PBS, C/HE-3D-SFnws samples underwent transversal cutting into $66 \times 23$ mm rectangular pieces. The latter were sterilized at 55 °C in a vacuum oven for 3 h through exposure to an ethylene oxide/$CO_2$ (10/90 $v/v$) mixture under a pressure of 42 psi. Next, the specimens were kept for 24 h in an aeration room followed by an 8 h degassing at 50 °C in a vacuum oven. Before use, the sterilized 3D-SFnws pieces were checked for the absence of morphological changes and the upkeep of mechanical properties. Finally, the sterilized C/HE-3D-SFnws samples were aseptically transferred to 4-well, multi-dish w/lid sterile, polystyrene culture plates (Cat. No. 176597, Nalge Nunc International, Rochester, NY, USA). Sterilized steel rods kept the 3D-SFnws samples edges at the bottom of the plates.

### 2.3. Cells

Nontumorigenic primary coronary artery AHSMCs were from ATCC (USA). The supplier company guaranteed that they expressed smooth muscle $\alpha$-actin and were negative for human immune deficiency virus, hepatitis B virus, hepatitis C virus, mycoplasma, bacteria, yeasts, and fungi. For first expansion purposes, the Vascular Cells Basal Medium (ATCC® Cat. no. PCS-100-030) with the Vascular Smooth Muscle Growth Kit (ATCC® Cat. no. PCS-100-042) was used as suggested by the seller. The Growth Kit holds recombinant human (rh) basic FGF, rh insulin, rh EGF, ascorbic acid, L-glutamine, and foetal bovine serum (FBS). For the experiments, AHSMCs' culture medium was DMEM (89% $v/v$; Life Technologies Italia, Monza, Italy) supplemented with heat-inactivated (at 56 °C for 30 min) exosome-depleted FBS (10% $v/v$; Life Technologies Italia), and penicillin–streptomycin solution (1% $v/v$; Lonza, Rome, Italy).

Microvascular HECs isolated from adult skin capillaries were from Cell Applications, Inc. (Cat. no. 300-05a, San Diego, CA, USA). The seller pledged that such cells were free from bacteria, yeasts, fungi, and mycoplasma, and expressed the Factor VIII-related antigen. HECs were grown in Endothelial Cell Growth Medium (Cat. no. C-22010, PromoCell GmbH, Heidelberg, Germany). For the experiments, HECs were kept in Endothelial Cell Basal Medium (Cat. no. C-22210, PromoCell) supplemented with heat-inactivated (at 56 °C for 30 min) exosome-depleted FBS (10% $v/v$; Life Technologies Italia), and penicillin–streptomycin solution (1% $v/v$; Lonza, Italy).

To ensure that the exosomes under study came solely from the AHSMCs, FBS was first heat-inactivated (at 56 °C for 30 min) and next centrifuged twice at $100,000 \times g$ for 120 min at 4 °C in an Optima TLX ultracentrifuge (Beckman, Indianapolis, IN, USA) using the type TLA 100.3 minirotor before its addition to the experimental media [16].

### 2.4. Intravital AHSMCs Staining and In Vitro Culture

Prior to their experimental use, 3rd or 4th passage AHSMCs were counted using a Handheld Automated Cell Counter (Scepter™, Merck, Darmstadt, Germany) according to the seller's instructions. To highlight the C/HE-3D-SFnws-attached cells, just prior to seeding $2.0 \times 10^6$ AHSMCs were intravitally stained with the green fluorescent lipophilic membrane dye (tracer) $DiOC_{18}(3)$ (3,3′-Dioctadecyl oxacarbocyanine perchlorate (Thermo Fisher Scientific, Milan, Italy) with maximal fluorescence excitation at 484 nm and emission at 590 nm wavelengths. $DiOC_{18}(3)$ was dissolved in DMSO and used to intravitally stain AHSMCs according to the seller's instructions. For each experiment, four equal aliquots ($5 \times 10^5$ each) of pre-stained AHSMCs were carefully seeded onto as many C/HE-3D-SFnws scaffolds placed inside separate 4-well, multi-dish polystyrene culture plates (Cat. No. 176597, Nalge Nunc International). For comparative purposes, equal aliquots ($5 \times 10^5$ cells each) of prestained AHSMCs were seeded in parallel onto four identical polystyrene Petri dishes. The cell cultures of both groups were incubated for 15 days at 37 °C in a 95% $v/v$ air, 5% $v/v$ $CO_2$ atmosphere. AHSMCs were regularly seen under an inverted fluorescence microscope (IM35, Zeiss, Oberkochen, Germany) equipped with proper excitation and emission filters. All the next procedural steps were the same for the control group and the experimental group. Pictures were taken using a CP12 digital camera (Optika Microscopes, Società a responsabilità limitata, Ponteranica, BG, Italy).

### 2.5. AHSMCs' Total DNA Assay

To estimate the cell growth on C/HE-3D-SFnws scaffolds, DNA cellular contents were calculated by the Quant-iT PicoGreen dsDNA Kit (Thermo Fisher Scientific). Three specimens of AHSMCs cultured on C/HE-3D-SFnws were assessed at experimental days 3, and 15 (Figure 1e), respectively. After washing the cells in PBS, 8 mL deionized water was added to the wells to detach and lyse the cells. Next, repeated vortexing and twice freezing–thawing improved cells lysis. Then, the DNA amounts were fluorometrically measured at excitation 480 nm and emission 520 nm wavelengths, respectively. A standard double stranded (ds) DNA curve of known concentrations served to calibrate the fluorescence intensities in relation to cell numbers.

### 2.6. Assay of D-glucose Consumption

Cell $D$-glucose cumulative consumption was assessed in the conditioned growth medium samples from AHSMCs cultured on 3D-SFnws by a glucose oxidase assay using the Amplex® Red Glucose/Glucose Oxidase Assay Kit (Thermo Fisher Scientific) following the instructions manual. Glucose oxidase reacted with $D$-glucose to form $D$-gluconolactone and hydrogen peroxide in the presence of horseradish peroxidase. Hydrogen peroxide reacted with the Amplex Red reagent in a 1:1 stoichiometric ratio to generate the red-fluorescent product resorufin, whose intensity was recorded fluorometrically at excitation and emission wavelengths of 560 and 590 nm, respectively.

### 2.7. Phoshoproteins Array Analysis

To assess the activation of TGF-$\beta$ and NF-$\kappa$B intracellular signalling pathways, the phosphorylation status of selected proteins (see Table 1) in lysates from AHSMCs grown on C/HE-3D-SFnws scaffolds or on polystyrene was analysed using the C-Series RayBio™ Phosphorylation pathway profiling array (RayBiotech Inc., Peachtree Corners, GA, USA). For each experiment, AHSMCs ($5 \times 10^5$ cells) were carefully seeded onto as many C/HE-3D-SFnws scaffolds placed inside separate 4-well, multi-dish polystyrene culture plates (Cat. No. 176597, Nalge Nunc International) (experimental group). In parallel, equal aliquots were seeded for comparative purposes onto four polystyrene Petri dishes (diameter, 10 cm; Thermo Fisher Scientific). The cell cultures of both groups were incubated at 37 °C in a 95% $v/v$ air, 5% $v/v$ CO$_2$ atmosphere for 15 days in DMEM (89% $v/v$; Life Technologies Italia, Italy) supplemented with heat-inactivated (at 56 °C for 30 min) exosome-depleted FBS (10% $v/v$; Life Technologies Italia), and penicillin–streptomycin solution (1% $v/v$; Lonza, Rome, Italy).

According to the manufacturer's instructions, cell lysates were collected from both groups by solubilizing the cells in 1X Lysis buffer (RayBiotech Inc.) added with protease inhibitor and phosphatase inhibitor cocktails. The sample protein concentrations were assessed via Bradford's method. Briefly, each membrane supporting a different antibody array was blocked with Intercept® TBS-blocking buffer (LI-COR Biosciences GmbH, Bad Homburg vor der Hohe, Germany) for 60 min at room temperature and then incubated with 50 μg of protein lysate, overnight at 4 °C. After washing, the detection antibody cocktail was added during a 2 h incubation at room temperature, followed by 1 h incubation with IRDye® 800 CW-conjugated anti rabbit antibody (1:3000 in Intercept® TBS-blocking buffer (LI-COR Biosciences GmbH) plus 0.2% $v/v$ Tween-20). The positive signals of the phosphorylated proteins were acquired with an Odissey® (LI-COR Biosciences GmbH) scanner and their densitometric values quantified using the Image Studio® (version 5.2) software package (LI-COR Biosciences GmbH). Finally, the results were (i) processed as integrated intensity absolute values; or (ii) normalized to the maximal integrated density obtained in each single array. In keeping with Neradil et al. [49], these two processing modes resulted in identical phosphorylation profiles for each of the examined proteins. Hence, the results were expressed as the means $\pm$ SDs of three distinct experiments.

**Table 1.** Phosphorylated proteins investigated via membrane-based double antibody arrays.

| Abbreviations | Names in Extenso | Phosphorylation Site (s) | References * |
|---|---|---|---|
| Akt/PKB | Akt/Protein Kinase B | Ser473 | 102–104 |
| ATF-2 | Activating Transcription Factor-2 | Thr69/Thr71 | 81 |
| ATM | Ataxia-Telangiectasia Mutated Ser/Thr kinase | Ser1981 | 82 |
| c-Fos | Protein of transcription factor AP1 | Thr232 | |
| c-Jun | Protein of transcription factor AP1 | Ser73 | 82 |
| CREB | cyclic AMP Response Element Binding protein | Ser133 | 77 |
| eIF-2$\alpha$ | eukaryotic translation Initiation Factor-2$\alpha$ | Ser51 | |
| HDAC-2 | Histone Deacetylase-2 | Ser394 | |
| HDAC-4 | Histone Deacetylase-4 | Ser632 | |
| I$\kappa$B$\alpha$ | NF-$\kappa$B Inhibitor $\alpha$ | Ser32 | 90 |
| MSK-1 | Mitogen- and Stress-activated protein Kinase | Ser376 | |
| NF-$\kappa$B | Nuclear Factor-$\kappa$B | Ser536 | 85 |
| SMAD-1 | Small Mother Against Decapentaplegic homolog-1 | Ser463/Ser465 | |
| SMAD-2 | Small Mother Against Decapentaplegic homolog-2 | Ser245/Ser250/Ser255 | 78–80 |
| SMAD-4 | Small Mother Against Decapentaplegic homolog-4 | Thr277 | |
| SMAD-5 | Small Mother Against Decapentaplegic homolog-5 | Ser463/Ser465 | |
| TAK-1 | TGF-$\beta$-Activated Kinase 1 | Ser412 | 76 |
| ZAP-70 | ZAP-70 Tyrosine protein kinase | Tyr292 | |

* These references relate only to proteins whose phosphorylation levels were significantly changed.

### 2.8. Isolation, Characterization, and Quantification of Exosomes

AHSMCs-conditioned media of both experimental and control groups were collected at 72 h intervals between day 3 and 15 and centrifuged at 2000$\times$ *g* for 30 min at 4 °C to remove cells and debris. The resulting supernatants were stored at −80 °C for later analysis. After thawing, the supernatants belonging to each group were pooled together and the corresponding total exosomal fractions were extracted using the Total Exosome Isolation Reagent No. 4478359 for cell culture media (provided by Thermo Fisher-Invitrogen, USA) according to the supplier's protocol with some modifications. Briefly, the supernatants were centrifuged at 15,000$\times$ *g* for 30 min and next mixed with the proprietary reagent, incubated overnight at 4 °C, and afterward centrifuged again at 10,000$\times$ *g* for 90 min at 4 °C. The final pellets held the exosome fractions. This procedure has been compared with others and its validity confirmed [50]. The total proteins of the exosome fractions were quantified via Bradford's method. The marker-based assessments of the exosomal preparations were performed using ELISA kits detecting the CD9 (ExoTEST$^{TM}$, HansaBio Med, Tallinn, Estonia) and CD81 markers (ExoELISA-ULTRA CD81, System Biosciences, Palo Alto, CA, USA). Notably, CD9 and CD81 are members of the transmembrane-4 superfamily proteins intensely expressed also by the exosomes of vascular SMCs [51]. Thereafter, equal exosomal particle numbers (i.e., $1.04 \times 10^{11}$) quantified via ELISA kits for CD9 and CD81 markers from the experimental (C/HE-3D-SFnws) and the control (polystyrene) groups were used in parallel for further processing.

### 2.9. AGFs Carried by AHSMC-Released Exosomes

The AGFs carried by exosomes were quantified with the Human Angiogenesis Antibody Array C1000 (RayBiotech) according to the manufacturer's protocols. Briefly, equal amounts of exosomal proteins of the control and experimental group were diluted in 2.0 mL PBS and next incubated with the antibody arrays, which had been pre-treated for 30 min with Intercept® TBS-blocking buffer (LI-COR). After an overnight incubation at 4 °C and a thorough washing, the array membranes were incubated for 2 h with 1.0 mL of a mix of array-specific biotin-conjugated primary antibodies, diluted 1:250 in Intercept® TBS-blocking buffer. Finally, the membranes were incubated at room temperature for

1 h with 2.0 mL of DyLight800-conjugated streptavidin (LGC Clinical Diagnostics' KPL, Gaithersburg, MD, USA), diluted 1:7500 in Intercept® TBS-blocking buffer. The positive signals of the various AGFs were acquired with an Odissey® scanner (LI-COR Biosciences GmbH) and their densitometric values quantified by using the Image Studio® software package (version 5.2, LI-COR Biosciences GmbH). The intensity values of the positive signals from each array were normalized via comparisons to corresponding positive controls. The results from three independent experiments were averaged and expressed as mean values ± SDs. This technology provided several advantages: (i) it allowed performing high-content screening using about the same sample volume as traditional ELISAs require; (ii) it improved the chances of discovering key factors while maintaining an ELISA-like sensitivity; (iii) it had a wider detection range, i.e., 10,000-fold, than typical ELISA assays (which is 100-to-1000-fold); and (iv) it owed a lower inter-array coefficient of variation of spot signal intensities (i.e., ~5–10%) than typical ELISAs do (i.e., ~10–15%).

### 2.10. In Vitro HECs' Cultures

HECs were seeded into 24-well plates at $15 \times 10^3$ cells/well and kept in Endothelial Cell Basal medium (Cat. no. C-22210, PromoCell) plus 10% *v/v* exosome-depleted FBS. A fluorescence CellTiter-Blue® cell viability assay (Promega, Madison, WI, USA) served to calculate out HECs numbers. This assay measures the conversion of resazurin into fluorescent resorufin by metabolically active cells. The fluorescence intensity produced is directly proportional to the number of viable cells [52]. Thus, at the devised time points, HECs were incubated for 1 h at 37 °C in 500 μL of culture medium added with 50 μL of CellTiter-Blue® reagent. The resulting resazurin was fluorometrically recorded using FP 6200 fluorometer (Jasco, Cremella (LC), Italy), with excitation and emission wavelengths of 560 nm and 590 nm, respectively. Twenty-four hours after plating (i.e., at experimental 0 h), a first CellTiter-Blue® test was conducted to obtain baseline values of cell numbers. After obtaining baseline values (0 h), half of the wells, i.e., the experimental group, were randomly selected to be added with a medium enriched with exosomes (at a final concentration, 2 μg mL$^{-1}$) that had been collected and quantified as detailed above. The remaining wells served as controls. The same amount of AHSMCs' exosomes were added to the experimental group again 24 h later. Then, to assess changes in cell numbers, at 72 h the CellTiter-Blue® test was performed again in wells of both groups. The fluorescence values gained were transformed into corresponding HECs numbers using an ad hoc constructed standard curve.

### 2.11. In Vitro HECs Migration Test

For migration studies, HECs were pre-labelled with fluorescent CellBrite® NIR 680 dye (1 μM; Biotium, Inc., Fremont, CA, USA) and ~$35 \times 10^3$ HECs were seeded into a silicone Culture Insert-2 Well (Ibidi GmbH, Graefelfing, Germany) inserted inside a 12-well plate. HECs were cultured in Endothelial Cell Basal medium (Cat. no. C-22210, Promocell) fortified with 10% *v/v* exosome-depleted FBS and kept at 37 °C and in 5% *v/v* $CO_2$ in air for at least 24 h to permit cell adhesion and the formation of a confluent monolayer. Then, the Culture Insert-2 Well removal left two defined cell patches, separated by a 500 μm wide gap. The culture medium was at once removed to be replaced either with a fresh medium holding exosomes (5 μg mL$^{-1}$) released from C/HE-3D-SFnws-adhering AHSMCs or with basal medium containing 10% *v/v* exosome-depleted FBS (control wells). The culture plate was incubated at 37 °C and at various time points the fluorescence signals due to the cells migration into the gap were measured using an Odyssey® Imager (LI-COR Biosciences GmbH). The fluorescence intensities were quantified using Image Studio® software (version 5.2, LI-COR Biosciences GmbH). To this aim, at time 0 h, a rectangular shape (corresponding to the area of the gap) was set to define the detection zone for fluorescent signals and then, at various time points, the fluorescence intensity into the rectangular shape was quantified in real-time as the sum of the pixel values within the shape's boundary. The HECs migration was expressed as percentage fluorescence values with respect to the experimental 0 h.

The migration assays were conducted in triplicate and the mean $\pm$ SD values served to construct curves reflecting the time-related gap reduction in the two groups.

### 2.12. In Vitro HECs Tubes and Nodes Formation Assay

The proangiogenic properties of exosomes released from AHSMCs grown on C/HE-3D-SFnws scaffolds were assessed using HECs and the PromoKine Angiogenesis Assay Kit (Cat. No. PK-CA577-K905; PromoCell) according to the manufacturer's instructions. Briefly, HECs were grown at 37 °C in air with $CO_2$ 5% $v/v$ until reaching about 90% confluency in Endothelial Cell Basal Medium (Cat. No. C-22210; PromoCell) added with the Supplement Mix (Cat. No C-39215; PromoCell). Next, HECs were harvested using trypsin 0.025% $v/v$ and resuspended in Endothelial Cell Basal Medium (Cat. No. C-22210; PromoCell) fortified with 2.5% $v/v$ exosome-depleted FBS. Then, an aliquot (50 µL) of Extracellular Matrix (ECM) solution was added to each well of a 96-well sterile cell culture plate kept on ice. Thereafter the plate was incubated at 37 °C to form a gel. After that, $20 \times 10^3$ HECs suspended in 100 µL culture medium were mixed with different concentrations (1, 2, 5, 10, and 20 µg mL$^{-1}$) of exosomes from C/HE-3D-SFnws-attached AHSMCs and directly added to each well. Controls on ECM gel received no exosomes. Finally, the plates were incubated for 5 h at 37 °C in air with $CO_2$ 5% $v/v$. Thereafter, HECs were checked and photographed at 100× magnification under a Zeiss IM35 microscope with a CP12 digital camera (Optika Microscopes). The total mean tube lengths (in µm) and nodes numbers per microscopic field at 100× magnification were quantified via morphometric methods [53]. Triplicate results were averaged and graphed as bars showing the means $\pm$ SDs.

### 2.13. Statistical Analysis

Data were expressed as mean values $\pm$ SDs. Descriptive statistical analyses were conducted using the *Analyse-it*™ software package (www.analyse-it.com, accessed on 10 January 2022). Shapiro–Wilk's test revealed that the data groups had a normal distribution. A one-sided Student's *t* test served to assess the level of statistical significance differences of data of AHSMCs cultured on 3D-SFnws vs. AHSMCs cultured on polystyrene. A one-way ANOVA with post hoc Tukey's test served for multiple comparisons of the results from endothelial tubes/nodes formation assays. Statistical significance was set at *p* value < 0.05.

## 3. Results

### 3.1. Carded/Hydroentangled (C/HE)-3D-SFnws

An earlier work from our laboratory detailed the physicochemical characteristics of the C/HE-3D-SFnws used here [16]. Briefly, the native SF microfibers are first mainly longitudinally oriented via carding and next transversally twisted via hydroentanglement technology. The upshot is a 3D nonwoven whose fabrication avoids the use of any gluing chemical (e.g., FA) while abiding by the biomechanical requirements of human soft tissues [54] (Figure 1a,b). Interestingly, from an applicative standpoint, according to the criteria of Wang et al. [55] and Thurber et al. [56], the properties of C/HE-3D-SFnws are consistent with a slow, i.e., medium-to-long term, biological breakdown that could advance the progressive repair/reconstruction of an injured/lost tissue in vivo.

### 3.2. Growth and Metabolism of AHSMCs on C/HE-3D-SFnws vs. Polystyrene

Observations under the fluorescence microscope revealed that within 3 h of careful seeding, about 80% of the intravitally stained AHSMCs adhered to the C/HE-3D-SFnws microfibers (data not shown). These cells increased in number with time and moved not only along the SF microfibers but also into the inter-fibre voids after secreting extracellular matrix (ECM) (Figure 1c,d). In keeping with this, between day 3 and 15 of staying in vitro, the double strand (ds) DNA amount of polystyrene-stuck AHSMCs rose by 7.8-fold, while that of the C/HE-3D-SFnws-attached AHSMCs increased by 10.9-fold (+39.7%, *p* < 0.05) (Figure 1e). Additionally, during the same time lag, the cumulative *D*-glucose uptake from the growth medium on the part of polystyrene-attached AHSMCs increased by 6.1-fold while that of C/HE-3D-SFnws-bound AHSMCs rose by 8.4-fold (+35.5%, $p < 0.05$) (Figure 1f). These significant increases in growth and metabolic activities of the C/HE-3D-SFnws-bound vs. polystyrene-stuck AHSMCs happened despite both experimental groups being kept in the same incubator and nourished with equal amounts of the same exosome-depleted 10% $v/v$ FBS growth medium. In fact, the two groups differed only in the substrate to which they adhered.

### 3.3. Signalling Pathways Activated in C/HE-3D-SFnws- vs. Polystyrene-Attached AHSMCs

The activation of intracellular signalling pathways, as revealed by the phosphorylation of specific amino acidic sites, crucially drives cells' functions. Thus, by using membrane-based double-antibody arrays, we assessed the relative phosphorylation—hence activity— levels of various proteins belonging to the TGF-β and NF-κB (also known as nuclear factor 'kappa-light-chain-enhancer' of activated B-cells) pathways in AHSMCs grown on either 3D-SFnws or polystyrene [57]. Table 1 lists the abbreviated names and the corresponding specific phosphorylation sites of the investigated proteins. Typical couples of developed array membranes for either experimental group are shown in Figure S1.

Conversely, Figure 2a,b show the integrated intensity values for each couple of specific protein spots and their statistical significance.

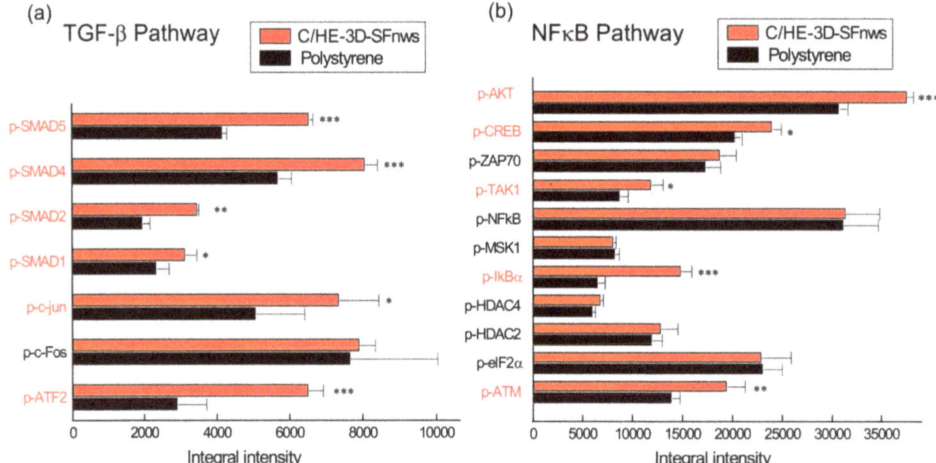



**Figure 2.** Signalling pathways activated in C/HE-3D-SFnws-attached vs. polystyrene-stuck AHSMCs. Equal amounts of total protein lysates from AHSMCs grown on either C/HE-3D-SFnws or polystyrene for 15 days in vitro were subjected to signalling pathway-specific membrane-based double-antibody arrays to assess any difference in specific sites phosphorylation of different proteins. The adhesion to C/HE-3D-SFnws significantly affects the activation of (**a**) TGF-β and (**b**) NF-κB signalling pathways in AHSMCs. Abbreviations in red highlight the increased phosphoproteins. The integrated intensity values for each couple of specific protein spots and their levels of statistical significance are shown as the mean values ± standard deviations (SDs). *, $p < 0.03$; **, $p < 0.002$; and ***, $p < 0.0005$. For technical details, consult the Section 2.7. Abbreviations: Akt/PKB, Akt/Protein kinase B; ATF-2, Activating Transcription Factor 2; ATM, Ataxia-Telangiectasia Mutated Ser/Thr kinase; c-Fos, Protoncogene c-Fos; c-Jun, Transcription factor AP1; CREB, cAMP Response Element Binding protein; eIF2α, eukaryotic translation Initiation Factor-2α; HDAC2, HDAC4, Histone DeACetylase 2/4; IκBα, NF-κB inhibitor α; MSK1, Mitogen- and Stress-activated protein Kinase 1; NF-κB, Nuclear Factor κB; SMAD1, SMAD2, SMAD4, SMAD5, Small Mother Against Decapentaplegic homolog; TAK-1, TGF-β-activated kinase 1; and ZAP70, Tyrosine protein kinase ZAP70.

As the results show, the adhesion to neatly different polymeric substrates did affect the two signalling pathways investigated. In more detail:

### 3.3.1. TGF-β Signalling Pathway

The TGF-β signalling pathway was activated more strongly in C/HE-3D-SFnws-grown than in polystyrene-stuck AHSMCs. In fact, significant increases in the phosphorylation of specific functional sites were exhibited by the downstream TGF-β signalling mediators SMAD-1 (+34%, $p$ = 0.0122); SMAD-2 (+76%, $p$ = 0.0026); SMAD-4 (+42%, $p$ = 0.0002); SMAD-5 (+56%, $p$ = 0.0001); and also by ATF-2 (+122%, $p$ = 0.0005); and c-Jun (+45%, $p$ = 0.03). Conversely, the specific site phosphorylation levels of c-Fos were alike ($p$ > 0.05) in the two groups (Figure 2a).

### 3.3.2. NF-κB Signalling Pathway

Five members of this pathway showed higher phosphorylation levels of specific functional sites in C/HE-3D-SFnws-grown than in polystyrene-stuck AHSMCs, i.e., TAK-1 (+36%, $p$ = 0.0174); TBK-1 (+31%, $p$ = 0.0092); IκBα (+126%, $p$ = 0.0001); ATM (+39%, $p$ = 0.002); Akt/PKB (+21%, $p$ = 0.0007); and CREB (+23%, $p$ = 0.0158) (Figure 2b). By contrast, the phosphorylation levels of specific sites concerning NF-κB; eIF-2α; HDAC-2; HDAC-4; and MSK-1 did not differ ($p$ > 0.05) between the two groups (Figure 2b).

Altogether, the two activated signalling pathways underlay the more intense proliferation and metabolism occurring in the C/HE-3D-SFnws-attached than in the polystyrene-stuck AHSMCs.

### *3.4. AGFs Released via Exosomes from AHSMCs Grown on C/HE-3D-SFnws vs. Polystyrene*

The AGFs present in equal amounts of the pooled CD9$^+$/CD81$^+$ exosomes released between day 3 and 15 from AHSMCs grown on either C/HE-3D-SFnws or polystyrene were identified and quantified by means of specific membrane-based double-antibody arrays [57]. Figure S2a,b shows corresponding couples of typical array membranes of the two groups. The quantitative and statistical analysis of equivalent spots revealed that the amounts of 8 out of the 43 potentially discoverable AGFs were significantly higher ($p$ < 0.05) in the exosomes released from C/HE-3D-SFnws-attached than from polystyrene-stuck AHSMCs (Figure 3). Interestingly, TGF-β was transported in alike amounts by the exosomes released from both groups (Figure S2a,b). In the C/HE-3D-SFnws group of exosomal proteins, the highest per cent increases vs. their polystyrene counterparts were those of Interleukin-6 (IL-6; +442%, $p$ < 0.0001); Interleukin-8 (IL-8; +117%, $p$ < 0.0001); and Growth-Regulated Oncogene (GRO)-α/-β/-γ chemokines (+100%, $p$ < 0.0001). Lesser but still significant increases were observed for Interleukin-4 (IL-4; +87%, $p$ < 0.0002); Interleukin-2 (IL-2; +71%, $p$ <0.001); Interleukin-1α (IL-1α; +48%, $p$ < 0.001); Granulocyte-Macrophage Colony-Stimulating Factor (GM-CSF; +41%, $p$ < 0.0087); and basic Fibroblast Growth Factor (bFGF; +33.3%, $p$ <0.033) (Figure 3). Conversely, the exosomal amounts of eight more known angiogenic compounds—Plasminogen/Angiostatin; ANGPT-2 (Angiopoietin-2); Tie-2 (Angiopoietin-1 receptor); MCP-1 (Monocyte Chemoattractant Protein-1); VEGF-D (Vascular Endothelial Growth Factor-D); VEGF-R3 (VEGF receptor 3); and TIMP-1 and TIMP-2 (Tissue Inhibitor of Metalloproteinase-1/-2)—did not significantly ($p$ > 0.05) differ between the C/HE-3D-SFnws and polystyrene groups. Finally, another 27 agents were identified in similarly low amounts in the exosomes from both groups (Figure S2a,b).

These findings are consistent with the reported angiogenic and growth-promoting effects of the more intensely expressed AGFs carried by the exosomes released from C/HE-3D-SFnws-attached AHSMCs (see Table 2 and for more details the Supplementary Materials).

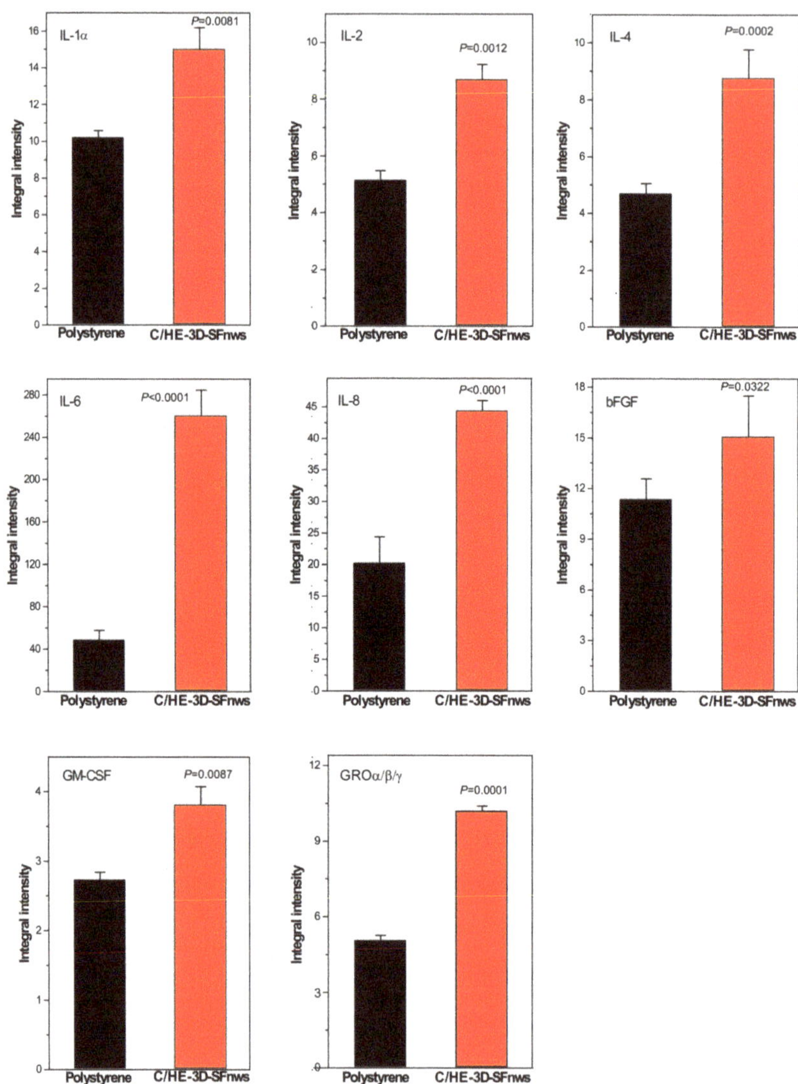



**Figure 3.** The angiogenic and trophic factors (AGFs) carried by the exosomes released from AHSMCs grown on C/HE-3D-SFnws vs. polystyrene. Equal amounts of exosomal proteins isolated from AHSMCs-conditioned pooled media samples of the two experimental groups were subjected to membrane-based double-antibody arrays, detecting the relative levels of multiple AGFs. The adhesion to C/HE-3D-SFnws significantly affected the levels of the eight exosomally carried AGFs shown in Figure. The bars are integral intensity values expressed as mean values ± standard deviations (SDs) from three distinct experiments, each conducted in duplicate. The corresponding *p* values of the differences between each couple of bars are also shown over the top of the right bars. For technical details, consult the Section 2.9.

**Table 2.** The main trophic and angiogenic features of each of the enriched AGFs conveyed by the exosomes released from 3D-SFnws-adhering AHSMCs.

| AGF | Trophic and Angiogenic Actions |
| --- | --- |
| IL-1α | • promotes angiogenesis in vivo by inducing VEGF's synthesis [58];<br>• activates the VEGF·VEGFR-2 signalling pathway;<br>• stimulates Platelet-Derived Growth Factor (PDGF)'s A chain [59], and bFGF expression [60];<br>• induces its own expression in vascular SMCs, exerting autocrine growth-stimulatory effects [61]. |
| IL-2 | • affects the permeability of ECs [62] and promotes ECs angiogenesis through the α and β IL-2 receptors;<br>• stimulates angiogenesis in animals and tube formation in HUVECs [63];<br>• enhances SMCs responsiveness to angiotensin II [64];<br>• in cooperation with IL-1α potentiates human SMCs proliferation [65]. |
| IL-4 | • increases the expression of vascular cell adhesion molecule (VCAM)-1, IL-6, and MCP-1 [66];<br>• induces cytoskeletal rearrangements both in HUVECs and in human coronary artery ECs [67];<br>• acts as a mild mitogen for both macro- and microvascular ECs [68–70]. |
| IL-6 | • exerts autocrine growth-stimulating effects on vascular SMCs inducing endogenous PDGF's production [71];<br>• support ECs' cell proliferation and mobility [72]. |
| IL-8 GRO-α/β/γ | • by sharing the evolutionary 'ELR' motif they all powerfully promote angiogenesis even in the absence of inflammation [73,74]. |
| bFGF | • regulates both angiogenesis and arteriogenesis (reviewed in [75]);<br>• enhances ECs and SMCs proliferation [76];<br>• regulates vascular remodelling and the proliferation of human dermal microvascular ECs [77,78]. |
| GM-CSF | • supports ECs leading to the formation of endothelial capillaries [79];<br>• stimulates the migratory repair of mechanically wounded ECs monolayers [80];<br>• stimulates angiogenesis, neovascularization, and arteriogenesis [81]. |

*3.5. Exosomes from C/HE-3D-SFnws-Stuck AHSMCs Powerfully Stimulate HECs to Grow, Migrate, and Make Endothelial Tubes/Nodes*

Next, we assessed whether the AGFs carried by the exosomes released from the C/HE-3D-SFnws-bound AHSMCs would have any real angiogenic power by inducing cultured HECs to proliferate, migrate, and/or de novo assemble into endothelial tubes and establish nodes in vitro.

First, adding exosomes (2 μg mL$^{-1}$) to HECs cultured in exosome-depleted 10% $v/v$ FBS medium significantly increased the numbers of viable, i.e., metabolically active cells 72 h later, as revealed by the CellTiter-Blue® assay vs. their untreated counterparts (Figure 4a).

Second, adding exosomes (5 μg mL$^{-1}$) to HECs cultured as above significantly advanced their migration into the "wound's area" between 24 h and 72 h as compared to untreated HECs. By 72 h, the gap's space covered by the exosome-treated HECs was about double ($p < 0.05$) that overlain by the untreated (control) cells (Figure 4b). Moreover, the mobilization of exosome-treated HECs toward other directions was also discernible in the experimental model used (not shown).

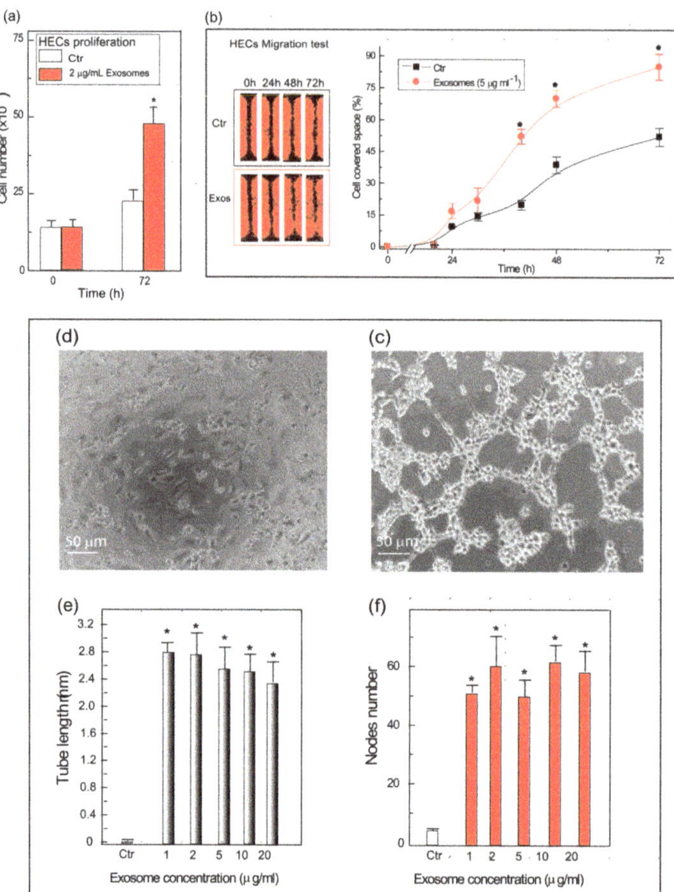

**Figure 4.** Exosomes released from AHSMCs grown on C/HE-3D-SFnws induce HECs' proliferation, migration, and tubes and nodes formation in vitro. (**a**) Stimulation of HECs' growth: exosomes (2 µg mL$^{-1}$) released from C/HD-3D-SFnws-adhering AHSMCs when added to HECs cultured in Endothelial Cell Basal medium plus 10% *v/v* exosome-depleted FBS strongly increased the numbers of viable cells 72 h later, as revealed by CellTiter-Blue® assay vs. the untreated cells (Ctr). *, $p < 0.05$ vs. Ctr. See Materials and Methods for technical details. (**b**) Stimulation of HECs' migration: adding exosomes (5 µg mL$^{-1}$) released from the C/HE-3D-SFnws-bound AHSMCs to HECs cultured in Endothelial Cell Growth Basal Medium fortified with 2.5% *v/v* exosome-depleted FBS significantly advanced their migration into the "wound area" between 24 h and 72 h as compared to untreated HECs (Ctr), as the gap's space covered by the exosome-treated HECs was, by 72 h, about double ($p < 0.05$) that occupied by the untreated cells (Ctr). For technical details consult the Section 2.11. (**c,d**) Tube and nodes formation: micrograph (**c**) showing plain attached HECs 5 h after seeding into a 96-well plate onto Extracellular Matrix (ECM) gel. In each of these wells 20 × 10$^3$ HECs were incubated at 37 °C in Endothelial Cell Growth Basal Medium fortified with 2.5% *v/v* exosome depleted FBS. No AHSMC-released exosomes were added. Micrograph (**d**) shows HECs 5 h after seeding onto ECM gel while being simultaneously exposed to increasing concentrations of exosomes released from C/HD-3D-SFnws-adhering AHSMCs, starting from 1 µg mL$^{-1}$ to 20 µg mL$^{-1}$; all the other conditions as in (**c**). Endothelial tube formation was strongly induced by the AHSMCs exosomes. (**c,d**), Phase contrast microscopy. Original magnification, 100×. (**e,f**) Bar graph showing the total length (in mm) of newly formed endothelial tubes (e) and the number of nodes (f) per microscopic

field under the conditions of the test, i.e., (i) control HECs cultured on ECM gel with no addition of exosomes set free from C/HE-3D-SFnws-attached AHSMCs; and (ii) HECs cultured on ECM gel exposed to increasing concentrations of exosomes released from C/HE-3D-SFnws-attached AHSMCs. Tube formation assay was performed as detailed in the Section 2.12. The number of nodes and the total tube length (in mm)/microscopic field of 332,667 $\mu m^2$ area were found via morphometric methods [53] on pictures taken at 100× magnification of five microscopic fields for each exosomal concentration. Triplicate results were averaged, and the bars show the means ± SDs. As compared to control HECs on Extracellular Matrix gel (Ctr), in the total absence of exosomes, the HECs treated with exosomes released from C/HE-3D-SFnws-attached AHSMCs showed, by 5 h, huge increases in endothelial tube lengths and in numbers of nodes that were dose-independent in the range evaluated. Pairwise one-tailed Student's *t* test and one-way analysis of variance (ANOVA) with post hoc Tukey's test were used for statistical analysis. *, $p < 0.001$ vs. Ctr. Conversely, no statistical difference ($p > 0.05$) in tube length and number of nodes at 5 h occurred within the several doses of exosomes evaluated.

Thirdly, control (no exosomes added) HECs plated on ECM gel in 2.5% $v/v$ exosome-depleted FBS medium formed very few endothelial tubes/nodes. Conversely, after 5 h exposure to different doses (1, 2, 5, 10, and 20 $\mu g$ $mL^{-1}$) of exosomes, the ECM gel-plated HECs formed quite extensive tubular networks, interconnected by a considerable number of nodes. The increases in lengths of the endothelial tubes (by 35-fold to 42-fold vs. untreated controls, $p < 0.001$) and in numbers of nodes (by 11-fold to 13-fold vs. untreated controls, $p < 0.001$) per microscopic field were quite conspicuous and alike ($p > 0.05$) for each of the doses assessed (Figure 4c–f).

Therefore, the AGFs carried by the exosomes released from 3D-SFnws-attached AHSMCs exerted effective mitogenic, mobilizing, and angiogenic activities when added to HECs cultured in vitro.

## 4. Discussion

The C/HE-3D-SFnws scaffolds we presently used consist of native SF microfibers in β-sheet form isolated from domesticated *Bombyx mori* silkworms. These C/HE-3D-SFnws [16] mark a technological evolution with respect to the earlier FA-crosslinked [10] and C/N-3D-SFnws [11]. Indeed, both these earlier prototypes performed quite well in terms of biocompatibility, host response, and the engineering/regeneration of a reticular connective tissue (see also below) [10,11]. However, from a biomechanical standpoint, both earlier kinds of SFnws were unsatisfactory, which blunted their potential clinical application. While the FA-crosslinked SFnws remained somehow stiff, even after a long-lasting hydration, the C/N-3D-SFnws were thin (i.e., 130 $\mu m$ thick) as their production required short (<25 mm) SF fibres as the starting material and hence they could be misshaped easily when handled [11,16]. Conversely, the carding/hydroentanglement processing of the present scaffolds required longer SF fibres (> 50 mm) thereby producing thicker and mechanically more robust structures. Therefore, citing Hu et al. [16] the novel C/HE-3D-SFnws *"maintain all the most appreciable characteristics of the carded-needled prototype (i.e., softness, lightness, interconnected porosity); display an outstanding handling stability (can be safely cut to realize any required size and shape); and at the same time provide the opportunity for modulating the biomechanical responses over a wider range of stress and strain values"*. Moreover, Hu et al. [16] also stated that being anisotropic, the C/HE-3D-SFnws scaffolds *"are particularly suited to fulfil* human *soft tissues mechanical needs engendered by directionally projected lines of force"*. Therefore, the C/HE-3D-SFnws characteristics appoint them as useful candidates for the guided engineering/regeneration and repair of injuries suffered by human soft tissues, hollow viscera walls and vessels included.

To be successful, an implanted biomaterial scaffold crucially requires an efficient neovascularization. When it misses this target, it is inexorably bound to fail. In earlier works, we showed that once grafted into the subcutaneous tissue of C57/BL6 mice, both the FA- and C/N-3D-SFnws underwent a fast neovascularization [10,11]. A more recent study using the same C/HE-3D-SFnws as used here showed that the scaffold-attached HDFs released exosomes loaded with surpluses of a dozen distinct AGFs, which powerfully stimulated HECs to abundantly produce endothelial tubes in vitro [16].

Fibroblasts are important, but not the sole constituents of the dermal and subcutaneous layers of the skin and of other connective tissue types. An added cell kind, the SMCs, populates the contractile layers or *tunicae* of hollow viscera, vessels included. Under normal conditions, adult SMCs express sets of specific proteins by which they regulate visceral and vascular contractile tone; secrete ECM components; and keep a quite low proliferation rate. However, SMCs never undergo terminal differentiation, as they own a distinct phenotypic plasticity [82]. In the adult vessel walls, SMCs of the epithelioid/synthetic-secretory and potentially migratory/proliferative phenotype prevail in the *tunica intima*, while SMCs of the spindle-shaped contractile phenotype hold sway in the *tunica media*; however, a certain number of SMCs of the less abundant phenotype is always present in either *tunica* [83]. In vitro and in physiological and pathological or harmful conditions in vivo, SMCs shift from their quiescent/spindle-shaped/contractile phenotype to their epithelioid/synthetic-secretory/migratory/proliferative phenotype. At variance with the former, the latter phenotype drives a vascular and/or visceral wall remodelling and is more prone to apoptosis [82,84,85]. In vivo, such a phenotype shift can result in neointima formation, stent occlusion, atherosclerosis, thrombosis, and asthma [86–89].

Wild-type or genetically modified rodent SMCs have often served as (mostly vascular) disease models [90,91]. However, animal SMCs, although undeniably useful, do not perfectly model human SMCs due to species-specific discrepancies in macroscopic anatomy, chromosome complement, genomic function, biochemistry, metabolism, and mechanical factors [30,91]. These limitations fully justify experimental studies into human SMCs as the most relevant models of human pathobiology. However, as human SMCs have different embryological origins, one should be wary of extrapolating the findings gained from SMCs of one origin to all the other SMCs [29,30]. Although the coronary artery AHSMCs used in this study have the same embryological origin as other visceral SMCs, the translatability of the present results to the latter cells requires further assessments.

The results of our earlier and present in vitro studies have repeatedly confirmed the high biocompatibility of 3D SFnws in relation, not only to HDFs and HECs, but also to AHSMCs. As well consistent with our earlier observations, SF was superior to polystyrene as an AHSMCs' adhesion substrate [10,11,15,16,38,92]. Various reports have proven that biomaterial substrates owning patterned or microtopographic structures advanced HECs' adhesion and proliferation more than nonpatterned ones did [93,94]. We recall here that 3D-SFnws have a patterned structure made of microfiber wisps separated by voids or grooves. Such patterning favoured the first adhesion of the AHSMCs, HDFs, and HECs. Subsequently, the same 3D-SFnws microfibers functioned as guides along which the proliferating AHSMCs first migrated and later invaded and colonized the intervening voids [10,11,15,16,92]. Our results revealed that higher increases in dsDNA (i.e., cell numbers) and glucose consumption occurred between day 3 and 15 in the AHSMCs attached to C/HE-3D-SFnws than to polystyrene. The same AHSMCs also showed higher growth and metabolic rates when stuck on a 3D SF nanofiber-microfiber-based vessel model [38]. In sharp contrast, unidirectionally aligned topologically patterned 2D SF films were reported to decrease SMCs' mitotic rate and to promote their transition from a synthetic/highly proliferative to a contractile/lowly proliferative phenotype [95,96]. At the root of these discordant findings might be differences in SF purity, age of SMCs donors, SMCs embryological origins, and passage numbers in vitro. Moreover, we recall here that when FBS, which carries various survival and growth factors, is added to the culture medium, the mechanism of contact inhibition of growth no longer regulates AHSMCs' proliferation [88]. At any rate, we used the very same growth medium fortified with identical per cent fractions of the same batch of exosome-depleted FBS to cultivate AHSMCs on C/HE-3D-SFnws or on polystyrene. Therefore, the exosome-depleted FBS we added did not affect the observed differences proliferation rates between AHSMCs grown on SFnws or polystyrene. Instead, the environmental context, i.e., the direct AHSMCs contact with SF or polystyrene, and/or the unlike autocrine effects brought about by the different AGFs amounts carried by AHSMC-released exosomes significantly affected the cell growth and metabolism of either experimental group.

Consistent with such a view, we found that AHSMCs produced and released discrete amounts of TGF-β, part of which was carried by exosomes (Figure S2, panel [a]). This TGF-β may have been one of the agents activating TAK-1 (also known as TGF-β-activated kinase 1) an evolutionarily conserved MAPK kinase kinase (MAPKKK) family member placed upstream the TGF-β intracellular signalling pathway [97,98]. However, as TGF-β levels in the exosomes and the growth medium (data not shown) of the two experimental groups were alike ($p > 0.05$), other agents must have played a part in TAK-1 activation. Besides TGF-β, TAK-1 is also activated by TNF, IL-1, Toll-like receptors (TLRs) ligands, and various kinases (see [99] for further details). The observed increase in the $Ser^{412}$ phosphorylation is indeed pivotal for TAK-1 activation [100]. It might have been related via an autocrine mechanism to the increased IL-1α content in the exosomes from SFnws-cultured AHSMCs (see also below). Moreover, both the catalytic subunit α of PKA (also known as cyclic AMP-dependent protein kinase A), or the cyclic AMP-controlled PRKX (or serine/threonine X-linked protein kinase) phosphorylate TAK-1 at $Ser^{412}$ activating it [100]. That an increased activity of cyclic AMP-related kinases was going on in the SF-stuck AHSMCs is consistent with the heightened $Ser^{133}$ phosphorylation and activation of the CREB (also known as cyclic AMP response element-binding) protein (Figure S1, panel (b)), which could be related to the increased proliferative activity of these cells [101]. Going back to the TGF-β pathway, the higher phosphorylation/activation of SMAD-1 and SMAD-5, two of the downstream TGF-β signalling mediators, suggests the involvement of both in the intensified growth and metabolism proper of the SF-attached AHSMCs [102]. The heightened phosphorylation/activation of SMAD-4 too—the cofactor forming complexes with any other SMAD homo- and heterodimers to transfer them into the nucleus and accordingly change gene expression [103]—supports an ongoing, more intense activity of the TGF-β signalling pathway in the SF-stuck AHSMCs. On the other hand, the less intense increase in phosphorylation/activation of the SMAD-2 mediator suggests that a fraction of the AHSMCs might have started shifting from the proliferative/secretory to the quiescent/contractile phenotype. This view agrees with the report that vascular SMCs cultured as multi-layered aggregates eventually entered a resting phenotype [104]. Moreover, the increased phosphorylation/activation of ATF-2, a protein encoded by a SMAD-targeted gene [105] belonging to the AP-1 family of transcription factors, which may also bind SMADs, further confirms that an intensification of various TGF-β pathway-related activities was occurring in the SFnws-adhering AHSMCs. AP-1 transcription factors and SMAD proteins do reciprocally interact thus modulating their respective effects in complex ways [106].

The said phosphorylation/activation of TAK-1 may also reveal a crosstalk between the TGF-β and NF-κB signalling pathways in the SFnws-adhering AHSMCs. In fact, TAK-1 signalling is all-important for the NF-κB's canonical activation mediated by signals from various receptors including IL-1 receptor I (IL-1RI·IL-1α; see also above) [107–109], and by stimuli linked to adaptive immunity [110–113]. Working TAK-1 phosphorylates at $Ser^{177}$ and $Ser^{181}$ and activates the IKKβ kinase. In turn, the activated IKKβ phosphorylates at $Ser^{32}$ (and $Ser^{36}$) IκBα. When not phosphorylated at such sites, IκBα binds and masks the nuclear localization signals (NLS) of the NF-κB factors Thus, the inactive IκBα·NF-κB transcription factor complexes stay sequestered in the cytoplasm [114]. Additionally, inside the nucleus unphosphorylated IκBα hinders the binding of the dimeric NF-κB transcription factors to their specific target gene sequences [115]. However, the phosphorylation of IκBα at Ser$^{32}$ (and Ser$^{36}$) by IKKβ targets IκBα to the S26 proteasome for degradation while letting the cytoplasmic NF-κB transcription factors to enter the nucleus regulating cell proliferation, adhesion, mobility, and reactive oxygen species (ROS) elimination [116,117] in the C/HE-3D-SFnws-stuck AHSMCs. In summary, our results suggest the existence of a circular positive interaction by which exosomally released cytokines such as IL-1α and IL-8 supported the more vigorous growth- and migration-related activities sustained by TAK-1/NF-kB/CREB signalling that in turn increased the expression of IL-1α and IL-8.

Interestingly, the present results also revealed a more intense phosphorylation/activation of the ubiquitously expressed ATM (also known as Ataxia-Telangiectasia Mutated), a serine/threonine kinase member of the PIKK (phosphatidylinositol-3 kinase-related kinases) protein family [118]. In the absence of any DNA damage [119,120] an increased ATM activity importantly participates in multiple cellular events (detailed in [121–125]). Moreover, ATM also phosphorylates and activates Akt/PKB at Ser$^{473}$, an intensified event that occurred in the SF-stuck AHSMCs, which may have positively affected their glucose uptake, protein synthesis, survival, and proliferation/differentiation [126–128].

Within the walls of vessels ECs and SMCs are close neighbours and reciprocally interact under physiological and pathological circumstances [129–131]. To build functioning blood vessels, tube forming ECs secrete TGF-β and PDGF to recruit SMCs [132]. Although Tang et al. [133] reported discrepant data, our results clearly show that 3D-SFnws-stuck AHSMCs did advance vasculogenesis by releasing exosomes that conveyed significant amounts of multiple angiogenic/growth factors. These findings consist with reports showing that the exosomes released from 3D cell culture models improved angiogenesis more effectively than those from 2D cultures [134]. However, we wish to stress that it is the *combination* of all the exosome- transported AGFs in their proper ratios that will optimally advance the three processes proper of neovascularization, i.e., vasculogenesis, angiogenesis, and arteriogenesis.

AHSMCs can be induced by a variety of stimuli to synthesize several cytokines and chemokines [135]. Therefore, a further potential benefit of our earlier and present results is that, after proper purification and standardization, the exosomes produced by 3D-SFnws bioreactors hosting AHSMCs or fibroblasts could find beneficial therapeutic applications as vascularization and cell growth stimulants, thus advancing injury healing and repair in clinical settings [16,26].

## 5. Conclusions and Future Perspectives

The present results reveal that AHSMCs cultured on the same C/HE-3D-SFnws as above released exosomes carrying 15 AGFs, of which 8 were significantly enriched. The AHSMCs exosomes also powerfully stimulated HECs to grow, migrate, and form dense tubes and nodes in vitro. Interestingly, only six of these AGFs, i.e., GRO-α/CXCL1, GRO-β/CXCL2, GRO-γ/CXCL3, IL-8/CXCL8, IL-4, and IL-1α, were significantly enriched in the exosomes released from both AHSMCs and HDFs [16] cultured on C/HE-3D-SFnws. However, it is worth noting that the exosomes released from the two cell types also transported discrete amounts of 10 other AGFs, e.g., MCP-1 (Monocyte chemoattractant protein-1), VEGF-D, ANGPT-1/-2, Tie2 (ANGPT-1 receptor), Angiostatin, uPAR (CD87), MMP-1/-9 (Matrix Metalloproteinase-1/-9), and TIMP-1/-2.

Finally, another score of compounds—i.e., TGF-β1, IGF-1, PDGF-BB, GM-CSF, I-309 (CCL-1), IL-10, IL-1β, Endostatin, I-TAC (Interferon-inducible T cell Alpha Chemoattractant), Leptin, PLGF (Placental Growth Factor), RANTES (Regulated upon Activation, Normal T Cell Expressed and Presumably Secreted chemokine), TNF-α, TPO (Thrombopoietin), VEGF-A, MCP-2, MCP-4, and PECAM-1 (Platelet Endothelial Cell Adhesion Molecule-1)—were conveyed solely by the exosomes released from the C/HE-3D-SFnws-adhering AHSMCs (Figure S2) being undetectable in those released from their HDFs counterparts [16].

Altogether, these observations show that patterns of enriched or not enriched AGFs released via exosomes by SF-stuck AHSMCs and HDFs exhibit not only substrate-type- but also cell type-specificity. The latter could affect the characteristics of the neovascularization AHSMCs and HDFs, respectively induced—a topic well worth investigating further. In this regard, D'Amore and Smith [136] showed that, according to their origin in small or large vessels, HECs' responses to a set of growth factors exhibited quantitative and qualitative differences.

Altogether, our results strengthen the view that once grafted in vivo 3D-SFnws can decidedly advance their own vascularization by inducing the colonizing human fibroblasts and SMCs to release loads of enriched AGFs via exosomes. We posit that an alike mechanism operated when we grafted FA- and C/N-3D SFnws into the subcutaneous tissues of mice in vivo [10,11].

Altogether, our results stress the importance of the AGFs in protein form, transported by the exosomes released from 3D-SFnws-attached AHSMCs and from HDFs [16]. The role(s) of any angiogenic RNAs carried by exosomes from the same sources will be addressed by future studies. The crucial insights into the interactions between SFnws and nontumorigenic AHSMCs, HDFs, and HECs we brought to light bode well for prospective applications of SF-based properly structured scaffolds in human (and even veterinary) clinical settings.

**Supplementary Materials:** The following supporting information can be downloaded at: https://www.mdpi.com/article/10.3390/polym14040697/s1, Figure S1: Developed double-antibody array membranes and array maps of human phosphorylation signalling pathways. Figure S2: Developed double-antibody array membranes. Details on the trophic and angiogenic actions of the AGFs conveyed in increased amounts by the exosomes released from C/HE-3D-SFnws-stuck AHSMCs.

**Author Contributions:** All the authors contributed to design the study, to perform the experiments, and to write the manuscript. U.A., A.C. and I.D.P. conceived the experiments with the 3D-SF nonwovens. P.H., A.C. and I.D.P. performed the in vitro cell culture and biochemical experiments. P.H., A.C. and I.D.P. also collected the results under the supervision of U.A., J.W., Z.W., A.C. and I.D.P. U.A. collected relevant literature, statistically analysed the results, and critically edited the manuscript. All authors have read and agreed to the published version of the manuscript.

**Funding:** The annual allotment of the FUR 2019 from the Ministry of Italian University and Research (MUR) given to A.C. and I.D.P. funded this research work. No funding from private or commercial sources supported it. Peng Hu holds the triannual Doctorate Student fellowship No. 754345 from the INVITE project funded by the European Un-ion's Horizon 2020 Research and Innovation Program under the Marie Sklodowska-Curie Grant Agreement.

**Data Availability Statement:** The datasets of this study are available upon request to the corresponding authors.

**Acknowledgments:** The authors gratefully thank Giuliano Freddi for assessing the physical and chemical properties of the SF-based 3D nonwoven scaffolds.

**Conflicts of Interest:** The authors declare no conflict of interest.

## Abbreviations

| | |
|---|---|
| 3D-SFnws | three-dimensional silk fibroin nonwovens |
| AGFs | angiogenic and growth factors |
| AHSMCs | adult human smooth muscle cells |
| Akt | protein kinase B or PKB |
| ANGPT-1/2 | angiopoietin-1/2 |
| ANOVA | analysis of variance |
| ATM | ataxia-telangiectasia mutated Ser/Thr kinase |
| bFGF | basic Fibroblast Growth factor |
| c-Fos | component of AP1 transcription factors |
| c-Jun | component of AP1 transcription factors |
| CREB | cyclic AMP response element binding protein |
| C/HE | carding/hydroentanglement |
| C/N | carding/needling |
| DiOC18(3) | (3,3′-Dioctadecyl oxacarbocyanine perchlorate or DiO |
| (ds)DNA | double strand DNA |
| ECM | extracellular matrix |
| ECs | endothelial cells |
| eIF-2$\alpha$ | eukaryotic translation initiation factor-2$\alpha$ |
| ELISA | enzyme-linked immunosorbent assay |
| FA | formic acid |
| FBS | foetal bovine serum |
| GM-CFS | granulocyte macrophage-colony stimulating factor |
| GRO-$\alpha$/-$\beta$/-$\gamma$ | growth-regulated oncogene-$\alpha$/-$\beta$/-$\gamma$ (or CXCL1/2/3) |
| HADAC | histone deacetylase |
| HDFs | human dermal fibroblasts |
| HECs | human ECs |
| IKK | IkappaB kinase |
| I$\kappa$B$\alpha$ | nuclear factor of kappa light polypeptide gene enhancer in B-cells inhibitor, alpha |
| IL- | interleukin- |
| MAPK | mitogen-activated protein kinase |
| MCP-1 | monocyte chemoattractant protein-1 (or CCL2) |
| MMP | matrix metalloprotease |
| MSK-1 | mitogen, and stress-activated protein kinase |
| NF-$\kappa$B | nuclear factor $\kappa$B |
| rh | recombinant human |
| SD | standard deviation |
| SF | silk fibroin |
| SMAD | small mother against decapentaplegic homolog |
| SMCs | smooth muscle cells |
| TAK-1 | TGF-$\beta$-activated kinase-1 |
| TCF/LEF | T-cell factor/Lymphoid enhancer factor |
| TGF-$\beta$ | transforming growth factor-beta |
| Tie-2 | ANGPT-1 receptor |
| TIMP-1/2 | tissue inhibitor of metalloproteinases-1/2 |
| TLR | Toll-like receptor |
| uPA | urokinase-like plasminogen activator |
| uPAR | uPA surface receptor |
| VEGF | vascular endothelial growth factor |
| VEGF-R | VEGF receptor |

# References


1. Guo, C.; Zhang, J.; Jordan, J.S.; Wang, X.; Henning, R.W.; Yarger, J.L. Structural Comparison of Various Silkworm Silks: An Insight into the Structure-Property Relationship. *Biomacromolecules* **2018**, *19*, 906–917. [CrossRef]
2. Zhou, C.Z.; Confalonieri, F.; Jacquet, M.; Perasso, R.; Li, Z.G.; Janin, J. Silk fibroin: Structural implications of a remarkable amino acid sequence. *Proteins* **2001**, *44*, 119–122. [CrossRef]
3. Sun, W.; Gregory, D.A.; Tomeh, M.A.; Zhao, X. Silk Fibroin as a Functional Biomaterial for Tissue Engineering. *Int. J. Mol. Sci.* **2021**, *22*, 1499. [CrossRef]
4. Kaewpirom, S.; Boonsang, S. Influence of alcohol treatments on properties of silk-fibroin-based films for highly optocally transparent coating applications. *RSC Adv.* **2020**, *10*, 15913–15923. [CrossRef]
5. Rockwood, D.N.; Preda, R.C.; Yücel, T.; Wang, X.; Lovett, M.L.; Kaplan, D.L. Materials fabrication from Bombyx mori silk fibroin. *Nat. Protoc.* **2011**, *6*, 1612–1631. [CrossRef]
6. Li, Y.; Yang, Y.; Yang, L.; Zeng, Y.; Gao, X.; Xu, H. Poly(ethylene glycol)-modified silk fibroin membrane as a carrier for limbal epithelial stem cell transplantation in a rabbit LSCD model. *Stem Cell Res. Ther.* **2017**, *8*, 256. [CrossRef]
7. Cheng, G.; Davoudi, Z.; Xing, X.; Yu, X.; Cheng, X.; Li, Z.; Deng, H.; Wang, Q. Advanced Silk Fibroin Biomaterials for Cartilage Regeneration. *ACS Biomater. Sci. Eng.* **2018**, *4*, 2704–2715. [CrossRef]
8. Zhang, Y.; Lu, L.; Chen, Y.; Wang, J.; Chen, Y.; Mao, C.; Yang, M. Polydopamine modification of silk fibroin membranes significantly promotes their wound healing effect. *Biomater. Sci.* **2019**, *7*, 5232–5237. [CrossRef]
9. Li, C.; Guo, C.; Fitzpatrick, V.; Ibrahim, A.; Zwierstra, M.J.; Hanna, P.; Lechtig, A.; Nazarian, A.; Lin, S.J.; Kaplan, D.L. Design of biodegradable, implantable devices towards clinical translation. *Nat. Rev. Mater.* **2020**, *5*, 61–81. [CrossRef]
10. Dal Prà, I.; Freddi, G.; Minic, J.; Chiarini, A.; Armato, U. De novo engineering of reticular connective tissue in vivo by silk fibroin nonwoven materials. *Biomaterials* **2005**, *26*, 1987–1999. [CrossRef]
11. Chiarini, A.; Freddi, G.; Liu, D.; Armato, U.; Dal Prà, I. Biocompatible silk noil-based three-dimensional carded-needled nonwoven scaffolds guide the engineering of novel skin connective tissue. *Tissue Eng. Part A* **2016**, *22*, 1047–1060. [CrossRef]
12. Armato, U.; Dal Pra, I.; Chiarini, A.; Freddi, G. Will silk fibroin nanofiber scaffolds ever hold a useful place in Translational Regenerative Medicine? *Int. J. Burns Trauma* **2011**, *1*, 27–33.
13. Holland, C.; Numata, K.; Rnjak-Kovacina, J.; Seib, F.P. The biomedical use of silk: Past, present, future. *Adv. Healthc. Mater.* **2019**, *8*, 1800465. [CrossRef]
14. ISO and CEN Definition of Nonwovens. Available online: https://www.edana.org/docs/default-source/edana-nonwovens/iso-and-cen-definition-of-nonwovens.pdf?sfvrsn=21822973_221 (accessed on 10 January 2022).
15. Dal Prà, I.; Chiarini, A.; Boschi, A.; Freddi, G.; Armato, U. Novel dermo-epidermal equivalents on silk fibroin-based formic acid-crosslinked three-dimensional nonwoven devices with prospective applications in human tissue engineering/regeneration/repair. *Int. J. Mol. Med.* **2006**, *18*, 241–247. [CrossRef]
16. Hu, P.; Chiarini, A.; Wu, J.; Freddi, G.; Nie, K.; Armato, U.; Dal Prà, I. Exosomes of adult human fibroblasts cultured on 3D silk fibroin nonwovens intensely stimulate neoangiogenesis. *Burns Trauma* **2021**, *9*, tkab003. [CrossRef]
17. Shabbir, A.; Cox, A.; Rodriguez-Menocal, L.; Salgado, M.; Van Badiavas, E. Mesenchymal stem cell exosomes induce proliferation and migration of normal and chronic wound fibroblasts and enhance angiogenesis in vitro. *Stem Cells Dev.* **2015**, *24*, 1635–1647. [CrossRef]
18. Baruah, J.; Wary, K.K. Exosomes in the Regulation of Vascular Endothelial Cell Regeneration. *Front. Cell. Dev. Biol.* **2020**, *7*, 353. [CrossRef]
19. Ribeiro, M.F.; Zhu, H.; Millard, R.W.; Fan, G.C. Exosomes Function in Pro- and Anti-Angiogenesis. *Curr. Angiogenes* **2013**, *2*, 54–59. [CrossRef]
20. Boulanger, C.M.; Loyer, X.; Rautou, P.E.; Amabile, N. Extracellular vesicles in coronary artery disease. *Nat. Rev Cardiol.* **2017**, *14*, 259–272. [CrossRef]
21. Théry, C.; Witwer, K.W.; Aikawa, E.; Alcaraz, M.J.; Anderson, J.D.; Andriantsitohaina, R.; Antoniou, A.; Arab, T.; Archer, F.; Atkin-Smith, G.K. Minimal information for studies of extracellular vesicles 2018 (MISEV2018): A position statement of the International Society for Extracellular Vesicles and update of the MISEV2014 guidelines. *J. Extracell. Vesicles* **2018**, *7*, 1535750. [CrossRef]
22. Caruso, S.; Poon, I. Apoptotic Cell-Derived Extracellular Vesicles: More Than Just Debris. *Front. Immunol.* **2018**, *9*, 1486. [CrossRef]
23. Tokarz, A.; Szuścik, I.; Kuśnierz-Cabala, B.; Kapusta, M.; Konkolewska, M.; Żurakowski, A.; Georgescu, A.; Stępień, E. Extracellular vesicles participate in the transport of cytokines and angiogenic factors in diabetic patients with ocular complications. *Folia Med. Cracov.* **2015**, *55*, 35–48.
24. Yan, L.; Wu, X. Exosomes produced from 3D cultures of umbilical cord mesenchymal stem cells in a hollow-fiber bioreactor show improved osteochondral regeneration activity. *Cell Biol. Toxicol.* **2020**, *36*, 165–178. [CrossRef]
25. Gurunathan, S.; Kang, M.H.; Jeyaraj, M.; Qasim, M.; Kim, J.H. Review of the isolation, characterization, biological function, and multifarious therapeutic approaches of exosomes. *Cells* **2019**, *8*, 307. [CrossRef]
26. Hu, P.; Yang, Q.; Wang, Q.; Shi, C.; Wang, D.; Armato, U.; Dal Prà, I.; Chiarini, A. Mesenchymal stromal cells-exosomes: A promising cell free therapeutic tool for wound healing and cutaneous regeneration. *Burns Trauma* **2019**, *7*, 38. [CrossRef]
27. Barnes, B.J.; Somerville, C.C. Modulating cytokine production via select packaging and secretion from extracellular vesicles. *Front. Immunol.* **2020**, *11*, 1040. [CrossRef]

28. Mikawa, T.; Gourdie, R.G. Pericardial mesoderm generates a population of coronary smooth muscle cells migrating into the heart along with ingrowth of the epicardial organ. *Dev. Biol.* **1996**, *174*, 221–232. [CrossRef]

29. Gadson, P.F., Jr.; Dalton, M.L.; Patterson, E.; Svoboda, D.D.; Hutchinson, L.; Schram, D.; Rosenquist, T.H. Differential response of mesoderm- and neural crest-derived smooth muscle to TGF-β1: Regulation of c-myb and α1 (I) procollagen genes. *Exp. Cell Res.* **1997**, *230*, 169–180. [CrossRef]

30. Topouzis, S.; Majesky, M.W. Smooth muscle lineage diversity in the chick embryo: Two types of aortic smooth muscle cell differ in growth and receptor-mediated transcriptional responses to transforming growth factor-β. *Dev. Biol.* **1996**, *178*, 430–445. [CrossRef]

31. Majesky, M.W. Developmental basis of vascular smooth muscle diversity. *Arterioscler. Thromb. Vasc. Biol.* **2007**, *27*, 1248–1258. [CrossRef]

32. Sinha, S.; Iyer, D.; Granata, A. Embryonic origins of human vascular smooth muscle cells: Implications for in vitro modeling and clinical application. *Cell Mol. Life Sci.* **2014**, *71*, 2271–2288. [CrossRef]

33. DeRuiter, M.C.; Poelmann, R.E.; VanMunsteren, J.C.; Mironov, V.; Markwald, R.R.; Gittenberger-de Groot, A.C. Embryonic endothelial cells transdifferentiate into mesenchymal cells expressing smooth muscle actins in vivo and in vitro. *Circ. Res.* **1997**, *80*, 444–451. [CrossRef]

34. Iyemere, V.P.; Proudfoot, D.; Weissberg, P.L.; Shanahan, C.M. Vascular smooth muscle cell phenotypic plasticity and the regulation of vascular calcification. *J. Intern. Med.* **2006**, *260*, 192–210. [CrossRef]

35. Olson, L.E.; Soriano, P. PDGFRβ signaling regulates mural cell plasticity and inhibits fat development. *Dev. Cell.* **2011**, *20*, 815–826. [CrossRef]

36. Rensen, S.S.; Doevendans, P.A.; van Eys, G.J. Regulation and characteristics of vascular smooth muscle cell phenotypic diversity. *Neth. Heart J.* **2007**, *15*, 100–108. [CrossRef]

37. Bargehr, J.; Low, L.; Cheung, C.; Bernard, W.G.; Iyer, D.; Bennett, M.R.; Gambardella, L.; Sinha, S. Embryological Origin of Human Smooth Muscle Cells Influences Their Ability to Support Endothelial Network Formation. *Stem Cells Transl. Med.* **2016**, *5*, 946–959. [CrossRef]

38. Alessandrino, A.; Chiarini, A.; Biagiotti, M.; Dal Prà, I.; Bassani, G.A.; Vincoli, V.; Settembrini, P.; Pierimarchi, P.; Freddi, G.; Armato, U. Three-Layered Silk Fibroin Tubular Scaffold for the Repair and Regeneration of Small Caliber Blood Vessels: From Design to in vivo Pilot Tests. *Front. Bioeng. Biotechnol.* **2019**, *7*, 356. [CrossRef]

39. Tanaka, K.; Fukuda, D.; Higashikuni, Y.; Hirata, Y.; Komuro, I.; Saotome, T.; Yamashita, Y.; Asakura, T.; Sata, M. Biodegradable Extremely-Small-Diameter Vascular Graft Made of Silk Fibroin can be Implanted in Mice. *J. Atheroscler Thromb.* **2020**, *27*, 1299–1309. [CrossRef]

40. Gays, D.; Hess, C.; Camporeale, A.; Ala, U.; Provero, P.; Mosimann, C.; Santoro, M.M. An exclusive cellular and molecular network governs intestinal smooth muscle cell differentiation in vertebrates. *Development* **2017**, *144*, 464–478. [CrossRef]

41. Yap, H.M.; Israf, D.A.; Harith, H.H.; Tham, C.L.; Sulaiman, M.R. Crosstalk between Signaling Pathways Involved in the Regulation of Airway Smooth Muscle Cell Hyperplasia. *Front. Pharmacol.* **2019**, *10*, 1148. [CrossRef]

42. Oszajca, K.; Szemraj, J. Assessment of the correlation between oxidative stress and expression of *MMP-2*, *TIMP-1* and *COX-2* in human aortic smooth muscle cells. *Arch. Med. Sci. Atheroscler. Dis.* **2021**, *6*, e158–e165. [CrossRef]

43. Clifford, R.L.; John, A.E.; Brightling, C.E.; Knox, A.J. Abnormal histone methylation is responsible for increased vascular endothelial growth factor 165a secretion from airway smooth muscle cells in asthma. *J. Immunol.* **2012**, *189*, 819–831. [CrossRef]

44. Clauser, S.; Meilhac, O.; Bièche, I.; Raynal, P.; Bruneval, P.; Michel, J.B.; Borgel, D. Increased secretion of Gas6 by smooth muscle cells in human atherosclerotic carotid plaques. *Thromb. Haemost.* **2012**, *107*, 140–149. [CrossRef]

45. Alagappan, V.K.; McKay, S.; Widyastuti, A.; Garrelds, I.M.; Bogers, A.J.; Hoogsteden, H.C.; Hirst, S.J.; Sharma, H.S. Proinflammatory cytokines upregulate mRNA expression and secretion of vascular endothelial growth factor in cultured human airway smooth muscle cells. *Cell Biochem. Biophys.* **2005**, *43*, 119–129. [CrossRef]

46. Feng, P.H.; Hsiung, T.C.; Kuo, H.P.; Huang, C.D. Cross-talk between bradykinin and epidermal growth factor in regulating IL-6 production in human airway smooth muscle cells. *Chang Gung Med. J.* **2010**, *33*, 92–99.

47. Seol, H.J.; Oh, M.J.; Kim, H.J. Endothelin-1 expression by vascular endothelial growth factor in human umbilical vein endothelial cells and aortic smooth muscle cells. *Hypertens. Pregnancy* **2011**, *30*, 295–301. [CrossRef]

48. Comelli, L.; Rocchiccioli, S.; Smirni, S.; Salvetti, A.; Signore, G.; Citti, L.; Trivella, M.G.; Cecchettini, A. Characterization of secreted vesicles from vascular smooth muscle cells. *Mol. Biosyst.* **2014**, *10*, 1146–1152. [CrossRef]

49. Neradil, J.; Kyr, M.; Polaskova, K.; Kren, L.; Macigova, P.; Skoda, J.; Sterba, J.; Veselska, R. Phospho-Protein Arrays as Effective Tools for Screening Possible Targets for Kinase Inhibitors and Their Use in Precision Pediatric Oncology. *Front. Oncol.* **2019**, *9*, 930. [CrossRef]

50. Patel, G.K.; Khan, M.A.; Zubair, H.; Srivastava, S.K.; Khushman, M.; Singh, S.; Singh, A.P. Comparative analysis of exosome isolation methods using culture supernatant for optimum yield, purity and downstream applications. *Sci. Rep.* **2019**, *9*, 5335. [CrossRef]

51. Kapustin, A.N.; Chatrou, M.L.; Drozdov, I.; Zheng, Y.; Davidson, S.M.; Soong, D.; Furmanik, M.; Sanchis, P.; De Rosales, R.T.; Alvarez-Hernandez, D.; et al. Vascular smooth muscle cell calcification is mediated by regulated exosome secretion. *Circ. Res.* **2015**, *116*, 1312–1323. [CrossRef]

52. Nakayama, G.R.; Caton, M.C.; Nova, M.P.; Parandoosh, Z. Assessment of the Alamar Blue Assay for Cellular Growth and Viability In Vitro. *J. Immunol. Methods* **1997**, *204*, 205–208. [CrossRef]
53. Armato, U.; Romano, F.; Andreis, P.G.; Paccagnella, L.; Marchesini, C. Growth stimulation and apoptosis induced in cultures of neonatal rat liver cells by repeated exposures to epidermal growth factor/urogastrone with or without associated pancreatic hormones. *Cell Tissue Res.* **1986**, *245*, 471–480. [CrossRef] [PubMed]
54. Holzapfel, G.A. Biomechanics of soft tissue. In *Handbook of Materials Behavior Models*; Lemaitre, J., Ed.; Section 10.11; Academic Press: Cambridge, MS, USA, 2001; pp. 1057–1071. ISBN 978-0-12-443341-0.
55. Wang, Y.; Rudym, D.D.; Walsh, A.; Abrahamsen, L.; Kim, H.J.; Kim, H.S.; Kirker-Head, C.; Kaplan, D.L. In vivo degradation of three-dimensional silk fibroin scaffolds. *Biomaterials* **2008**, *29*, 3415–3428. [CrossRef] [PubMed]
56. Thurber, A.E.; Omenetto, F.G.; Kaplan, D.L. In vivo bioresponses to silk proteins. *Biomaterials* **2015**, *71*, 145–157. [CrossRef]
57. Huang, R.; Jiang, W.; Yang, J.; Mao, Y.Q.; Zhang, Y.; Yang, W.; Yang, D.; Burkholder, B.; Huang, R.F.; Huang, R.P. A biotin label-based antibody array for high-content profiling of protein expression. *Cancer Genom. Proteom.* **2010**, *7*, 129–141.
58. Fahey, E.; Doyle, S.L. IL-1 Family Cytokine Regulation of Vascular Permeability and Angiogenesis. *Front. Immunol.* **2019**, *10*, 1426. [CrossRef]
59. Raines, E.W.; Dower, S.K.; Ross, R. Interleukin-1 mitogenic activity for fibroblasts and smooth muscle cells is due to PDGF-AA. *Science* **1989**, *243*, 393–396. [CrossRef]
60. Gay, C.G.; Winkles, J.A. Interleukin 1 regulates heparin-binding growth factor 2 gene expression in vascular smooth muscle cells. *Proc. Natl. Acad. Sci. USA* **1991**, *88*, 296–300. [CrossRef]
61. Beasley, D.; Cooper, A.L. Constitutive expression of interleukin-1α precursor promotes human vascular smooth muscle cell proliferation. *Am. J. Physiol. Heart Circ. Physiol.* **1999**, *276*, H901–H912. [CrossRef]
62. Downie, G.H.; Ryan, U.S.; Hayes, B.A.; Friedman, M. Interleukin-2 directly increases albumin permeability of bovine and human vascular endothelium in vitro. *Am. J. Respir. Cell Mol. Biol.* **1992**, *7*, 58–65. [CrossRef]
63. Bae, J.; Park, D.; Lee, Y.S.; Young, D. Interleukin-2 promotes angiogenesis by activation of Akt and increase of ROS. *J. Microbiol. Biotechnol.* **2008**, *18*, 377–382. [PubMed]
64. Nabata, T.; Fukuo, K.; Morimoto, S.; Kitano, S.; Momose, N.; Hirotani, A.; Nakahashi, T.; Nishibe, A.; Hata, S.; Niinobu, T.; et al. Interleukin-2 modulates the responsiveness to angiotensin II in cultured vascular smooth muscle cells. *Atherosclerosis* **1997**, *133*, 23–30. [CrossRef]
65. Schultz, K.; Murthy, V.; Tatro, J.B.; Beasley, D. Endogenous interleukin-1α promotes a proliferative and proinflammatory phenotype in human vascular smooth muscle cells. *Am. J. Physiol. Heart Circ. Physiol.* **2007**, *292*, H2927–H2934. [CrossRef] [PubMed]
66. Lee, Y.W.; Eum, S.Y.; Chen, K.C.; Hennig, B.; Toborek, M. Gene expression profile in interleukin-4-stimulated human vascular endothelial cells. *Mol. Med.* **2004**, *10*, 19–27. [CrossRef]
67. Skaria, T.; Burgener, J.; Bachli, E.; Schoedon, G. IL-4 Causes Hyperpermeability of Vascular Endothelial Cells through Wnt5A Signaling. *PLoS ONE* **2016**, *11*, e0156002. [CrossRef]
68. Klein, N.J.; Rigley, K.P.; Callard, R.E. IL-4 regulates the morphology, cytoskeleton, and proliferation of human umbilical vein endothelial cells: Relationship between vimentin and CD23. *Int. Immunol.* **1993**, *5*, 293. [CrossRef]
69. Toi, M.; Harris, A.L.; Bicknell, R. Interleukin-4 is a potent mitogen for capillary endothelium. *Biochem. Biophys. Res. Commun.* **1991**, *174*, 1287–1293. [CrossRef]
70. Fukushi, J.; Morisaki, T.; Shono, T.; Nishie, A.; Torisu, H.; Ono, M.; Kuwano, M. Novel biological functions of interleukin-4: Formation of tube-like structures by vascular endothelial cells in vitro and angiogenesis in vivo. *Biochem. Biophys. Res. Commun.* **1998**, *250*, 444–448. [CrossRef]
71. Ikeda, U.; Ikeda, M.; Oohara, T.; Oguchi, A.; Kamitani, T.; Tsuruya, Y.; Kano, S. Interleukin 6 stimulates growth of vascular smooth muscle cells in a PDGF-dependent manner. *Am. J. Physiol.* **1991**, *260*, H1713–H1717. [CrossRef]
72. Ljungberg, L.U.; Zegeye, M.M.; Kardeby, C.; Fälker, K.; Repsilber, D.; Sirsjö, A. Global Transcriptional Profiling Reveals Novel Autocrine Functions of Interleukin 6 in Human Vascular Endothelial Cells. *Mediators Inflamm.* **2020**, *2020*, 4623107. [CrossRef]
73. Keeley, E.C.; Mehrad, B.; Strieter, R.M. Chemokines as mediators of neovascularization. *Arterioscler. Thromb. Vasc. Biol.* **2008**, *28*, 1928–1936. [CrossRef] [PubMed]
74. Strieter, R.M.; Polverini, P.J.; Kunkel, S.L.; Arenberg, D.A.; Burdick, M.D.; Kasper, J.; Dzuiba, J.; Van Damme, J.; Walz, A.; Marriott, D. The functional role of the ELR motif in CXC chemokine-mediated angiogenesis. *J. Biol. Chem.* **1995**, *270*, 27348–27357. [CrossRef] [PubMed]
75. Presta, M.; Dell'Era, P.; Mitola, S.; Moroni, E.; Ronca, R.; Rusnati, M. Fibroblast growth factor/fibroblast growth factor receptor system in angiogenesis. *Cytokine Growth Factor Rev.* **2005**, *16*, 159–178. [CrossRef]
76. Tomanek, R.J.; Hansen, H.K.; Christensen, L.P. Temporally expressed PDGF and FGF-2 regulate embryonic coronary artery formation and growth. *Arterioscler. Thromb. Vasc. Biol.* **2008**, *28*, 1237–1243. [CrossRef] [PubMed]
77. Holnthoner, W.; Pillinger, M.; Groger, M.; Wolff, K.; Ashton, A.W.; Albanese, C.; Neumeister, P.; Pestell, R.G.; Petzelbauer, P. Fibroblast growth factor-2 induces Lef/Tcf-dependent transcription in human endothelial cells. *J. Biol. Chem.* **2002**, *277*, 45847–45853. [CrossRef] [PubMed]
78. Wang, X.; Xiao, Y.; Mou, Y.; Zhao, Y.; Blankesteijn, W.M.; Hall, J.L. A role for the β-catenin/T-cell factor signaling cascade in vascular remodeling. *Circ. Res.* **2002**, *90*, 340–347. [CrossRef]

79. Krubasik, D.; Eisenach, P.A.; Kunz-Schughart, L.A. Granulocyte-macrophage colony stimulating factor induces endothelial capillary formation through induction of membrane-type 1 matrix metalloproteinase expression in vitro. *Int. J. Cancer* **2008**, *122*, 1261–1272. [CrossRef]
80. Bussolino, F.; Ziche, M.; Wang, J.M.; Alessi, D.; Morbidelli, L.; Cremona, O.; Bosia, A.; Marchisio, P.C.; Mantovani, A. In vitro and in vivo activation of endothelial cells by colony-stimulating factors. *J. Clin. Investig.* **1991**, *87*, 986–995. [CrossRef]
81. Buschmann, I.R.; Busch, H.J.; Mies, G.; Hossmann, K.A. Therapeutic induction of arteriogenesis in hypoperfused rat brain via granulocyte-macrophage colony-stimulating factor. *Circulation* **2003**, *108*, 610–615. [CrossRef]
82. Frismantiene, A.; Philippova, M.; Erne, P.; Resink, T.J. Smooth muscle cell-driven vascular diseases and molecular mechanisms of VSMC plasticity. *Cell Signal.* **2018**, *52*, 48–64. [CrossRef]
83. Bochaton-Piallat, M.L.; Ropraz, P.; Gabbiani, F.; Gabbiani, G. Phenotypic heterogeneity of rat arterial smooth muscle cell clones: Implications for the development of experimental intimal thickening. *Arterioscler. Thromb. Vasc. Biol.* **1996**, *16*, 815–820. [CrossRef] [PubMed]
84. Bochaton-Piallat, M.L.; Gabbiani, F.; Redard, M.; Desmoulière, A.; Gabbiani, G. Apoptosis takes part in cellularity regulation during rat intimal thickening. *Am. J. Pathol.* **1995**, *146*, 1059. [PubMed]
85. Slomp, J.; Gittenberger-de Groot, A.C.; Glukhova, M.A.; van Munsteren, J.C.; Kockx, M.M.; Schwartz, S.M.; Koteliansky, V.E. Differentiation, dedifferentiation, and apoptosis of smooth muscle cells during the development of the human ductus arteriosus. *Arterioscler. Thromb. Vasc. Biol.* **1997**, *17*, 1003–1009. [CrossRef] [PubMed]
86. Schwartz, S.M.; Majesky, M.W.; Murry, C.E. The intima: Development and monoclonal responses to injury. *Atherosclerosis* **1995**, *118*, S125–S140. [CrossRef]
87. Campbell, J.H.; Campbell, G.R. Smooth muscle phenotypic modulation—A personal experience. *Arterioscler. Thromb. Vasc. Biol.* **2012**, *32*, 1784–1789. [CrossRef] [PubMed]
88. Chamley-Campbell, J.; Campbell, G.R.; Ross, R. The smooth muscle cell in culture. *Physiol. Rev.* **1979**, *59*, 1–61. [CrossRef] [PubMed]
89. Liu, M.; Gomez, D. Smooth Muscle Cell Phenotypic Diversity. *Arterioscler. Thromb. Vasc. Biol.* **2019**, *39*, 1715–1723. [CrossRef] [PubMed]
90. Getz, G.S.; Reardon, C.A. Animal models of atherosclerosis. *Arterioscler. Thromb. Vasc. Biol.* **2012**, *32*, 1104–1115. [CrossRef] [PubMed]
91. Bruemmer, D.; Daugherty, A.; Lu, H.; Rateri, D.L. Relevance of angiotensin II-induced aortic pathologies in mice to human aortic aneurysms. *Ann. N. Y. Acad. Sci.* **2011**, *1245*, 7–10. [CrossRef]
92. Chiarini, A.; Petrini, P.; Bozzini, S.; Dal Pra, I.; Armato, U. Silk fibroin/poly(carbonate)-urethane as a substrate for cell growth: In vitro interactions with human cells. *Biomaterials* **2003**, *24*, 789–799. [CrossRef]
93. Tan, J.Y.; Wen, J.C.; Shi, W.H.; He, Q.; Zhu, L.; Liang, K.; Shao, Z.Z.; Yu, B. Effect of microtopographic structures of silk fibroin on endothelial cell behavior. *Mol. Med. Rep.* **2013**, *7*, 292–298. [CrossRef] [PubMed]
94. Roca-Cusachs, P.; Alcaraz, J.; Sunyer, R.; Samitier, J.; Farré, R.; Navajas, D. Micropatterning of single endothelial cell shape reveals a tight coupling between nuclear volume in G1 and proliferation. *Biophys. J.* **2008**, *94*, 4984–4995. [CrossRef] [PubMed]
95. Beamish, J.A.; He, P.; Kottke-Marchant, K.; Marchant, R.E. Molecular regulation of contractile smooth muscle cell phenotype: Implications for vascular tissue engineering. *Tissue Eng. Part B Rev.* **2010**, *16*, 467–491. [CrossRef] [PubMed]
96. Gupta, P.; Moses, J.C.; Mandal, B.B. Surface Patterning and Innate Physicochemical Attributes of Silk Films Concomitantly Govern Vascular Cell Dynamics. *ACS Biomater. Sci. Eng.* **2019**, *5*, 933–949. [CrossRef]
97. Hata, A.; Chen, Y.G. TGF-β Signaling from Receptors to Smads. *Cold Spring Harb. Perspect. Biol.* **2016**, *18*, a022061. [CrossRef] [PubMed]
98. Tzavlaki, K.; Moustakas, A. TGF-β Signaling. *Biomolecules* **2020**, *10*, 487. [CrossRef]
99. Xu, Y.R.; Lei, C.Q. TAK1-TABs Complex: A Central Signalosome in Inflammatory Responses. *Front. Immunol.* **2021**, *11*, 608976. [CrossRef]
100. Ouyang, C.; Nie, L.; Gu, M.; Wu, A.; Han, X.; Wang, X.; Shao, J.; Xia, Z. Transforming growth factor (TGF)-β-activated kinase 1 (TAK1) activation requires phosphorylation of serine 412 by protein kinase A catalytic subunit α (PKACalpha) and X-linked protein kinase (PRKX). *J. Biol. Chem.* **2014**, *289*, 24226–24237. [CrossRef] [PubMed]
101. Molnar, P.; Perrault, R.; Louis, S.; Zahradka, P. The cyclic AMP response element-binding protein (CREB) mediates smooth muscle cell proliferation in response to angiotensin II. *J. Cell Commun. Signal.* **2014**, *8*, 29–37. [CrossRef]
102. Daly, A.C.; Randall, R.A.; Hill, C.S. Transforming growth factor β-induced Smad1/5 phosphorylation in epithelial cells is mediated by novel receptor complexes and is essential for anchorage-independent growth. *Mol. Cell. Biol.* **2008**, *28*, 6889–6902. [CrossRef]
103. Roelen, B.A.; Cohen, O.S.; Raychowdhury, M.K.; Chadee, D.N.; Zhang, Y.; Kyriakis, J.M.; Alessandrini, A.A.; Lin, H.Y. Phosphorylation of threonine 276 in Smad4 is involved in transforming growth factor-β-induced nuclear accumulation. *Am. J. Physiol. Cell Physiol.* **2003**, *285*, C823–C830. [CrossRef] [PubMed]
104. Jäger, M.A.; De La Torre, C.; Arnold, C.; Kohlhaas, J.; Kappert, L.; Hecker, M.; Feldner, A.; Korff, T. Assembly of vascular smooth muscle cells in 3D aggregates provokes cellular quiescence. *Exp. Cell. Res.* **2020**, *388*, 111782. [CrossRef] [PubMed]
105. Sano, Y.; Harada, J.; Tashiro, S.; Gotoh-Mandeville, R.; Maekawa, T.; Ishii, S. ATF-2 is a common nuclear target of Smad and TAK1 pathways in transforming growth factor-β signaling. *J. Biol. Chem.* **1999**, *274*, 8949–8957. [CrossRef]

106. Verrecchia, F.; Tacheau, C.; Schorpp-Kistner, M.; Angel, P.; Mauviel, A. Induction of the AP-1 members c-Jun and JunB by TGF-β/Smad suppresses early Smad-driven gene activation. *Oncogene* **2001**, *20*, 2205–2211. [CrossRef] [PubMed]
107. Hayden, M.S.; Ghosh, S. Shared principles in NF-κB signaling. *Cell* **2008**, *132*, 344–362. [CrossRef]
108. Shim, J.H.; Xiao, C.; Paschal, A.E.; Bailey, S.T.; Rao, P.; Hayden, M.S.; Lee, K.Y.; Bussey, C.; Steckel, M.; Tanaka, N.; et al. TAK1, but not TAB1 or TAB2, plays an essential role in multiple signaling pathways in vivo. *Genes Dev.* **2005**, *19*, 2668–2681. [CrossRef]
109. Takaesu, G.; Surabhi, R.M.; Park, K.J.; Ninomiya-Tsuji, J.; Matsumoto, K.; Gaynor, R.B. TAK1 is critical for IκB kinase-mediated activation of the NF-κB pathway. *J. Mol. Biol.* **2003**, *326*, 105–115. [CrossRef]
110. Dai, L.; Aye Thu, C.; Liu, X.Y.; Xi, J.; Cheung, P.C. TAK1, more than just innate immunity. *IUBMB Life* **2012**, *64*, 825–834. [CrossRef]
111. Schmid, J.A.; Birbach, A. IκB kinase β (IKKβ/IKK2/IKBKB)—A key molecule in signaling to the transcription factor NF-κB. *Cytokine Growth Factor Rev.* **2008**, *19*, 157–165. [CrossRef]
112. Perkins, N.D.; Gilmore, T.D. Good cop, bad cop: The different faces of NF-κB. *Cell Death Differ.* **2006**, *13*, 759–772. [CrossRef]
113. Hayden, M.S.; West, A.P.; Ghosh, S. NF-κB and the immune response. *Oncogene* **2006**, *25*, 6758–6780. [CrossRef] [PubMed]
114. Zandi, E.; Rothwarf, D.M.; Delhase, M.; Hayakawa, M.; Karin, M. The IκB kinase complex (IKK) contains two kinase subunits, IKKα and IKKβ, necessary for IκB phosphorylation and NF-κB activation. *Cell* **1997**, *91*, 243–252. [CrossRef]
115. Jacobs, M.D.; Harrison, S.C. Structure of an IκBα/NF-κB complex. *Cell* **1998**, *95*, 749–758. [CrossRef]
116. Lingappan, K. NF-κB in Oxidative Stress. *Curr. Opin. Toxicol.* **2018**, *7*, 81–86. [CrossRef]
117. Chen, J.; Chen, Y.; Chen, Y.; Yang, Z.; You, B.; Ruan, Y.C.; Peng, Y. Epidermal CFTR suppresses MAPK/NF-κB to promote cutaneous wound healing. *Cell. Physiol. Biochem.* **2016**, *39*, 2262–2274. [CrossRef]
118. McKinnon, P.J. ATM and ataxia telangiectasia. *EMBO Rep.* **2004**, *5*, 772–776. [CrossRef]
119. Lee, J.H.; Paull, T.T. ATM activation by DNA double-strand breaks through the Mre11-Rad50-Nbs1 complex. *Science* **2005**, *308*, 551–554, Erratum in *Science* **2005**, *308*, 1870. [CrossRef]
120. Lee, J.H.; Paull, T.T. Activation and regulation of ATM kinase activity in response to DNA double-strand breaks. *Oncogene* **2007**, *26*, 7741–7748. [CrossRef]
121. Yang, C.; Tang, X.; Guo, X.; Niikura, Y.; Kitagawa, K.; Cui, K.; Wong, S.T.; Fu, L.; Xu, B. Aurora-B mediated ATM serine 1403 phosphorylation is required for mitotic ATM activation and the spindle checkpoint. *Mol. Cell.* **2011**, *44*, 597–608. [CrossRef]
122. Sun, M.; Guo, X.; Qian, X.; Wang, H.; Yang, C.; Brinkman, K.L.; Serrano-Gonzalez, M.; Jope, R.S.; Zhou, B.; Engler, D.A.; et al. Activation of the ATM-Snail pathway promotes breast cancer metastasis. *J. Mol. Cell Biol.* **2012**, *4*, 304–315. [CrossRef]
123. Alexander, A.; Cai, S.L.; Kim, J.; Nanez, A.; Sahin, M.; MacLean, K.H.; Inoki, K.; Guan, K.L.; Shen, J.; Person, M.D.; et al. ATM signals to TSC2 in the cytoplasm to regulate mTORC1 in response to ROS. *Proc. Natl. Acad. Sci. USA* **2010**, *107*, 4153–4158, Erratum in *Proc. Natl. Acad. Sci. USA* **2012**, *109*, 8352. [CrossRef] [PubMed]
124. Krüger, A.; Ralser, M. ATM is a redox sensor linking genome stability and carbon metabolism. *Sci. Signal.* **2011**, *4*, pe17. [CrossRef] [PubMed]
125. Cosentino, C.; Grieco, D.; Costanzo, V. ATM activates the pentose phosphate pathway promoting anti-oxidant defence and DNA repair. *EMBO J.* **2011**, *30*, 546–555. [CrossRef] [PubMed]
126. López-Carballo, G.; Moreno, L.; Masiá, S.; Pérez, P.; Barettino, D. Activation of the phosphatidylinositol 3-kinase/Akt signaling pathway by retinoic acid is required for neural differentiation of SH-SY5Y human neuroblastoma cells. *J. Biol. Chem.* **2002**, *277*, 25297–25304. [CrossRef]
127. Boehrs, J.K.; He, J.; Halaby, M.J.; Yang, D.Q. Constitutive expression and cytoplasmic compartmentalization of ATM protein in differentiated human neuron-like SH-SY5Y cells. *J. Neurochem.* **2007**, *100*, 337–345. [CrossRef]
128. Li, Y.; Xiong, H.; Yang, D.Q. Functional switching of ATM: Sensor of DNA damage in proliferating cells and mediator of Akt survival signal in post-mitotic human neuron-like cells. *Chin. J. Cancer* **2012**, *31*, 364–372. [CrossRef]
129. Antonelli-Orlidge, A.; Saunders, K.B.; Smith, S.R.; D'Amore, P.A. An activated form of transforming growth factor beta is produced by cocultures of endothelial cells and pericytes. *Proc. Natl. Acad. Sci. USA* **1989**, *86*, 4544–4548. [CrossRef]
130. Fillinger, M.F.; Sampson, L.N.; Cronenwett, J.L.; Powell, R.J.; Wagner, R.J. Coculture of endothelial cells and smooth muscle cells in bilayer and conditioned media models. *J. Surg. Res.* **1997**, *67*, 169–178. [CrossRef]
131. Mach, F.; Schönbeck, U.; Sukhova, G.K.; Bourcier, T.; Bonnefoy, J.Y.; Pober, J.S.; Libby, P. Functional CD40 ligand is expressed on human vascular endothelial cells, smooth muscle cells, and macrophages: Implications for CD40-CD40 ligand signaling in atherosclerosis. *Proc. Natl. Acad. Sci. USA* **1997**, *94*, 1931–1936. [CrossRef]
132. Hirschi, K.K.; Rohovsky, S.A.; Beck, L.H.; Smith, S.R.; D'Amore, P.A. Endothelial cells modulate the proliferation of mural cell precursors via platelet-derived growth factor-BB and heterotypic cell contact. *Circ. Res.* **1999**, *84*, 298–305. [CrossRef]
133. Tang, R.; Zhang, G.; Chen, S.Y. Smooth Muscle Cell Proangiogenic Phenotype Induced by Cyclopentenyl Cytosine Promotes Endothelial Cell Proliferation and Migration. *J. Biol. Chem.* **2016**, *291*, 26913–26921. [CrossRef] [PubMed]
134. Zhang, Y.; Chopp, M.; Zhang, Z.G.; Katakowski, M.; Xin, H.; Qu, C.; Ali, M.; Mahmood, A.; Xiong, Y. Systemic administration of cell-free exosomes generated by human bone marrow derived mesenchymal stem cells cultured under 2D and 3D conditions improves functional recovery in rats after traumatic brain injury. *Neurochem. Int.* **2017**, *111*, 69–81. [CrossRef] [PubMed]
135. Gerthoffer, W.T.; Singer, C.A. Secretory Functions of Smooth Muscle: Cytokines and Growth Factors. *Mol. Interv.* **2002**, *2*, 447–456. [CrossRef] [PubMed]
136. D'Amore, P.A.; Smith, S.R. Growth factor effects on cells of the vascular wall: A survey. *Growth Factors* **1993**, *8*, 61–75. [CrossRef]

*Article*

# Highly Porous Composite Scaffolds Endowed with Antibacterial Activity for Multifunctional Grafts in Bone Repair

Ana S. Neto [1], Patrícia Pereira [2,3], Ana C. Fonseca [2,*], Carla Dias [4], Mariana C. Almeida [4], Inês Barros [5,6,7], Catarina O. Miranda [5,6,7], Luís P. de Almeida [5,6,7,8,9], Paula V. Morais [4], Jorge F. J. Coelho [2] and José M. F. Ferreira [1,*]

1   Department of Materials and Ceramic Engineering/CICECO—Aveiro Institute of Materials, University of Aveiro, 3810-193 Aveiro, Portugal; sofia.neto@ua.pt
2   Department of Chemical Engineering, CEMMPRE, University of Coimbra, 3030-790 Coimbra, Portugal; ppereira@ipn.pt (P.P.); jcoelho@eq.uc.pt (J.F.J.C.)
3   IPN, Instituto Pedro Nunes, Associação para a Inovação e Desenvolvimento em Ciência Tecnologia, Rua Pedro Nunes, 3030-199 Coimbra, Portugal
4   Department of Life Sciences, CEMMPRE, University of Coimbra, 3001-401 Coimbra, Portugal; carla_spd@hotmail.com (C.D.); mcalmeida@uc.pt (M.C.A.); pvmorais@ci.uc.pt (P.V.M.)
5   CNC—Center for Neuroscience and Cell Biology, University of Coimbra, 3004-504 Coimbra, Portugal; ines.barros2095@gmail.com (I.B.); csmiranda@cnc.uc.pt (C.O.M.); luispa@cnc.uc.pt (L.P.d.A.)
6   CIBB—Center for Innovative Biomedicine and Biotechnology, University of Coimbra, 3004-504 Coimbra, Portugal
7   IIIUC—Institute for Interdisciplinary Research, University of Coimbra, 3030-789 Coimbra, Portugal
8   Faculty of Pharmacy, University of Coimbra, 3000-548 Coimbra, Portugal
9   Viravector—Viral Vector for Gene Transfer Core Facility, University of Coimbra, 3004-504 Coimbra, Portugal
*   Correspondence: anafs@eq.uc.pt (A.C.F.); jmf@ua.pt (J.M.F.F.)

**Citation:** Neto, A.S.; Pereira, P.; Fonseca, A.C.; Dias, C.; Almeida, M.C.; Barros, I.; Miranda, C.O.; de Almeida, L.P.; Morais, P.V.; Coelho, J.F.J.; et al. Highly Porous Composite Scaffolds Endowed with Antibacterial Activity for Multifunctional Grafts in Bone Repair. *Polymers* 2021, 13, 4378. https://doi.org/10.3390/polym13244378

Academic Editors: Andrada Serafim and Stefan Ioan Voicu

Received: 1 November 2021
Accepted: 10 December 2021
Published: 14 December 2021

**Publisher's Note:** MDPI stays neutral with regard to jurisdictional claims in published maps and institutional affiliations.

**Abstract:** The present study deals with the development of multifunctional biphasic calcium phosphate (BCP) scaffolds coated with biopolymers—poly($\varepsilon$-caprolactone) (PCL) or poly(ester urea) (PEU)—loaded with an antibiotic drug, Rifampicin (RFP). The amounts of RFP incorporated into the PCL and PEU-coated scaffolds were $0.55 \pm 0.04$ and $0.45 \pm 0.02$ wt%, respectively. The in vitro drug release profiles in phosphate buffered saline over 6 days were characterized by a burst release within the first 8h, followed by a sustained release. The Korsmeyer–Peppas model showed that RFP release was controlled by polymer-specific non-Fickian diffusion. A faster burst release ($67.33 \pm 1.48\%$) was observed for the PCL-coated samples, in comparison to that measured ($47.23 \pm 0.31\%$) for the PEU-coated samples. The growth inhibitory activity against *Escherichia coli* and *Staphylococcus aureus* was evaluated. Although the RFP-loaded scaffolds were effective in reducing bacterial growth for both strains, their effectiveness depends on the particular bacterial strain, as well as on the type of polymer coating, since it rules the drug release behavior. The low antibacterial activity demonstrated by the BCP-PEU-RFP scaffold against *E. coli* could be a consequence of the lower amount of RFP that is released from this scaffold, when compared with BCP-PCL-RFP. In vitro studies showed excellent cytocompatibility, adherence, and proliferation of human mesenchymal stem cells on the BCP-PEU-RFP scaffold surface. The fabricated highly porous scaffolds that could act as an antibiotic delivery system have great potential for applications in bone regeneration and tissue engineering, while preventing bacterial infections.

**Keywords:** cuttlefish bone; biphasic calcium phosphate; polymeric coatings; rifampicin; drug delivery system

## 1. Introduction

Bacterial infections are one of the main problems associated with the implantation of conventional medical devices and can lead to increases in patient morbidity and mortality [1]. Bacterial infections are equally seen in tissue engineering approaches using scaffolds [2]. Irreversible adhesion of microorganisms creates a biofilm that protects bacteria from phagocytosis and antibiotics [3,4]. Thus, inhibiting irreversible bacterial adhesion is one of the most important prior steps for granting a successful implantation procedure. Bacterial infections could be prevented using scaffolds that allow early local antibiotic administration. The effectiveness of a delivery system is highly dependent on the control of the drug release profile. Since there is a high risk of infection immediately after implantation, the delivery system should promote an initial burst release of the antibiotic. This initial burst release should be followed by a sustained release to avoid a latent infection [3].

Calcium phosphates (CaP) are the most common biomaterials used in bone tissue engineering due to their similarity to the mineral component of bone and their excellent bioactivity. Nowadays, biphasic calcium phosphate (BCP) represents the gold standard of CaP biomaterials [5]. BCP usually combine relatively stable hydroxyapatite (HA) with a more soluble β-tricalcium phosphate (β-TCP) phase, thus allowing better control over the bioactivity and biodegradability of the scaffold. In this way, the stability of the material is ensured during the ingrowth of the bone [6]. Despite all the potential of CaP materials, they have some disadvantages, mainly their brittleness and low strength. These disadvantages can be mitigated by applying polymeric coatings to improve the robustness of the inorganic material [7]. In this context, synthetic polymers offer a number of advantageous over the natural ones, including the ability to easily adjust their physicochemical properties, and more reproducible synthesis and production processes [8]. The synergistic combinations between CaP and polymers have also been explored for drug delivery in bone tissue engineering [9–11]. Despite the ability of CaP scaffolds to incorporate pharmaceutical agents by surface adsorption, [12,13] they have low efficiency as sustained release systems. In addition, the relatively high temperatures required to obtain high skeletal density and suitable mechanical properties tend to reduce the surface area available for adsorption, making them potentially unsuitable for drug incorporation and release [14]. These drawbacks can be overcome by incorporating the desired active pharmaceutical ingredient into a polymeric coating [14]. Drug release from a polymeric coating is characterized by an initial burst release followed by a sustained release [15]. Several factors influence the drug release profile, namely, coating degradation, interaction with the polymer, and diffusion of the drug. Moreover, the properties of the scaffolds such as porosity, pore size and interconnectivity play a crucial role in the drug release profile [14].

Cuttlefish bone (CB) has a unique architecture with about 93% porosity [16]. The successful hydrothermal conversion (HT) of CB into BCP scaffolds has been reported previously [17]. Due to the brittleness and low strength of the hydrothermally converted HA scaffolds, polymeric reinforcement coatings have been reported elsewhere [18–20]. In our previous work, poly(ε-caprolactone) (PCL) or poly(ester urea) (PEU) coatings were used to improve the mechanical properties [17]. PCL is one of the most used synthetic polymers in bone tissue engineering. It is a biocompatible polymer with good mechanical properties and its degradation products are non-toxic [21,22]. On the other hand, PEU is a synthetic polymer with promising properties. The presence of α-amino acids improves cell-material interactions and enables a further functionalization that represents a powerful tool in the biomedical field [23,24]. In the present work, a further step was taken and the polymer-coated BCP scaffolds were investigated as a vehicle for the uptake and release of an antibiotic to avoid bacterial infection, thus obtaining multifunctional scaffolds. With this purpose, BCP scaffolds derived from CB were synthesized, heat treated and coated with PCL or PEU solutions, in which the antibiotic rifampicin (RFP), was dissolved (Figure 1). RFP has a broad-spectrum against Gram-positive and-negative bacterial strains. Moreover, it binds to the enzyme RNA-polymerase and blocks the bacterial DNA function [25]. As one of the most potent and broad-spectrum antibiotics, RFP has been explored as a pharmaceutical agent to prevent the formation of biofilm [26–28].

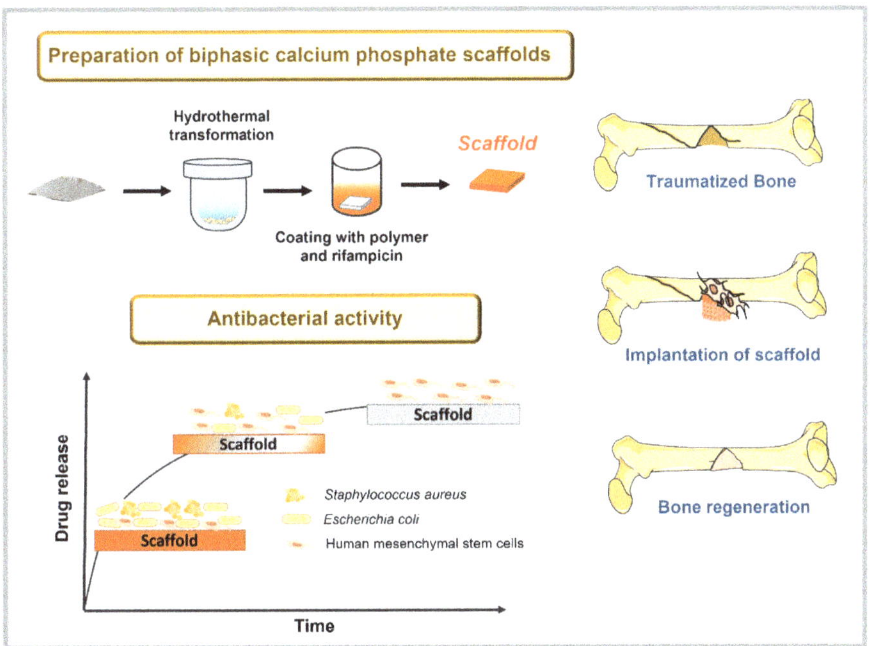



**Figure 1.** Overall strategy for the preparation of multifunctional biphasic calcium phosphate scaffolds coated with polymeric materials and with drug delivery properties.

## 2. Materials and Methods

### 2.1. Preparation of BCP Scaffolds

The bones of cuttlefish, *Sepia officinalis*, were cut into cylinders of approximately 6 mm in diameter and 3 mm in height. Using differential and gravimetric thermal analysis (DTA/TG, Labsys Setaram TG-DTA/DSC, Caluire-, France, heating rate of 10 °C min$^{-1}$), the exact amount of $CaCO_3$ in the CB (calcium precursor) was calculated. The CB cylinders were then subjected to hydrothermal transformation (HT) in the presence of a phosphorous precursor. Briefly, the samples were sealed in a stainless-steel autoclave lined with poly(tetrafluorethylene) (PTFE) with the required volume of an aqueous solution of $(NH_4)_2HPO_4$ (Panreac AppliChem, Castellar del Vallès, Spain) for 24 h at 200 °C. The obtained scaffolds were then subjected to heat treatment to remove the organic matter at 700 °C for 1 h at a heating rate of 0.5 °C min$^{-1}$, followed by a heating ramp of 2 °C min$^{-1}$ up to 1200 °C and a dwelling time of 2 h at this temperature for sintering.

### 2.2. Preparation of Polymeric Coated Scaffolds Loaded with RFP

The sintered BCP scaffolds were coated with PCL (Perstorp Specialty Chemicals AB, Perstorp, Sweden, CAPA$^{TM}$ 6800 $M_n$ = 80,000 g mol$^{-1}$) and PEU ($M_n$ = 63,000 g mol$^{-1}$). The polymer solutions were prepared at a concentration of 5% (*w/v*); PCL was dissolved in dichloromethane (Sigma, Darmstadt, Germany), while PEU was dissolved in chloroform (Fisher Scientific, Loughborough, UK). To improve the solubility of the PEU, 2% (*v/v*) of *N,N'*-dimethylformamide (Sigma-Aldrich, Darmstadt, Germany) was added to the chloroform solution. The scaffolds were coated by the dip coating method using a vacuum system for 20 min at the pressure of 0.4 bar. Two different samples were obtained: BCP coated with PCL (BCP-PCL) and BCP coated with PEU (BCP-PEU). To obtain RFP loaded samples, RFP powder (Panreac AppliChem, Castellar del Vallès, Spain) was dissolved in the polymer solution at a concentration of 1.5 mg.mL$^{-1}$. The scaffolds coated with RFP-containing PCL or PEU solutions, BCP-PCL-RFP or BCP-PEU-RFP, respectively, were obtained following the same procedure described above for preparing the BCP-PCL and the BCP-PEU samples. The incorporated RFP content was determined by immersing the scaffolds in dimethyl sulfoxide (DMSO), which allowed for the complete dissolution of the drug. Subsequently, the RFP content was measured by UV-Vis spectrophotometry at the wavelength of 338 nm. A calibration curve (Figure S1) was plotted within an appropriate RFP concentration range in DMSO.

### 2.3. Characterization of the Obtained Scaffolds

All of the scaffold samples (BCP, BCP-PCL, BCP-PEU, BCP-PCL-RFP and BCP-PEU-RFP) were characterized by Fourier transform infrared spectroscopy (FTIR), X-ray diffraction (XRD) and differential scanning calorimetry (DSC). FTIR spectra were acquired at room temperature (RT) using an Agilent Technologies Carey 630 spectrometer (Agilent Technologies, Inc., Santa Clara, CA, USA) equipped with a Golden Gate Single Reflection Diamond ATR. Data were recorded in a range from 650 to 4000 cm$^{-1}$ with a spectral resolution of 4 cm$^{-1}$ and 64 accumulations. XRD measurements were performed in a high-resolution X-ray diffractometer (PANalytical X'Pert Pro, Malvern Panalytical, Los Altos, CA, USA) with Cu K$\alpha$ radiation ($\lambda = 1.5406$ Å) using a step-scanning mode with a $2\theta$ angle from $10°$ to $100°$ and a step size of $0.0260°$ per second. DSC measurements were performed in a Netzsch DSC-214 (Netzsch, Selb, Germany) under a nitrogen atmosphere using an aluminium pan containing approximately 5 mg of sample. A heating rate of 10 °C min$^{-1}$ was used within a temperature range from $-40$ °C to 500 °C.

### 2.4. In Vitro RFP Release Study

In order to evaluate the in vitro RFP release profile, the scaffolds were placed in glass tubes containing 2 mL of PBS and incubated at 37 °C in a shaker for different time points (10 and 30 min; 1, 2, 6, 8, and 24 h; 2, 3, 4, 5 and 6 days). The amount of RFP released into the medium was measured by UV-Vis spectrophotometry at wavelength 338 nm (Figure S2). Measurements were performed in triplicate. The kinetic release of RFP from the polymer coated BCP scaffold was modelled using the Korsmeyer–Peppas model (Equation (1))

$$\frac{M_t}{M_\infty} = kt^n \left( \frac{M_t}{M_\infty} \leq 0.6 \right) \tag{1}$$

where $M_t$ and $M_\infty$ are the amounts of drug released at time $t$ and infinity, respectively; $k$ is the release constant and $n$ is the release exponent and is related to the release mechanism [29].

### 2.5. Antibacterial Activity Assay

Antibacterial assays of the RFP-loaded and unloaded scaffolds were performed against Gram-negative, *Escherichia coli* ATCC25922, and Gram-positive, *Staphylococcus aureus* ATCC25923. The bacterial strains were grown at 37 °C for 24 h in Luria-Bertani (LB) medium. A bacterial suspension was then prepared in 10 mL of PBS solution with turbidity adjusted to 0.5 according to the McFarland standard ($1.5 \times 10^8$ CFU mL$^{-1}$). The obtained suspension was diluted 10-fold and then 375 µL was mixed with 1125 µL of LB medium and added to the wells containing the different polymer (PCL or PEU) coated BCP scaffolds with or without RFP. The well without any scaffold was used as a positive control. The samples were incubated with an orbital shaker at 115 rpm and 37 °C. The number of viable cells remaining in the culture medium after 24, 48 and 72 h was analysed by the spread plate method. Briefly, 100 µL of bacterial suspension was taken from the wells and spread evenly on LB agar plates. At 72 h incubation, the scaffolds in contact with the bacterial suspension, were gently washed twice with PBS and the cells attached were detached by sonication for 10 min. The obtained bacterial suspension was also spread on LB agar plates. Finally, the LB dishes were incubated overnight at 37 °C and the number of colony-forming units (CFUs) was determined. The experiments were repeated in triplicate and with two replicates in each experiment.

*2.6. Isolation and Culture of Human Mesenchymal Stem Cells from Umbilical Cord Matrix*

Human umbilical cords from healthy donors after birth, with parental consent, were kindly donated by Crioestaminal Saúde e Tecnologia (Biocant Park, Cantanhede, Portugal). The umbilical cords were stored in sterile 50 mL tubes at RT between 12 and 48 h before tissue processing. The samples were cut into small pieces of approximately 5 cm. The pieces obtained were washed with sterile PBS to remove blood. The umbilical veins were also washed to remove blood and blood clots. To avoid contamination by endothelial cells, the umbilical veins and arteries were also removed. Samples were then dried in tissue culture plates to promote adhesion of the fragment to the polystyrene surface, and, after adhesion, human mesenchymal stem cells (hMSCs) proliferation medium (alpha-MEM without ribonucleosides and deoxyribonucleosides (GIBCO™ Invitrogen Corporation, Carlsbad, CA, USA) supplemented with 10% foetal bovine serum (Cytiva HyClone™ Fetal Bovine Serum (FBS) U.S. Origin, Fisher Scientific, Loughborough, UK), 1% penicillin/streptomycin and 1% amphotericin B (GIBCO™ Invitrogen Corporation, Carlsbad, CA, USA)) was added to the cell culture plate. Samples were cultured at 37 °C with 5% $CO_2$ and 95% humidity, for 10 days until hMSCs migrated from the umbilical cord matrix and defined colonies formed. Finally, fragments were removed from the umbilical matrix and cells were passaged. Passages 2–4 were used in all further experiments.

*2.7. In Vitro Cytocompatibility Assays*

Before seeding the cells, all scaffolds were sterilized by UV-irradiation on both sides for 15 min and pre-wetted in a culture medium for 6 h. The culture medium was removed and hMSCs were seeded at a density of $1 \times 10^5$ per scaffold onto the top of the scaffolds. To promote cell adhesion, the seeded scaffolds were incubated for 2 h followed by the addition of 1 mL of culture medium. The culture medium was changed every 2–3 days. Cells seeded into the scaffolds were cultured in an incubator at 37 °C and a humidified atmosphere (5% $CO_2$ and 95% air) for a period of 1–14 days.

*2.8. Cell Viability and Proliferation*

Cell viability and proliferation of seeded scaffolds with and without RFP were determined using the CellTiter 96® AQueous One Solution Cell Proliferation colorimetric assay (MTS assay, Promega Corporation, Madison, WI, USA) according to the manufacturer's instructions. After 1, 7 and 14 days of incubation, the culture medium was removed and 500 µL of serum-free culture medium containing MTS reagent (10%) was added to each well and incubated for 3 h at 37 °C and 5% $CO_2$. Then, 100 µL of each well (in quadruplicate) were transferred to a 96-well plate and the absorbance was measured at 490 nm. A negative control (untreated cells), i.e., cells cultured without being exposed to the scaffolds, was performed. Cell viability was calculated as the percentage of viable cells relative to the untreated control cells, which were considered to have 100% viability.

*2.9. Cell Attachment*

To evaluate cell adhesion and morphology of attached cells to scaffolds with and without RFP, SEM was used. After 14 days of culturing, the seeded scaffolds were rinsed with PBS and fixed with paraformaldehyde (4% in PBS 1×) for 1 h at RT. After fixation, samples were washed again with PBS 1× and were dehydrated stepwise through ethanol solutions (30, 50, 70 and 90%) for 15 min with a final dehydration in absolute ethanol for 30 min. The dried cell-seeded scaffolds were sputter-coated with a gold layer before visualization by SEM. SEM images were acquired at various magnifications, at accelerating voltage of 10 kV, using a high-resolution field emission scanning electron microscope, with EDS, WDS (STEM ZEISS, Merlin, Oberkochen, Germany).

*2.10. Statistical Analysis*

Each experiment was performed in triplicate, using two replicates in each experiment. Results are expressed as mean ± standard deviation (SD) and two-way-analysis of variance

(ANOVA) was used for statistical analysis. Results were considered statistically different when the *p*-value was less than 0.001. Data analysis and statistical tests were performed in GraphPad Prism 6.03 software (San Diego, CA, USA).

## 3. Results and Discussion

### 3.1. Characterization of the Scaffolds

The chemical groups present in the BCP scaffolds uncoated, and coated with PCL (BCP-PCL), PEU (BCP-PEU), and further RFP loading (BCP-PCL-RFP, BCP-PEU-RFP) were investigated by FTIR-ATR (Figure 2).

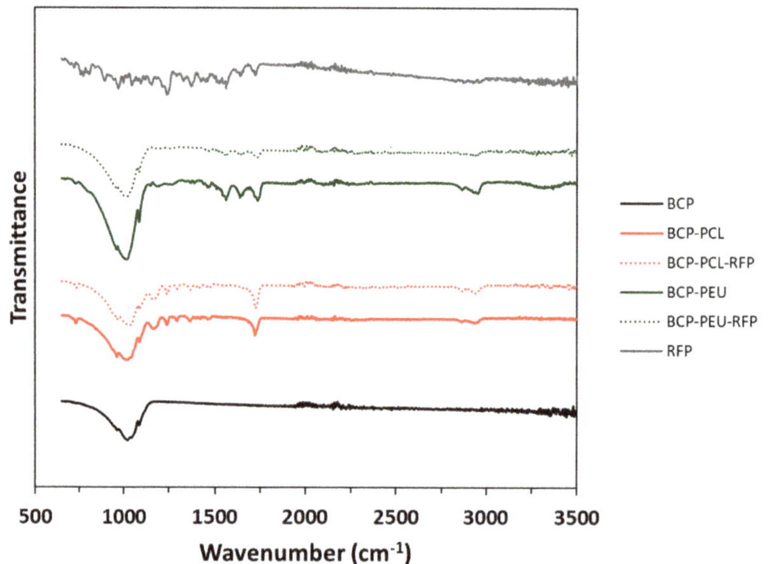


**Figure 2.** FTIR-ATR of BCP scaffolds, uncoated and after coating with different polymers (BCP-PCL and BCP-PEU) and further loaded with RFP (BCP-PCL-RFP and BCP-PEU-RFP).

The BCP scaffolds exhibit the characteristic bands of $-PO_4$ groups. The band at $960\ cm^{-1}$ is related to the $\nu_1$–$PO_4$ stretching, and the stretching $\nu_3$–$PO_4$ mode shows intense bands at 1028 and $1082\ cm^{-1}$. The characteristic –OH band at $3575\ cm^{-1}$ is not observed in the FTIR-ATR spectrum. This could be a consequence of the partial de-hydroxylation of HA at 1200 °C and its absence in the β-TCP phase. The polymer coated scaffolds show an overlap of the BCP spectrum with the characteristic spectra of the polymers. In the PCL coated scaffolds, the characteristic bands of C=O stretching vibrations (ester linkage) can be observed at $1720\ cm^{-1}$, $CH_2$ stretching modes at 2946 and $2896\ cm^{-1}$ and bending modes at 1362, 1399 and $1457\ cm^{-1}$. At 1239, 1041 and $1107\ cm^{-1}$ the bands associated with C–O–C stretching vibrations are observed. The C–O and C–C stretching of the amorphous and crystalline phases appear at 1166 and $1293\ cm^{-1}$, respectively. In turn, the PEU-coated samples also exhibit the characteristic PEU bands. The band at $1560\ cm^{-1}$ is associated with the bending of the N–H group and the stretching of the C–N group of the urea bond. The band at $3370\ cm^{-1}$ corresponds to the stretching vibration of the N–H group of the urea linkage. The C=O vibration of ester and urea is at 1735 and $1636\ cm^{-1}$, respectively. RFP has its characteristic bands of C=O at $1559\ cm^{-1}$, furanone (C=O) at $1638\ cm^{-1}$, acetyl (C=O) at $1720\ cm^{-1}$ and amide ($N–CH_3$) at $2890\ cm^{-1}$ [30]. The addition of RFP did not change the spectra of the samples. In fact, the spectra before and after incorporation of RFP were identical. Although the presence of RFP cannot be detected by FTIR, it can be clearly seen in Figure 3 where the samples with RFP change colour from white to burnt orange.

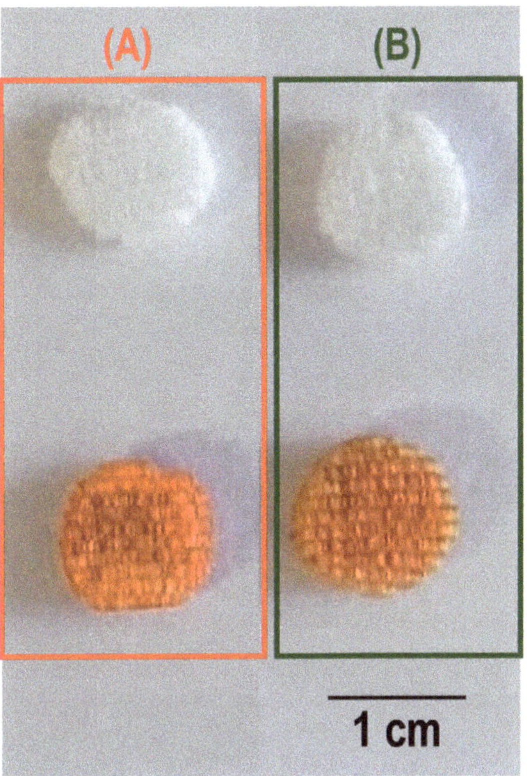



**Figure 3.** BCP scaffolds coated with different polymers (PCL (**A**) and PEU (**B**)—white), and further loaded with RFP (burnt orange).

### 3.2. In Vitro RFP Release Study

The RFP contents determined for PCL and PEU-coated samples were 0.55 ± 0.04% and 0.45 ± 0.02%, respectively. The loading capacity of the PCL coated scaffolds was significantly higher compared to that of PEU. The drug release profiles were determined over a period of 6 days and the results are shown in Figure 4. The cumulative release profiles for the different compositions are roughly similar, being characterized by two stages, an initial burst release (~8 h) followed by a sustained release period. Despite this similarity, the fraction of RFP released from the scaffolds revealed to be dependent on the polymer coating used. Indeed, the RFP release from BCP-PCL-RFP (67.33 ± 1.48%) was significantly higher during the first 8 h compared to the BCP-PEU-RFP systems, which released only 47.23 ± 0.31%. After 6 days, the total RFP content released reached 85.33 ± 0.36% and 61.73 ± 0.16% for BCP-PCL-RFP and BCP-PEU-RFP, respectively. The RFP release data revealed a good agreement with the Korsmeyer–Peppas model described in Equation (1), as deduced from the fitting parameters reported in Table 1. The values of release exponent ($n$) ranging from 0.46 to 1, indicate a non-Fickian diffusion for cylindrical samples. Moreover, the calculated values of constant ($k$) were 0.3058 and 0.1841 for BCP-PCL-RFP and BCP-PEU-RFP, respectively. The correlation coefficient was kept above 0.93 for all the scaffolds compositions.

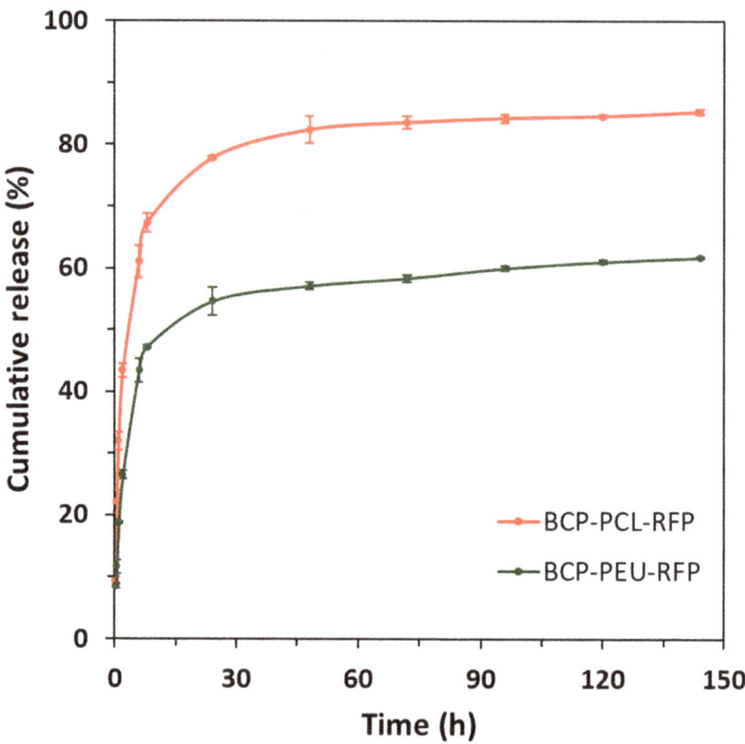



**Figure 4.** In vitro release profiles of RFP from the BCP scaffolds coated with PCL and PEU, in PBS (pH = 7.4), at 37 °C.

**Table 1.** Exponent ($n$), constant ($k$) and correlation coefficient ($R^2$) obtained from the Korsmeyer–Peppas model for the release of RFP from the BCP scaffolds coated with PCL and PEU.

| Scaffold | $n$ | $k$ | $R^2$ |
| --- | --- | --- | --- |
| BCP-PCL-RFP | 0.6215 | 0.3058 | 0.9807 |
| BCP-PEU-RFP | 0.4640 | 0.1841 | 0.9890 |

*3.3. Antibacterial Activity Assays*

The antibacterial activity data was assessed by determining the time-dependent reduction of cells of *E. coli* and *S. aureus* in the presence of the scaffolds, using the CFU determination by the spread plate method (Table 2). The polymer coated samples, without RFP, did not exhibit any antimicrobial activity. Regarding the polymer coated scaffolds, loaded with RFP, all were effective in reducing the growth of *S. aureus*. Nevertheless, with *E. coli* the BCP-PEU-RFP scaffold did not show the ability to inhibit bacterial growth, contrarily to what was observed with BCP-PEU-RFP. After 24 h of incubation with *E. coli*, the BCP-PCL-RFP samples showed higher reduction in bacterial growth compared to BCP-PEU-RFP samples. After 72 h of incubation, the BCP-PCL-RFP resulted in a 100% reduction in the number of *E. coli*. For the BCP-PCL-RPF few attached cells (both *E.coli* and *S. aureus*) were observed after the 72 h, evidenced the difficulty of both strains in colonize these scaffolds.

**Table 2.** Antibacterial activity of the different scaffolds. The antibacterial activity was determined by following the number of viable cells as CFU ml$^{-1}$.

| | | *E. coli* | | |
|---|---|---|---|---|
| **Scaffold** | **24 h** | **48 h** | **72 h** | **Attached Cells** |
| BCP-PCL | >3 × 10$^3$ | | | |
| BCP-PCL-RFP | 4.25 ± 0.25 | 0.00 ± 0.00 | 0.00 ± 0.00 | 6.00 ± 6.00 |
| BCP-PEU | >3 × 10$^3$ | | | |
| BCP-PEU-RFP | >3 × 10$^3$ | >3 × 10$^3$ | >3 × 10$^3$ | >3 × 10$^3$ |
| | | *S. aureus* | | |
| **Scaffold** | **24 h** | **48 h** | **72 h** | **Attached Cells** |
| BCP-PCL | >3 × 10$^3$ | | | |
| BCP-PCL-RFP | 0.00 ± 0.00 | 0.17 ± 0.24 | 0.00 ± 0.00 | 0.00 ± 0.00 |
| BCP-PEU | >3 × 10$^3$ | | | |
| BCP-PEU-RFP | 0.17 ± 0.24 | 0.00 ± 0.00 | 0.33 ± 0.47 | 10.00 ± 10.50 |

*3.4. Viability and Proliferation of Human Mesenchymal Stem Cells*

The cytocompatibility of the obtained scaffolds in the hMSCs (a precursor of osteoblastic-lineage cells [31]) was monitored using MTS assay (cell viability and proliferation), and by morphological/adhesion studies using light and electron microscopes. According to the inverted microscope images (Figure 5), hMSCs seeded on the surface of the BCP-PCL, BCP-PEU, and BCP-PEU-RFP scaffolds showed similar or better cell growth than cells cultured on plates (untreated cells), presenting the fusiform morphology characteristic of this type of cells, after 14 days of culture. It can be observed that the hMSCs tended to grow towards the scaffolds (Figure 5). However, the cell density on BCP-PCL-RFP scaffolds was decreased and the morphology showed an unhealthy appearance (spheroidal shape) compared to the control group.

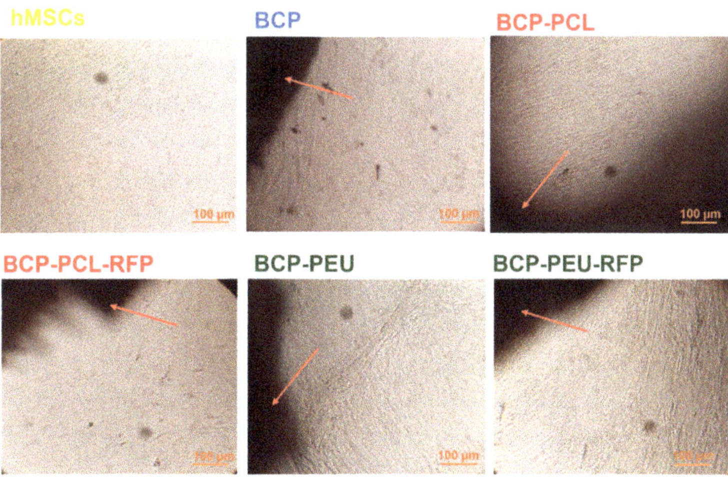


**Figure 5.** Images of hMSCs observed using an inverted microscope (scale bar, 100 µm), after 14 days of culture. Control group (cell culture plate), BCP scaffold, BCP-PCL scaffold, BCP-PCL-RFP scaffold, BCP-PEU scaffold, BCP-PEU-RFP scaffold.

The results of cell viability (% relative to control—untreated cells) and proliferation (the optical density at 490 nm) of hMSCs seeded into unloaded and RFP-loaded scaffolds are displayed in Figure 6. No cytotoxicity to hMSCs was exerted by the coated scaffolds after 1 day of incubation when compared to the control group, irrespective of RFP loading, as can be deduced from the cell viability values of ~84% (BCP-PCL), ~92% (BCP-PEU), ~81% (BCP-PCL-RFP), and ~90% (BCP-PEU-RFP). However, after 7 days of culture, a significant decrease in cell viability to ~23% was observed for the BCP-PCL-RFP scaffolds. Similar results were obtained after 14 days. Compared to untreated cells, low cell viability (~50%) was registered at all time points for BCP scaffolds, as shown in Figure 6. Our results suggest that the BCP-PEU and BCP-PEU-RFP scaffolds exhibit a good level of cytocompatibility.

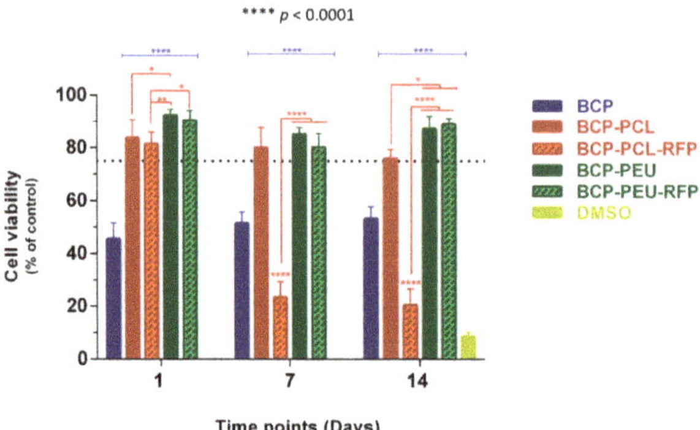


**Figure 6.** Cell viability of hMSCs onto BCP, BCP-PCL, BCP-PCL-RFP, BCP-PEU and BCP-PEU-RFP scaffolds. * $p = 0.0432$, ** $p = 0.0008$, **** $p < 0.0001$.

Regarding proliferation studies, the MTS data after 7 and 14 days of culture showed that cell density increased with culture time in all scaffolds, except for the BCP-PCL-RFP one, which inhibited proliferation of hMSCs. As shown in Figure 7, there were no significant differences in the proliferation rate of hMSCs between the BCP-PEU/BCP-PEU-RFP scaffolds and control group (untreated cells) at days 7 and 14. Compared to the BCP-PCL scaffolds, the hMSCs proliferation levels were better in the BCP-PEU scaffolds and significant differences were observed (**** $p < 0.0001$). Similar results were found in the BCP-PEU-RFP scaffolds, suggesting that PEU coating promotes good cell adhesion and sufficient proliferation.

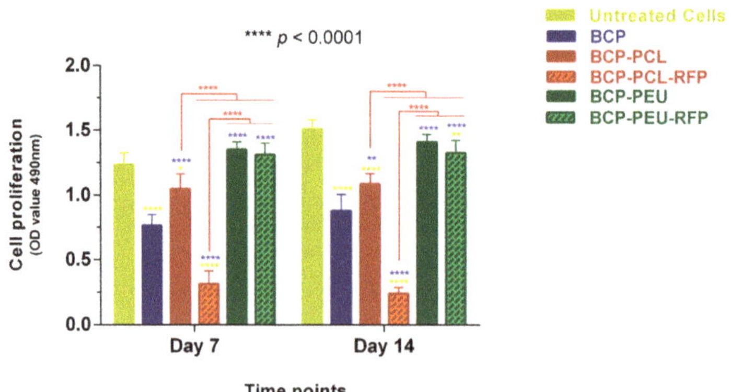


**Figure 7.** MTS assay for the proliferation of human mesenchymal stem cells (hMSCs) on the scaffolds cultured under different conditions on day 7 and 14. * $p < 0.05$; ** $p = 0.0026$, **** $p < 0.0001$.

### 3.5. Cell Attachment

Cell adhesion and morphology were examined by SEM. Figure 8 (without cells) shows SEM micrographs of the surface of the BCP-PCL, BCP-PEU, and BCP-PEU-RFP scaffolds after polymer impregnation, respectively, before the cell culture. In Figure 8 (with cells), SEM images of the surface of the BCP-PCL, BCP-PEU, and BCP-PEU-RFP scaffolds seeded with hMSCs cells, after 14 days of culture, are shown. SEM microscopic images (at low magnification) showed excellent adhesion and homogeneous distribution of hMSCs on the surface BCP-PCL, BCP-PEU, and BCP-PEU-RFP scaffolds, corroborating the MTS results. The surface of the scaffolds was almost completely covered by the cells and the extracellular matrix secreted by them. As can be seen in the high magnification images, a layer of cells and matrix on the surface of the scaffolds is observed. Regarding the morphology of the cells, it is difficult to withdraw conclusions.

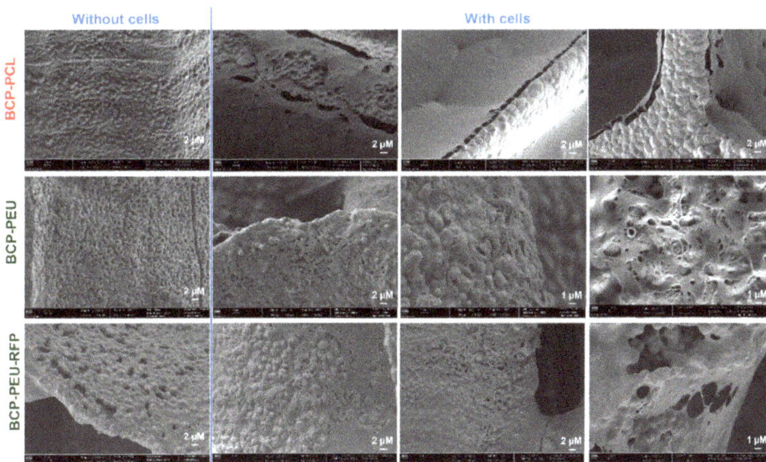


**Figure 8.** SEM micrographs of the surface of BCP-PCL, BCP-PEU and BCP-PEU-RFP scaffolds after polymer impregnation, without and with hMSCs after 14 days of culture.

## 4. Discussion

A complete conversion of CB into a BCP scaffold was obtained as reported in a previous work [17]. The unique CB architecture was retained after HT and sintering. Moreover, polymeric (PCL or PEU) coatings significantly improved the mechanical properties of the scaffolds without having a significant negative effect on porosity of the scaffold [17]. This means that the composite scaffolds skilfully combine the highly porous inorganic BCP with the polymeric coatings, improving the overall properties and expanding the application possibilities. Moreover, this synergistic effect can also be explored in the context of drug delivery. The main objective of this work was to develop a composite scaffold derived from CB that can act as a drug delivery system of an antibiotic, to avoid bacterial growth at the implantation site.

The discrepancy observed in the amounts of RFP incorporated into the scaffolds coated with the different polymers, 0.55 ± 0.04 wt% (PCL) and 0.45 ± 0.02 wt% (PEU), could be attributed to the higher viscosity of the PCL solution in comparison to that of PEU. This resulted in thicker PCL coatings [17], which are able to accommodate higher amounts of RFP. Due to the small amounts of RFP incorporated, it was not possible to detect differences between the coated samples with and without RFP using FTIR-ATR (Figure 2). It was also not possible to detect the presence of RFP in the samples using XRD and DSC (results in Supplementary Information, Figures S3 and S4). Nevertheless, its incorporation into the coated scaffolds can be easily deduced by the change of their colour from white to burnt orange (Figure 3). Moreover, the in vitro drug release studies (Figure 4) and the antimicrobial studies (Table 2) are other clear evidence of the presence of RFP in the coated samples.

The in vitro drug release studies (Figure 4) revealed that the applied polymeric coatings allowed for obtaining suitable RFP release profiles. All RFP loaded samples exhibited similar release profiles, characterized by an initial burst within the first 8 h, followed by a sustained release period. The relative amounts released during the burst period were 67.33 ± 1.48% and 47.23 ± 0.31% for PCL and PEU-coated samples, respectively. The incorporation of RFP into the scaffolds was aimed at preventing the risk of infection in the initial period after implantation. Therefore, the observed initial burst brings an important advantage to eliminate possible bacterial contamination during the medical procedure [32]. Moreover, it is important to highlight that RFP released during the burst period is still within the therapeutic limits, since a cytotoxic effect is observed only at a concentration of 100 $\mu$g.mL$^{-1}$ [33]. This initial burst release may be due to different mechanisms, such as pore diffusion, surface desorption and the absence of a diffusion barrier regulating the diffusion process [34].

The Korsmeyer–Peppas model is suitable for describing the kinetics of drug release from thin films and cylindrical or spherical samples such as those used in this study. The drug release mechanisms can be divided into Fickian diffusion, non-Fickian transport and zero-order release when n is 0.45, 0.45–1 and 1, respectively [29]. For the different coating compositions used in the present work, the n value ranged from 0.46 to 0.62, which is characteristic of a non-Fickian diffusion. This diffusion behaviour indicates a release that is simultaneously controlled by diffusion and by dissolution [35].

The release profile is strongly dependent on the type of polymer coating. Indeed, a faster RFP release was observed from the scaffolds coated with PCL, compared to the scaffolds coated with PEU. The two different polymers, PCL and PEU, have carboxyl groups that allow the formation of hydrogen bonds with RFP. Moreover, these groups have affinity for calcium present in BCP scaffolds. PEUs, in addition to carboxyl groups, have amide groups that can also form hydrogen bonds with RFP, while having affinity for phosphorus from BCP scaffolds. Therefore, PEU coating is expected to produce stronger interactions with both BCP scaffolds and RFP compared to PCL. The lower affinity between PCL and BCP scaffolds results in weaker and non-continuous interfaces that provide a larger surface area for contact with the PBS solution and desorption of the drug. These factors explain and determine drug release behaviour and consistently account for its higher amount released from PCL coated samples within the first 8 h during burst period.

RFP is a broad-spectrum antibiotic for both Gram-positive and Gram-negative bacteria and was therefore chosen as the antibiotic in this work. The antibacterial activity data of the RFP-loaded scaffolds against Gram-negative *E. coli* and Gram-positive *S. aureus* reported in Table 2 confirm its useful action against both types of bacteria. However, the efficacy depends on the particular bacterial strain and also on the type of polymer coating. When incubated with *E. coli*, a more rapid decrease in the number of viable bacterial cells was recorded for the BCP-PCL-RFP samples. The BCP-PEU-RFP scaffold did not show antibacterial activity towards *E. coli*. This result can be associated with the higher amount of RFP released from the PCL-coated scaffold in comparison with the PEU-coated counterpart (Figure 4). Most probably the amount of RFP released from the BCP-PEU-RFP during the time of incubation is not sufficient to exert its antibacterial effect against *E. coli*. On the other hand, all the RFP-loaded scaffolds exhibited stronger antibacterial activity against *S. aureus*. In fact, the reduction of bacterial growth reached almost 100% after 24 h of incubation for all compositions. The higher sensitivity of *S. aureus* to RFP is associated with the better permeability of RFP through Gram-positive cell walls than through Gram-negative ones [36].

The in vitro studies performed with hMSCs support the suitability of the prepared scaffolds for bone regeneration during local antibiotic therapy. MTS metabolic assay showed that the unloaded and RFP-loaded PEU scaffolds, as well as the BCP-PCL scaffolds, could provide a favourable platform for the survival/growth of hMSCs. However, a low proliferation rate (~50%) was found in the presence of the uncoated BCP scaffolds, which proved to be much more fragile as compared to the coated ones. Similar results with BCP scaffolds were obtained by Kim and co-workers [37], with human osteoblast-like MG-63 cells, and of Hongmim and co-workers [38] with murine mesenchymal stem cells. This implies that the highly porous inorganic BCP scaffolds together with the polymer coatings (PCL and PEU) play a synergistic role, by exhibiting less or no cytotoxicity and providing the suitable conditions for cell adhesion and proliferation. Compared to the BCP-PEU scaffolds, the PCL coated BCP scaffolds showed low proliferation and viability values over the experimental period (7 and 14 days). These findings could be due to a reduction in the porosity of the BCP-PCL scaffold, which consequently leads to a reduction in the total surface area available for cell adhesion [17]. Similar biological observations were reported by Milovac and co-workers [39] and Siddiqi and co-workers [40] for the PCL coated BCP scaffolds. Moreover, the BCP-PCL-RFP scaffolds were shown to negatively affect the viability and proliferation of hMSCs (i.e., higher cell toxicity and lower cell proliferation rate), resulting cell damage. This result, however, can be related with the higher amount of RFP released, after 1 day, from the PCL-coated scaffold ($\approx$80%) in comparison to the amount of drug released from PEU-coated scaffold ($\approx$50%). The higher amount of RFP released can induce toxicity on hMSCs, affecting their viability and proliferation. In contrast, cells proliferating on the surface of BCP-PEU-RFP scaffold were not negatively affected. The results revealed that BCP-PEU-RFP scaffolds have excellent properties for the intended applications, namely: cytocompatibilitys, and an optimal composition to support cell proliferation.

## 5. Conclusions

In this study, a drug delivery system using composite scaffolds, from BCP, coated with biopolymers and loaded with a pharmaceutical drug (RFP), was successfully developed. The polymeric (PCL or PEU) coatings bring synergistic benefits to the mechanical properties of the scaffolds [17], opening also the possibility of the scaffold to be used as a delivery system. The drug release kinetic mechanism was shown to be a non-Fickian diffusion. The RFP release from the BCP-coated scaffolds was similar for both coatings (PCL and PEU), being characterized by a burst release in the first 8 h, followed by a sustained release. A higher initial release of RFP from the PCL-coated samples could be attributed to the weaker interaction between the polymer and the drug and also between the polymer and the inorganic BCP scaffold. The antibacterial activity of the different scaffolds against *E. coli* and *S. aureus* showed that the samples without the antibiotic had no antimicrobial activity. Both RFP-loaded scaffolds were effective in reducing bacterial growth, particularly in the case of *S. aureus*. Regarding *E. coli*, only BCP-PCL-RFP was shown to have antibacterial activity. The in vitro cytotoxicity studies revealed that hMSCs adhered to the surface of BCP-PEU-RFP scaffolds, proliferated, and remained viable after 7 and 14 days of culture. Overall, the results show that the BCP-PEU-RFP scaffolds are the most promising for the application.

**Supplementary Materials:** The following are available online at https://www.mdpi.com/article/10.3390/polym13244378/s1, Figure S1. Calibration curve of RFP diluted in DMSO, Figure S2. Calibration curve of RFP diluted in PBS, Figure S3. XRD of uncoated BCP scaffolds, coated with different polymers (BCP-PCL, and BCP-PEU) and loaded with RFP (BCP-PCL-RFP and BCP-PEU-RFP), Figure S4. DCS of uncoated BCP scaffolds, coated with different polymers (BCP-PCL and BCP-PEU) and loaded with RFP (BCP-PCL-RFP and BCP-PEU-RFP).

**Author Contributions:** Conceptualization, J.F.J.C. and J.M.F.F.; Formal analysis, A.S.N., P.P., C.D., M.C.A., I.B. and C.O.M.; Funding acquisition, J.F.J.C. and J.M.F.F.; Investigation, A.S.N., P.P., A.C.F., C.D., M.C.A., I.B. and C.O.M.; Methodology, A.C.F., J.F.J.C. and J.M.F.F.; Resources, J.F.J.C., J.M.F.F., L.P.d.A. and P.V.M.; Supervision, L.P.d.A., P.V.M., J.F.J.C. and J.M.F.F.; Writing—original draft, A.S.N., P.P. and A.C.F.; Writing—review and editing, J.F.J.C. and J.M.F.F. All authors have read and agreed to the published version of the manuscript.

**Funding:** This work was developed within the scope of the project CICECO-Aveiro Institute of Materials, FCT Ref. UID/CTM/50011/2019, financed by national funds through Fundação para a Ciência e a Tecnologia (FCT/MCTES). This research was also partially sponsored by FEDER funds through the program COMPETE2020—Programa Operacional Factores de Competitividade—and by national funds through FCT, under projects UIDB/00285/2020, UIBD/00511/2020 and UIDB/04539/2020. We would like to thank the financial support to European Regional Development Fund through the Regional Operational Program Center2020: Brain-Health2020 projects (CENTRO-01-0145-FEDER-000008), ViraVector (CENTRO-01-0145-FEDER-022095). Inês Barros was supported by FCT fellowship (SFRH/BD/148877/2019).

**Institutional Review Board Statement:** Not applicable.

**Informed Consent Statement:** Not applicable.

**Data Availability Statement:** The data presented in this study is available in the article.

**Acknowledgments:** The authors would also like to thank João Ramalho-Santos, Ana Sofia Rodrigues and Ana Branco for hMSCs isolation.

**Conflicts of Interest:** The authors declare no conflict of interest.

## References

1.  Percival, S.L.; Suleman, L.; Vuotto, C.; Donelli, G. Healthcare-associated infections, medical devices and biofilms: Risk, tolerance and control. *J. Med. Microbiol.* **2015**, *64*, 323–334. [CrossRef] [PubMed]
2.  Mouriño, V.; Boccaccini, A.R. Bone tissue engineering therapeutics: Controlled drug delivery in three-dimensional scaffolds. *J. R. Soc. Interface* **2010**, *7*, 209–227. [CrossRef] [PubMed]
3.  Zilberman, M.; Elsner, J.J. Antibiotic-eluting medical devices for various applications. *J. Control. Release* **2008**, *130*, 202–215. [CrossRef] [PubMed]
4.  Donlan, R.M. Biofilms and Device-Associated Infections. *Emerg. Infect. Dis.* **2001**, *7*, 277–281. [CrossRef]
5.  Bouler, J.M.; Pilet, P.; Gauthier, O.; Verron, E. Biphasic calcium phosphate ceramics for bone reconstruction: A review of biological response. *Acta Biomater.* **2017**, *53*, 1–12. [CrossRef] [PubMed]
6.  Lobo, S.E.; Arinzeh, T.L. Biphasic calcium phosphate ceramics for bone regeneration and tissue engineering applications. *Materials* **2010**, *3*, 815–826. [CrossRef]
7.  Motealleh, A.; Eqtesadi, S.; Pajares, A.; Miranda, P. Enhancing the mechanical and in vitro performance of robocast bioglass scaffolds by polymeric coatings: Effect of polymer composition. *J. Mech. Behav. Biomed. Mater.* **2018**, *84*, 35–45. [CrossRef]
8.  Cortizo, M.S.; Belluzo, M.S. Biodegradable Polymers for Bone Tissue Engineering. In *Industrial Applications of Renewable Biomass Products*; Goyanes, S.N., D'Accorso, N.B., Eds.; Springer: Cham, Switzerland, 2017; pp. 47–74.
9.  Araújo, M.; Viveiros, R.; Philippart, A.; Miola, M.; Doumett, S.; Baldi, G.; Perez, J.; Boccaccini, A.R.; Aguiar-Ricardo, A.; Verné, E. Bioactivity, mechanical properties and drug delivery ability of bioactive glass-ceramic scaffolds coated with a natural-derived polymer. *Mater. Sci. Eng. C Mater. Biol. Appl.* **2017**, *77*, 342–351. [CrossRef]
10. Kim, H.W.; Knowles, J.C.; Kim, H.E. Hydroxyapatite/poly($\varepsilon$-caprolactone) composite coatings on hydroxyapatite porous bone scaffold for drug delivery. *Biomaterials* **2004**, *25*, 1279–1287. [CrossRef] [PubMed]
11. Bose, S.; Sarkar, N.; Banerjee, D. Effects of PCL, PEG and PLGA polymers on curcumin release from calcium phosphate matrix for in vitro and in vivo bone regeneration. *Mater. Today Chem.* **2018**, *8*, 110–120. [CrossRef]
12. Al-Sokanee, Z.N.; Toabi, A.A.H.; Al-Assadi, M.J.; Alassadi, E.A.S. The drug release study of ceftriaxone from porous hydroxyapatite scaffolds. *AAPS PharmSciTech* **2009**, *10*, 772–779. [CrossRef] [PubMed]
13. Chai, F.; Hornez, J.C.; Blanchemain, N.; Neut, C.; Descamps, M.; Hildebrand, H.F. Antibacterial activation of hydroxyapatite (HA) with controlled porosity by different antibiotics. *Biomol. Eng.* **2007**, *24*, 510–514. [CrossRef] [PubMed]
14. Bose, S.; Tarafder, S. Calcium phosphate ceramic systems in growth factor and drug delivery for bone tissue engineering: A review. *Acta Biomater.* **2012**, *8*, 1401–1421. [CrossRef] [PubMed]
15. KKundu, B.; Soundrapandian, C.; Nandi, S.K.; Mukherjee, P.; Dandapat, N.; Roy, S.; Datta, B.K.; Mandal, T.K.; Basu, D.; Bhattacharya, R.N. Development of new localized drug delivery system based on ceftriaxone-sulbactam composite drug impregnated porous hydroxyapatite: A systematic approach for in vitro and in vivo animal trial. *Pharm. Res.* **2010**, *27*, 1659–1676. [CrossRef] [PubMed]
16. Birchall, J.D.; Thomas, N.L. On the architecture and function of cuttlefish bone. *J. Mater. Sci.* **1983**, *18*, 2081–2086. [CrossRef]
17. Neto, A.S.; Fonseca, A.C.; Abrantes, J.C.C.; Coelho, J.F.J.; Ferreira, J.M.F. Surface functionalization of cuttlefish bone-derived biphasic calcium phosphate scaffolds with polymeric coatings. *Mater. Sci. Eng. C Mater. Biol. Appl.* **2019**, *105*, 110014. [CrossRef] [PubMed]
18. Kim, B.S.; Kang, H.J.; Lee, J. Improvement of the compressive strength of a cuttlefish bone-derived porous hydroxyapatite scaffold via polycaprolactone coating. *J. Biomed. Mater. Res. B Appl. Biomater.* **2013**, *101*, 1302–1309. [CrossRef]

19. Milovac, D.; Gallego Ferrer, G.; Ivankovic, M.; Ivankovic, H. PCL-coated hydroxyapatite scaffold derived from cuttlefish bone: Morphology, mechanical properties and bioactivity. *Mater. Sci. Eng. C Mater. Biol. Appl.* **2014**, *34*, 437–445. [CrossRef]
20. Rogina, A.; Antunovic, M.; Milovac, D. Biomimetic design of bone substites based on cuttlefish bone-derived hydroxyapatite and biodegradable polymers. *J. Biomed. Mater. Res. B Appl. Biomater.* **2018**, *107*, 197–204. [CrossRef]
21. Sabir, M.I.; Xu, X.; Li, L. A review on biodegradable polymeric materials for bone tissue engineering applications. *J. Mater. Sci.* **2009**, *44*, 5713–5724. [CrossRef]
22. Saranya, N.; Saravanan, S.; Moorthi, A.; Ramyakrishna, B.; Selvamurugan, N. Enhanced Osteoblast Adhesion on Polymeric Nano-Scaffolds for Bone Tissue Engineering. *J. Biomed. Nanotechnol.* **2011**, *7*, 238–244. [CrossRef] [PubMed]
23. Fonseca, A.C.; Gil, M.H.; Simoes, P.N. Biodegradable poly(ester amide)s—A remarkable opportunity for the biomedical area: Review on the synthesis, characterization and applications. *Prog. Polym. Sci.* **2014**, *39*, 1291–1311. [CrossRef]
24. Li, S.; Xu, Y.; Yu, J.; Becker, M.L. Enhanced osteogenic activity of poly(ester urea) scaffolds using facile post-3D printing peptide functionalization strategies. *Biomaterials* **2017**, *141*, 176–187. [CrossRef]
25. Campbell, E.A.; Korzheva, N.; Mustaev, A.; Murakami, K.; Nair, S.; Goldfarb, A.; Darst, S.A. Structural mechanism for rifampicin inhibition of bacterial RNA polymerase. *Cell* **2001**, *104*, 901–912. [CrossRef]
26. Sandeep Kranthi Kiran, A.; Kizhakeyil, A.; Ramalingam, R.; Verma, N.K.; Lakshminarayanan, R.; Kumar, T.S.; Doble, M.; Ramakrishna, S. Drug loaded electrospun polymer/ceramic composite nanofibrous coatings on titanium for implant related infections. *Ceram. Int.* **2019**, *45*, 18710–18720. [CrossRef]
27. Aragón, J.; Feoli, S.; Irusta, S.; Mendoza, G. Composite scaffold obtained by electro-hydrodynamic technique for infection prevention and treatment in bone repair. *Int. J. Pharm.* **2019**, *557*, 162–169. [CrossRef] [PubMed]
28. Ruckh, T.T.; Oldinski, R.A.; Carroll, D.A.; Mikhova, K.; Bryers, J.D.; Popat, K.C. Antimicrobial effects of nanofiber poly(caprolactone) tissue scaffolds releasing rifampicin. *J. Mater. Sci. Mater. Med.* **2012**, *23*, 1411–1420. [CrossRef]
29. Ritger, P.L.; Peppas, N.A. A simple equation for description of solute release I. Fickian ans non-Fikian release from non-swellable devices in the form of slabs, spheres, cylinders or discs. *J. Control. Release* **1987**, *5*, 23–36. [CrossRef]
30. Pati, R.; Sahu, R.; Panda, J.; Sonawane, A. Encapsulation of zinc-rifampicin complex into transferrin-conjugated silver quantum-dots improves its antimycobacterial activity and stability and facilitates drug delivery into macrophages. *Sci. Rep.* **2016**, *6*, 24184. [CrossRef]
31. Bianco, P.; Sacchetti, B.; Riminucci, M. Stem cells in skeletal physiology and endocrine diseases of bone. *Endocr. Dev.* **2011**, *21*, 91–101. [PubMed]
32. Martin, V.; Ribeiro, I.A.; Alves, M.M.; Gonçalves, L.; Claudio, R.A.; Grenho, L.; Fernandes, M.H.; Gomes, P.; Santos, C.F.; Bettencourt, A.F. Engineering a multifunctional 3D-printed PLA-collagen-minocycline-nanoHydroxyapatite scaffold with combined antimicrobial and osteogenic effects for bone regeneration. *Mater. Sci. Eng. C Mater. Biol. Appl.* **2019**, *101*, 15–26. [CrossRef]
33. Yuan, J.; Wang, B.; Han, C.; Lu, X.; Sun, W.; Wang, D.; Lu, J.; Zhao, J.; Zhang, C.; Xie, Y. In vitro comparison of three rifampicin loading methods in a reinforced porous β-tricalcium phosphate scaffold. *J. Mater. Sci. Mater. Med.* **2015**, *26*, 174. [CrossRef] [PubMed]
34. Huang, X.; Brazel, C.S. On the importance and mechanisms of burst release in matrix-controlled drug delivery systems. *J. Control. Release* **2001**, *73*, 121–136. [CrossRef]
35. Chandrasekaran, S.K.; Paul, D.R. Dissolution-Controlled transport from dispersed matrixes. *J. Pharm. Sci.* **1982**, *71*, 1399–1402. [CrossRef]
36. Wherli, W. Rifampin: Mechanisms of Action and Resistance. *Rev. Infect. Dis.* **1983**, *5* (Suppl. S3), S407–S411. [CrossRef]
37. Kim, B.S.; Kang, H.J.; Yang, S.S.; Lee, J. Comparison of in vitro and in vivo bioactivity: Cuttlefish-bone-derived hydroxyapatite and synthetic hydroxyapatite granules as a bone graft substitute. *Biomed. Mater.* **2014**, *9*, 025004. [CrossRef] [PubMed]
38. Hongmin, L.; Wei, Z.; Xingrong, Y.; Jing, W.; Wenxin, G.; Jihong, C.; Xin, X.; Fulin, C. Osteoinductive nanohydroxyapatite bone substitute prepared via in situ hydrothermal transformation of cuttlefish bone. *J. Biomed. Mater. Res. B Appl. Biomater.* **2015**, *103*, 816–824. [CrossRef]
39. Milovac, D.; Gamboa-Martínez, T.C.; Ivankovic, M.; Ferrer, G.G.; Ivankovic, H. PCL-coated hydroxyapatite scaffold derived from cuttlefish bone: In vitro cell culture studies. *Mater. Sci. Eng. C Mater. Biol. Appl.* **2014**, *42*, 264–272. [CrossRef] [PubMed]
40. Siddiqi, S.A.; Manzoor, F.; Jamal, A.; Tariq, M.; Ahmad, R.; Kamran, M.; Chaudhry, A.; Rehman, I.U. Mesenchymal stem cell (MSC) viability on PVA and PCL polymer coated hydroxyapatite scaffolds derived from cuttlefish. *RSC Adv.* **2016**, *6*, 32897–32904. [CrossRef]

Article

# Architecture and Composition Dictate Viscoelastic Properties of Organ-Derived Extracellular Matrix Hydrogels

Francisco Drusso Martinez-Garcia [1,2,†], Roderick Harold Jan de Hilster [1,3,†], Prashant Kumar Sharma [2,4], Theo Borghuis [1,3], Machteld Nelly Hylkema [1,3], Janette Kay Burgess [1,2,3] and Martin Conrad Harmsen [1,2,*]

[1] Department of Pathology and Medical Biology, University Medical Center Groningen, University of Groningen, Hanzeplein 1 (EA11), 9713 GZ Groningen, The Netherlands; f.d.martinez.garcia@umcg.nl (F.D.M.-G.); r.h.j.de.hilster@umcg.nl (R.H.J.d.H.); t.borghuis@umcg.nl (T.B.); m.n.hylkema@umcg.nl (M.N.H.); j.k.burgess@umcg.nl (J.K.B.)

[2] W.J. Kolff Institute for Biomedical Engineering and Materials Science-FB41, University Medical Center Groningen, University of Groningen, A. Deusinglaan 1, 9713 AV Groningen, The Netherlands; p.k.sharma@umcg.nl

[3] Groningen Research Institute for Asthma and COPD (GRIAC), University Medical Center Groningen, University of Groningen, Hanzeplein 1 (EA11), 9713 AV Groningen, The Netherlands

[4] Department of Biomedical Engineering-FB40, University Medical Center Groningen, University of Groningen, A. Deusinglaan 1, 9713 AV Groningen, The Netherlands

* Correspondence: m.c.harmsen@umcg.nl

† These authors contributed equally.

Citation: Martinez-Garcia, F.D.; de Hilster, R.H.J.; Sharma, P.K.; Borghuis, T.; Hylkema, M.N.; Burgess, J.K.; Harmsen, M.C. Architecture and Composition Dictate Viscoelastic Properties of Organ-Derived Extracellular Matrix Hydrogels. *Polymers* **2021**, *13*, 3113. https://doi.org/10.3390/polym13183113

Academic Editors: Stefan Ioan Voicu and Andrada Serafim

Received: 16 August 2021
Accepted: 13 September 2021
Published: 15 September 2021

**Publisher's Note:** MDPI stays neutral with regard to jurisdictional claims in published maps and institutional affiliations.

**Abstract:** The proteins and polysaccharides of the extracellular matrix (ECM) provide architectural support as well as biochemical and biophysical instruction to cells. Decellularized, ECM hydrogels replicate in vivo functions. The ECM's elasticity and water retention renders it viscoelastic. In this study, we compared the viscoelastic properties of ECM hydrogels derived from the skin, lung and (cardiac) left ventricle and mathematically modelled these data with a generalized Maxwell model. ECM hydrogels from the skin, lung and cardiac left ventricle (LV) were subjected to a stress relaxation test under uniaxial low-load compression at a 20%/s strain rate and the viscoelasticity determined. Stress relaxation data were modelled according to Maxwell. Physical data were compared with protein and sulfated GAGs composition and ultrastructure SEM. We show that the skin-ECM relaxed faster and had a lower elastic modulus than the lung-ECM and the LV-ECM. The skin-ECM had two Maxwell elements, the lung-ECM and the LV-ECM had three. The skin-ECM had a higher number of sulfated GAGs, and a highly porous surface, while both the LV-ECM and the lung-ECM had homogenous surfaces with localized porous regions. Our results show that the elasticity of ECM hydrogels, but also their viscoelastic relaxation and gelling behavior, was organ dependent. Part of these physical features correlated with their biochemical composition and ultrastructure.

**Keywords:** extracellular matrix; ECM hydrogel; viscoelasticity; decellularized organs; Maxwell model

## 1. Introduction

The extracellular matrix (ECM) is the acellular component of all organs and tissues: a three-dimensional (3D) mixture of proteins embedded in a gel of water-retaining negatively charged polysaccharides such as glycosaminoglycans (GAGs) [1]. While the ECM composition is tissue-specific, its components and organization can vary among structures within the same organ [2,3]. The ECM guides cell fate and provides mechanical support to the cells embedded within [4,5].

Historically, ECM mechanics were solely evaluated in terms of elasticity (i.e., elastic modulus often denoted by $E$, and also called stiffness)—the resistance of an object to undergo reversible deformation (strain, $\varepsilon$) in response to applied force (Stress, $\sigma$) [6,7]. In purely elastic materials, the mechanical energy is stored as strain and the elastic modulus remains strain rate independent [8]. Strain rate ($\dot{\varepsilon}$) is the speed with which a material is compressed. Due to variations in ECM composition and organization, the response to mechanical stress and strain differs among organs [9]. Nevertheless, more recent studies have shown that due to a large water content, the ECM is not elastic but viscoelastic in nature [7,10,11], where viscosity plays an active role in matrix mechanics [7]. Viscosity is a material property that arises from the resistance of a fluid to deformation. The combination of both elastic and viscous responses leads to a time-dependent stress dissipation (i.e., stress relaxation), a phenomenon known as viscoelasticity [8]. Unlike purely elastic materials that store and retain energy, a viscoelastic material will dissipate energy in the presence of stress over time, making the elastic modulus strain rate dependent. Thus, viscoelasticity is an inherent property of the ECM, that has only recently been recognized within biological systems [7,10,11].

The common in vitro systems to mimic in vivo ECM are hydrogels: water-swollen polymeric networks [12–15]. The viscoelasticity of hydrogels is tailored by varying the type and molecular weight of the constituent polymers, their concentration, and the crosslinking conditions (e.g., temperature, pH) [16–20]. Using hydrogels as ECM mimics illustrates that the cells perceive the surrounding viscoelasticity by applying a pushing or pulling force and sensing the time-dependent deformation response from the environment [10,21].

Hydrogels can be prepared from individual ECM components such as collagen, elastin or non-sulphated GAGs such as hyaluronic acid [18,19,22,23]. These homopolymeric hydrogels demonstrate the influence of individual matrix components in fundamental aspects of cell biology [18,19]. While collagen and elastin are the major load-bearing and elastic ECM components, other molecules, such as GAGs, hold on to water and play a role in matrix mechanics by offering resistance to the mobility of water within the ECM [24]. The main role of GAGs in ECM mechanics is that of lubrication and stress absorption. While homopolymeric networks demonstrate the contribution of individual components in matrix-cell biology, a factual representation of matrix viscoelasticity requires the presence of the native, heterogeneous components from the ECM. Thus, using organ-derived ECM might provide a biomimetic model that, due to their source, retains native ECM components involved in matrix mechanics [25].

We set out to compare three organs that are continuously subjected to mechanical forces but differ in function, i.e., the skin, lung and left ventricle (LV) of the heart. Skin is pliable and deformed due to body movement, while the lungs undergo inflation/deflation cycles via the action of the diaphragm and compression of the chest. Finally, the heart is a continuously beating muscle. The molecular composition of the ECM shares the presence of collagen type I, while organ-specific differences exist that relate to the function of the skin, lung and LV. We hypothesized that the composition and architecture of the ECM of skin, lung and LV dictates their mechanical properties in particular viscoelasticity. We set out to test this hypothesis using hydrogels from decellularized skin, lung and LV.

## 2. Materials and Methods

### 2.1. Decellularization

Porcine skin, lungs and hearts (~6-month, female) were purchased from a local slaughterhouse (Kroon Vlees, Groningen, the Netherlands). The heart and skin were decellularized as described previously [26]. The lung was decellularized as described by Pouliot et al. [27] with the exception that in this case the lungs were finely blended prior to decellularization.

The dissected LVs from pig heart and skin were dissected into 1 cm$^3$ cubes. The tissues were washed with 1x Dulbecco's phosphate-buffered saline (DPBS; BioWhittaker®, Walkersville, MD, USA) at room temperature (RT) and then minced in a commercial Bourgini 21.3001 blender (Bourgini, Breda, the Netherlands) with DPBS until the pieces were ~1 mm$^3$ in size. After a second DPBS wash, the tissue homogenate was sonicated for 1 min at 100% power. After sonication, the tissue homogenate was washed with DPBS again and incubated in 0.05% trypsin (Thermo Fisher Scientific, Waltham, MA, USA) at 37 °C for 3 h under constant shaking. After trypsin treatment, tissue material was washed again with DPBS and frozen at −20 °C for at least 24 h.

The homogenate was thawed, washed with Milli-Q® water for 3 h and then sequentially treated with saturated NaCl (6 M) solution, 70% ethanol, 1% SDS solution (Sigma-Aldrich, St. Louis, MO, USA), 1% Triton X-100 (Sigma-Aldrich), 1% sodium deoxycholate (Sigma-Aldrich) and 30 μg/mL DNase (Sigma-Aldrich) (in $MgSO_4$ 1.3 mM and $CaCl_2$ 2 mM), with three washes with Milli-Q® water between treatments, 24 h each at RT with constant shaking, except for the enzymatic treatments, which were at 37 °C while shaking. The resultant decellularized ECM was stored in sterile DPBS containing 1% penicillin-streptomycin (Gibco Invitrogen, Carlsbad, CA, USA) at 4 °C.

The lung was dissected, and the cartilaginous airways removed before cutting into ~1 $cm^3$ cubes and minced until it was ~1 $mm^3$ in size with a commercial blender. The resulting tissue homogenate was then repeatedly washed with Milli-Q® water and spun down at 3000× *g* until the supernatant cleared completely. The remaining tissue homogenate went through two rounds of sequential treatment with 0.1% Triton X-100, 2% sodium deoxycholate, 1 M NaCl solution and 30 μg/mL DNase in 1.3 mM $MgSO_4$ and 2 mM $CaCl_2$, 10 mM Tris pH 8 (Sigma-Aldrich) solution each for 24 h at 4 °C with constant shaking, except for the enzymatic treatments, which were at 37 °C with shaking. Between treatments, the homogenate was washed three times with Milli-Q® water, centrifuged at 3000× *g* between washes. After two cycles of decellularization, the tissue homogenate was sterilized by adding 0.18% peracetic acid and 4.8% ETOH, and left shaking at 4 °C for 24 h. After tissue sterilization, the resultant decellularized ECM was stored in sterile DPBS containing 1% penicillin-streptomycin at 4 °C (Figure 1a).

### 2.2. Hydrogel Preparation

The blended, decellularized skin-ECM, lung-ECM and LV-ECM were snap-frozen in liquid nitrogen and lyophilized with a FreeZone Plus lyophilizer (Labconco, Kansas City, MO, USA) and then ground into a powder with an A11 Analytical mill (IKA, Staufen, Germany). Then, 20 mg/mL of ECM powder was digested with 2 mg/mL porcine pepsin (Sigma-Aldrich) in 0.01 M HCl under constant agitation at RT either for 8 h (LV-ECM) or 48 h (lung-ECM and skin-ECM) (Figure 1b). After digestion, the pH was neutralized with 0.1 M NaOH and brought to 1× DPBS with one-tenth volume 10× DPBS to generate the so-called ECM pre-gel solution. For hydrogel formation, pre-gel from each organ-derived ECM was poured in a mold and incubated at 37 °C for 1 h (Figure 1a). After gelation, the hydrogels were equilibrated in HBSS medium (Lonza, Bazel, Switzerland) and incubated for 24 h prior to experiments.

### 2.3. Turbidity Assay

The gelling kinetics of skin-ECM, lung-ECM and LV-ECM hydrogels were analyzed with a turbidimetric assay [28–30]. The ECM pre-gel solutions were pipetted (150 μL) into a precooled (4 °C) 96-well plate (Corning Inc., Corning, NY, USA). The cooled 96-well plate containing the pre-gels was loaded into a pre-heated (37 °C) CLARIOstar Plus multi-mode microplate Reader (BMG Labtech, Ortenberg, Germany), and the absorbance measured at 405 nm with 30-second intervals for 2 h. Absorbance values were normalized with the following formula:

$$NA = \frac{(A - A_{min})}{(A_{max} - A_{min})} \times 100\% \tag{1}$$

where *NA* is the normalized absorbance, *A* is the absorbance at any given time, $A_{min}$ is the lowest observed absorbance and $A_{max}$ is the maximal absorbance. The normalized curves were plotted to start from gelation, omitting the lag time. From the sigmoidal-shaped turbidity curves, we calculated the following kinetic parameters: $A_{min}$ and $A_{max}$, $T_{lag}$ (the time value at which the normalized absorbance is 0), $T_{1/2}$ (the time at which the normalized absorbance is 50%), $T_{end}$ (the time at which the normalized absorbance is 100%) and *S* (the slope of the linear portion of the curve), indicating the speed of gelation. Three independent turbidity measurements were performed with three replicates each (*n* = 3).

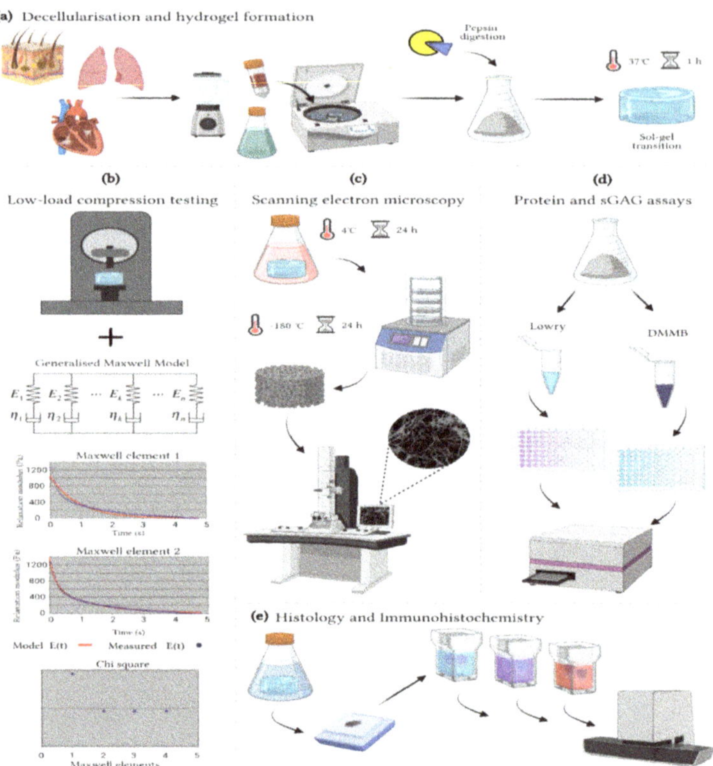

**Figure 1.** Methods. (**a**) Decellularization and hydrogel formation of skin-ECM, lung-ECM and LV-ECM (**b**) Low-load compression testing at 20% strain for 100 s. The data analyzed with a generalized Maxwell model of viscoelasticity. The number of Maxwell elements were determined based on curve-fitting the stress relaxation data (Relaxation modulus; Pa). The figure shows skin-ECM data, where two Maxwell elements were sufficient to explain their viscoelasticity, confirmed by the decrease in $Chi^2$. (**c**) Scanning electron microscopy (SEM). All hydrogels were fixed for 24 h in 2% glutaraldehyde and 2% paraformaldehyde (1:1 ratio) at 4 °C, freeze-dried for 24 h, metal coated and visualized with SEM. (**d**) Protein and sulphated glycosaminoglycans (sGAGs) quantification with Lowry and DMMB assays. (**e**) Histology and Immunohistochemistry. Hydrogels were fixed for 24 h in 2% formalin, processed conventionally with a graded ethanol series for dehydration, paraffin embedded and sectioned. Sections (5 μm) were stained with Alcian Blue, Picrosirius Red (PSR) and Masson's Trichrome (MTC) as well as immune stained for collagen type I (COL1A1) and Elastin and scanned with a Hamamatsu section scanner. (Figure created with BioRender).

### 2.4. Viscoelastic Properties

The elastic modulus and stress relaxation properties of skin-ECM, lung-ECM and LV-ECM hydrogels were evaluated on the Low-Load Compression Tester (LLCT) as described previously [26,31,32]. Data were acquired with LabVIEW 7.1, and subsequently analyzed with MatLab 2018 (MathWorks® Inc., Natick, USA). Hydrogels (300 μL) underwent uniaxial compression with a 2.5-millimeter diameter plunger at three different locations, leaving 2 mm from the edge and 2.5 mm between each compression site. When compressed, each hydrogel reached 80% of its original thickness (i.e., strain, $\varepsilon$ = 0.2) at a strain rate ($\dot{\varepsilon}$) of 0.2 s$^{-1}$ (or a deformation rate of 20%·s$^{-1}$) at room temperature ($\cong$25 °C). The elastic modulus was determined as the slope between the stress–strain curve.

After compression, the strain was kept constant (0.2) for 100 s, and the stress was continuously monitored. The time to reach 50% stress relaxation was determined by comparing the stress relaxation percentage at $t = 0$ s and $t = 100$ s. The relaxing stress as a function of time ($\sigma(t)$) was divided by the constant strain of 0.2 to obtain the value of continuously decreasing modulus $E(t)$. Data were acquired according to a generalized Maxwell model, Equation (2), to calculate the values of $E_i$ and $\tau_i$ for individual Maxwell elements, where $i$ varies from 1 to $n$. The $\tau_i$ is the relaxation time constant for each individual Maxwell element and is the ratio of $\eta_i$ (dashpot) and $E_i$ (spring) for that element (Figure 1). The number of Maxwell elements necessary to fit the experimental data were determined by visually fitting a plot that shows the decrease in $X^2$ value with the addition of every extra Maxwell element (Figure 1b). The required number of Maxwell elements were chosen when no further decrease in Chi$^2$ was observed (Figure 1b).

$$E(t) = E_1 e^{-t/\tau_1} + E_2 e^{-t/\tau_2} + E_3 e^{-t/\tau_3} + \ldots E_n e^{-t/\tau_n} \tag{2}$$

The relative importance ($R_i$) of each Maxwell element in terms of percentage within the relaxation process was expressed as the proportion of its spring constant, $E_i$, to the sum of all spring constants Equation (3).

$$R_i = 100 \times \frac{E_i}{\sum_{i=1}^{n} E_i} \tag{3}$$

### 2.5. Protein Quantification

Total protein content was quantified with a Pierce™ Modified Lowry Protein Assay Kit (Thermo Fisher Scientific). For this, 2 µL of pre-gel solution was diluted in 38 µL of 1x DPBS and transferred to a well in a non-adhesive Costar® 96-well plate (Corning Inc., Kennebunk, ME, USA). Next, 200 µL of modified Lowry protein assay was added per well before incubation at RT for 10 min. Fresh 1 N Folin-Ciocalteu's phenol reagent was prepared by diluting 2 N Folin-Ciocalteu's phenol reagent with an equal volume of Milli-Q® water and 20 µL of this solution was added per well and incubated at RT for 30 min. The absorbance was read at 750 nm with Benchmark Plus™ microplate spectrophotometer system (Bio-Rad, Hercules, CA, USA). The protein concentration was determined based on a calibration curve derived from a dilution series of bovine serum albumin (Thermo Fisher Scientific). DPBS served as absorbance blanks. The protein concentration (µg/mL) from each organ-derived ECM was calculated from four independent experiments each performed in triplicate (Figure 1d).

### 2.6. Sulphated Glycosaminoglycans (sGAGs) Quantification

Total sGAGs content was quantified with a 1,9-Dimethyl-Methylene Blue zinc chloride double salt (DMMB) assay based on reported protocols [26,33,34]. For this, 20–25 mg of ECM powder was incubated in 300 µL of 75 mM NaCl, 25 mM EDTA, 50 µL of 10% SDS, and 5 µL of proteinase K (19.9 mg/mL) (Thermo Scientific) at 60 °C overnight. Next, 20 µL of digested organ-derived ECM was mixed with 200 µL of DMMB solution (comprising 19 mg DMMB in 40 mM Glycine, 38 mM NaCl, 100 mM acetic acid pH 3) and the absorbance read at 525 nm immediately. Serial dilutions of chondroitin sulphate (Sigma-Aldrich) were used for the calibration curve, and the absorbance blank corrected with DMMB solution. The total sGAGs content (µg/mL) from each organ-derived ECM was calculated from four independent experiments each performed in triplicate (Figure 1d).

### 2.7. Histological Characterisation

Hydrogels were fixed with 2% formalin for 24 h at 4 °C. All samples were conventionally processed for histology using a graded alcohol to dehydrate and paraffin embedded. Sections of 1 µm thickness were deparaffinized and stained with Alcian Blue (pH 2.5) to visualize GAGs [35] and Masson's Trichrome (MTC) and Picrosirius Red (PSR) to visualize collagens [36,37], following the previously cited protocols. Slides were covered with Permount™ Mounting Medium (Fisher Chemical™, Waltham, MA, USA) (Figure 1e). Analyses derived from histology staining are detailed in Section 2.9.

### 2.8. Immunohistochemistry

Sections (4 µm) were deparaffinized and incubated with citrate buffer for antigen retrieval. After blocking to prevent non-specific background staining, sections were incubated with 1 µg/mL of mouse anti-human COL1A1 (Abcam, Cambridge, UK) or 1 µg/mL of goat anti-human elastin (Cedarlane, Burlington, VT, USA), respectively, at 4 °C, overnight. After 3 DPBS washes, sections were incubated with a 1:100 dilution of an anti-mouse horse radish peroxidase conjugate (Dako, Santa Clara, CA, USA) or a 1:100 dilution of an anti-goat horse radish peroxidase conjugate (Dako) at RT for 1 h. The staining was then visualized with Vector® NovaRED™ (Vector Laboratories, Burlingame, USA). Slides were counterstained with Hematoxylin and covered with Permount® mounting media (Figure 1e).

### 2.9. Imaging and Image Analysis

All stained sections were scanned with a Slide Scanner (Hamamatsu Photonics K.K., Herrsching, Germany) at 20× magnification (Figure 1e). PSR fluorescent images (PSR-fluo) were generated with Zeiss LSM 780 CLSM confocal microscope (Carl Zeiss NTS GmbH, Oberkochen, Germany), $\lambda_{ex}$ 561 nm/$\lambda_{em}$ 566/670 nm at 40× magnification [38]. COL1A1, Elastin scans and PSR-fluo images were analyzed with TWOMBLI, an ImageJ/Fiji [39] plugin to quantify patterns in ECM [40]. Before analyzing the COL1A1 and Elastin, the Vector® NovaRED™ color was isolated from the images using a color deconvolution plugin [41]. The images with only Vector® NovaRED™ color were subsequently used for the analysis.

TWOMBLI was used to determine the number of fibers, end points, branching points, total fiber length and alignment, lacunarity (number and size of gaps in the matrix) and high-density matrix proportion (measure for compactness of matrix).

### 2.10. Hydrogel Ultrastructure

Hydrogel ultrastructure was visualized with scanning electron microscopy (SEM). First, all hydrogels were fixed with a 1% paraformaldehyde, 1% formalin at 4 °C for 24 h. Then, the hydrogels were washed three times with DPBS and once with Milli-Q® water to remove any remaining fixatives and salts. The hydrogels were plunged in liquid nitrogen and freeze-dried. Dried samples were glued on top of 0.5″ SEM pin stubs (Agar Scientific, Stansted, UK) and Au-Pd coated after rinsing with Argon with Leica EM SCD050 sputter coater device (Leica Microsystems B.V., Amsterdam, Netherlands). Hydrogels were visualized at 5000× and 10,000× magnification, at 3.0 kV with Zeiss Supra 55 STEM (Carl Zeiss NTS GmbH) (Figure 1c).

### 2.11. Statistical Analysis

All statistical analyzes were performed using GraphPad Prism v9.1.0 (GraphPad Company, San Diego, USA). All data were scrutinized for outliers using the robust regression and outlier removal (ROUT) test and analyzed for normality using Shapiro–Wilk and D'Agostino and Pearson tests [42–44]. Based on this, LLCT data were analyzed with Kruskal–Wallis and Dunn's post hoc test and with one-way ANOVA and Tukey's post hoc test. Lowry, TWOMBLI and turbidity data were also analyzed with one-way ANOVA and Tukey. DMMB data were analyzed with Student's *t*-test. Graphs are presented as median with quartiles or mean values with standard deviation (SD). All *p* values below * 0.05; ** 0.01; *** 0.001 and **** 0.0001 were considered statistically significant.

### 3. Results

### 3.1. Turbidity

Measuring the gelation kinetics of ECM hydrogels using turbidimetric analysis is based on the increased turbidity during gelling, i.e., increased absorbance. The quantitative breakdown of the following parameters from these curves: $A_{min}$, $A_{max}$, $T_{lag}$, $T_{1/2}$, $T_{end}$ and $S$, is shown in Table 1.

**Table 1.** Summary of Gelation kinetics parameters for organ derived ECM hydrogels [1].

| Organ ECM | $A_{min}$ | $A_{max}$ | $T_{lag}$ (min) | $T_{1/2}$ (min) | $T_{end}$ (min) | S (%/ min) |
|---|---|---|---|---|---|---|
| Skin | 2.3 ± 0.2 ****b | 3.5 ± 0.0 ****b | 13.1 ± 2.9 | 12.2 ± 2.9 ****b | 16.9 ± 3.2 ****b | 5.3 ± 0.8 ****b |
| Lung | 1.0 ± 0.1 ****c | 1.4 ± 0.2 ****c | 10.4 ± 4.7 | 43.2 ± 16.8 ****c | 94.1 ± 11.7 *c | 1.2 ± 0.3 |
| LV | 1.5 ± 0.1 ****a | 1.8 ± 0.2 ****a | 10.9 ± 6.5 | 20.9 ± 4.3 **a | 75.7 ± 17.5 ****a | 1.5 ± 0.6 ****a |

[1] All data is shown as Mean ± standard deviation (S.D.). Statistical differences between (a) skin-ECM, (b) lung-ECM and (c) left ventricle (LV)-ECM are highlighted and their significance shown: * $p < 0.05$, ** $p < 0.01$ and **** $p < 0.0001$; according to one-way ANOVA and Tukey after robust regression and outlier removal (ROUT).

All the organ-derived ECMs gelated in a sigmoidal pattern that started after a lag period of 10–13 min (Figure 2). The minimum and maximum absorbance ($A_{min}$ and $A_{max}$) remained highest in the skin-ECM, while the lung-ECM and the LV-ECM were lower and comparable. The skin-ECM sol-gel transition was faster than the lung-ECM and the LV-ECM. The total gelling time ($T_{end}$) was, therefore, the shortest in the skin-ECM, followed by the LV-ECM and finally the lung-ECM (Figure 2, Table 1).

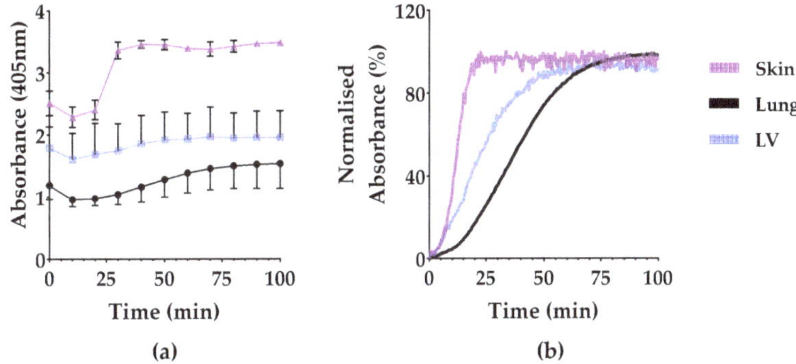


**Figure 2.** Turbidity of skin-ECM, lung-ECM and LV-ECM hydrogels. (**a**) Turbidity data (mean and S.D.) at every 10 min; (**b**) Normalized absorbance from the start of gelation. All data derived from a minimum of three independent experiments performed in triplicates. Data are presented as mean with SD.

### 3.2. Elastic Modulus and Stress Relaxation

The elastic modulus of the skin-ECM (1.66 ± 0.82 kPa) was lower than the lung-ECM (4.98 ± 1.81 kPa) and the LV-ECM (4.38 ± 1.73 kPa) (Figure 3a).

The time to reach 50% stress relaxation was fastest in the skin-ECM (5.16 ± 4.57 s), compared to the lung-ECM (49.40 ± 4.35 s) and lastly the LV-ECM (51.63 ± 1.18 s). The elastic modulus or stress relaxation of the LV-ECM and the lung-ECM did not differ (Figure 3b,c).

### 3.3. Maxwell Analysis

Maxwell analysis showed differences in the relative importance ($R_i$) and the time each Maxwell element remains active (tau; $\tau$) among organ-derived ECM hydrogels.

The fastest or first (1st) Maxwell element had a greater $R_i$ in the skin-ECM (66.42 ± 9.07%), than both the lung-ECM (53.02 ± 3.39%) and the LV-ECM (53.48 ± 4.72%). The intermediate or second (2nd) Maxwell element, had also a greater $R_i$ in the skin-ECM (33.58 ± 9.07%) than in the LV-ECM (33.58 ± 9.07%). The slow or third (3rd) Maxwell element was not detected in the skin-ECM but had a lower $R_i$ in the lung-ECM (16.29 ± 2.15 s) than in the LV-ECM (19.05 ± 3.29 s) (Figure 4a).

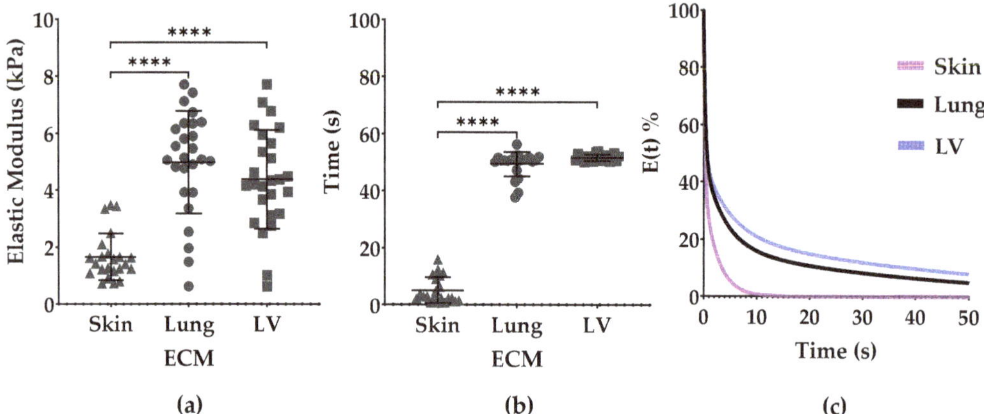

**Figure 3.** Viscoelastic properties of skin-ECM, lung-ECM and LV-ECM. (**a**) Elastic modulus; (**b**) Time to reach 50% stress relaxation; (**c**) Average stress relaxation normalized to start from 100%. All data derived from a minimum of three independent experiments performed in triplicates from low-load compression testing at 0.2 strain. Data are presented as mean with SD. Statistical differences according to one-way ANOVA and Dunn's post hoc test **** $p < 0.0001$.

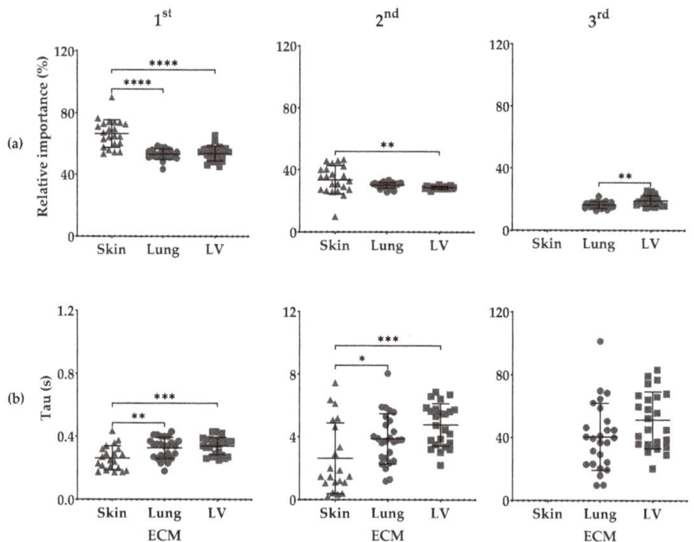

**Figure 4.** Maxwell analysis of skin-ECM, lung-ECM and LV-ECM viscoelasticity. (**a**) Relative importance of 1st, 2nd and 3rd Maxwell elements; (**b**) Tau ($\tau$) of 1st, 2nd and 3rd Maxwell elements reported in seconds (s). All data derived from a minimum of three independent experiments performed in triplicates. Data are presented as mean with SD. Statistical differences according to one-way ANOVA and Tukey (1st and 2nd Maxwell Elements) and Student's $t$-test (3rd Maxwell Element) * $p < 0.05$; ** $p < 0.01$; *** $p < 0.001$ and **** $p < 0.0001$.

The $\tau_1$ from first 1st Maxwell element remained active for less time in the skin-ECM ($0.26 \pm 0.08$ s) than in the lung-ECM ($0.33 \pm 0.07$ s) and the LV-ECM ($0.34 \pm 0.06$ s). The $\tau_2$ was also active for a shorter time in the skin-ECM ($2.63 \pm 2.25$ s) than in the lung-ECM ($3.88 \pm 1.60$ s) and the LV-ECM ($4.78 \pm 1.35$ s). No differences were found between the $\tau_3$ of the lung-ECM ($40.41 \pm 21.33$ s) and the LV-ECM ($51.24 \pm 18.14$ s) (Figure 4b).

### 3.4. Protein and sGAGs Content

The protein content of the pre-gels of the skin-ECM (973 ± 207 µg/mL), lung-ECM (1029 ± 154 µg/mL) and LV-ECM (912 ± 98 µg/mL) did not differ (Figure 5a). The pre-gels of the skin-ECM (202 ± 39 µg/mL) had significantly higher sGAGs contents than the LV-ECM (11 ± 10 µg/mL ****). In the lung-ECM, the sGAGs were below the detection limit of the DMMB assay Figure 5b).

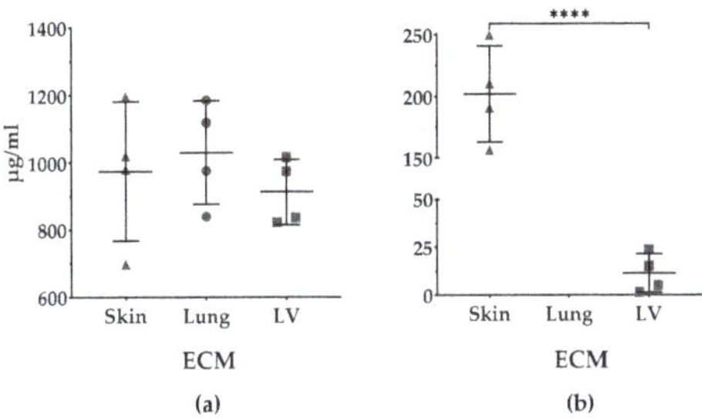

**Figure 5.** Quantification of protein and sulfated GAGs (sGAGs) of organ-derived ECM. (**a**) Protein content (µg/mL) of skin-ECM, lung-ECM and LV-ECM according to Lowry assay; (**b**) sGAGs content (µg/mL) skin-ECM and LV-ECM and according to DMMB assay. No sGAGs were detected in lung-ECM. All data derived from four independent experiments performed in triplicates. Data are presented as mean with SD. Mean values per experiment (n = 4) are also shown. Lowry data analyzed with one-way ANOVA (*p* = ns). DMMB data analyzed with Student's *t*-test **** *p* < 0.0001.

### 3.5. Histologic Assessment

Histological staining of the skin-ECM, the lung-ECM and the LV-ECM with Alcian Blue Masson's Trichrome (MTC) and Picrosirius Red (PSR) is shown in Figure 6.

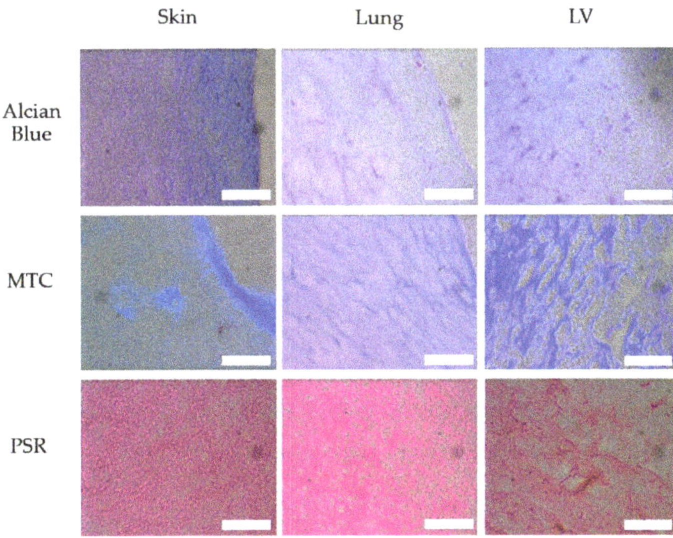


**Figure 6.** Alcian Blue, MTC and PSR, stains in sections of skin-ECM, lung-ECM and LV-ECM. Scale bars: 50 µm.

All the hydrogels contained detectable levels of proteins and sGAGs that were arranged as fibrous meshworks. The skin-ECM had markedly higher levels of sGAGs by Alcian Blue compared to the lung-EMC and the LV-ECM, which had comparable levels.

The collagen networks in the EMC hydrogels, as visualized with MTC and PSR, showed differences. Collagen was observed as dense, condensed fibrous networks in the LV-ECM (MTC and PSR) and the skin-ECM (PSR). The lung-ECM showed a more finely distributed network of collagen fibers that was intermediate of the skin -ECM and the LV-ECM, which had also bound less dye both in the MTC and PSR stains. The LV-ECM was heterogeneous with large interfibrous areas of irregular size, while both the lung-ECM and the skin-ECM appeared more homogeneously organized. Interestingly, this architecture showed more prominently in the skin-ECM after PSR staining because MTC staining showed a more irregular binding.

### 3.6. Matrix Organisation

Both MTC and PSR predominantly stain collagen-type fibers. This was corroborated by the immunohistochemistry for COL1A1, a component of the major tissue collagen, i.e., type I (Figure 7). The collagen I architecture of the LV-ECM showed condensed fibers surrounding large irregularly shaped voids. In contrast, in the lung-ECM and the skin-ECM, the collagen I architecture was comprised of a fine reticular meshwork. The skin-ECM had a higher collagen I content than the lung-ECM. The elastin was distributed in condensed patches in all the ECM hydrogels (Figure 7). It would appear that the LV-ECM contained higher levels of elastin than the skin-ECM and the lung-ECM.

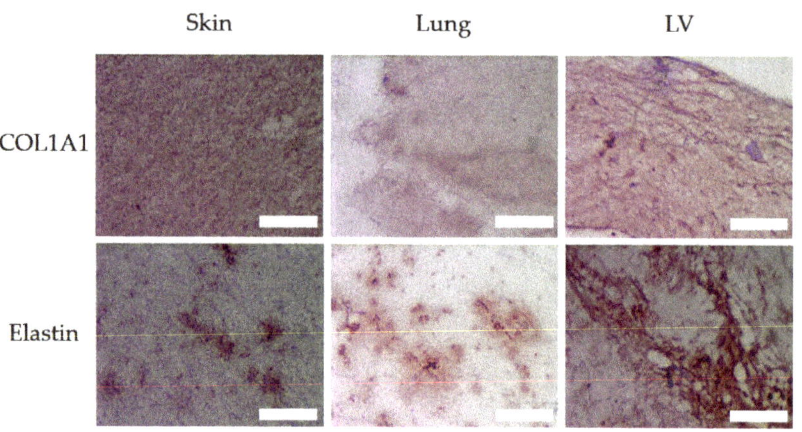


**Figure 7.** Immunohistochemistry (red) of COL1A1 and Elastin of skin, lung and LV ECM hydrogels. Scale bars represent 50 μm.

Fluorescent imaging of PSR-stained sections (Figure 8) was used to run detailed analyses of the collagen fibers with respect to size, shape and organization (Twombli, Table 2).

### 3.6.1. Number of Fibers and Length (Mean and Total)

The mean and total number and length of the immuno-stained collagen type I fibers differed between all three organ-derived ECM hydrogels (Figure 7, Table 2). The skin-ECM had less elastin fibers, that were also shorter compared to the LV-ECM. The lung-ECM had shorter elastin fibers than the LV-ECM (Figure 7, Table 2). Histochemical picrosirius fluoro-micrographs revealed less fibers in the skin-ECM and the LV-ECM compared to the lung-ECM (Figure 8, Table 2). The mean fiber length of the lung-ECM was longer compared to the skin-ECM and the LV-ECM (Figure 7, Table 2).

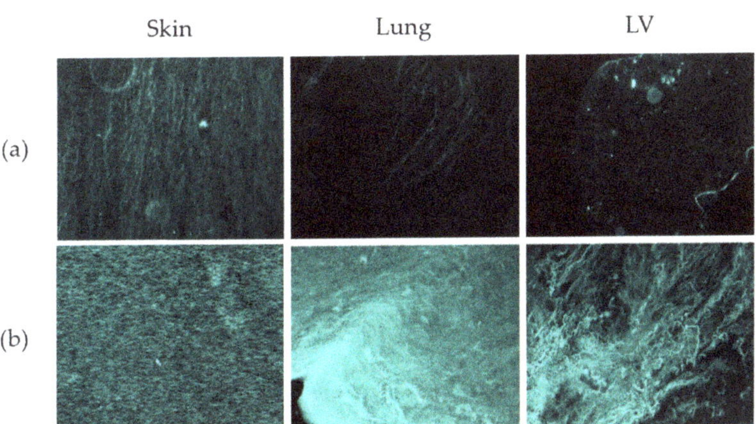


**Figure 8.** Fluorescent micrographs of PSR stained skin, lung and LV ECM hydrogels. (**a**) Unstained sections (autofluorescence); (**b**) PSR-stained sections. Original objective magnification 40×.

**Table 2.** Results from TWOMBLI Analysis.

| Organ ECM | Number of Fibers | [1] Mean Fiber Length | [1] Total Fiber Length | Fiber Alignment | Number of Branch Points | Number of End Points | Lacunarity | [2] HDMˆ1000% |
|---|---|---|---|---|---|---|---|---|
| **COL1A1** | | | | | | | | |
| Skin | $15.3 \pm 1.6$ | $25.3 \pm 0.5$ | $8504 \pm 2064$ | $0.19 \pm 0.04$ | $121.3 \pm 81.8$ | $552 \pm 95$ | $14.7 \pm 0.8$ | $82.7 \pm 44.0$ |
| Lung | $14.9 \pm 0.7$ | $25.4 \pm 1.2$ | $28{,}316 \pm 17{,}989$ | $0.11 \pm 0.07$ | $350.7 \pm 254.0$ | $1902 \pm 1179$ | $16.7 \pm 3.1$ | $0.9 \pm 0.1$ *a |
| LV | $13.9 \pm 1.6$ | $24.3 \pm 1.2$ | $15{,}337 \pm 785$ | $0.11 \pm 0.01$ | $158.7 \pm 39.0$ | $1106 \pm 45$ | $21.6 \pm 4.7$ | $4.2 \pm 2.7$ *a |
| **Elastin** | | | | | | | | |
| Skin | $10.7 \pm 1.7$ *c | $19.3 \pm 2.3$ *c | $2852 \pm 270$ **c | $0.04 \pm 0.01$ | $27.0 \pm 7.8$ *c | $273 \pm 55c$ ***c | $95.2 \pm 15.6$ ***c | $9.1 \pm 6.1$ **c |
| Lung | $11.7 \pm 0.7$ | $26.0 \pm 2.7$ | $9349 \pm 1330$ *c | $0.04 \pm 0.04$ | $95.7 \pm 23.2$ *c | $808 \pm 153$ **c | $98.5 \pm 16.0$ ***c | $11.2 \pm 0.9$ **c |
| LV | $16.09 \pm 2.60$ | $26.0 \pm 2.7$ *a | $17{,}869 \pm 5099$ | $0.09 \pm 0.12$ | $259.0 \pm 109.7$ | $1101 \pm 169$ | $18.5 \pm 1.7$ | $89.5 \pm 33.0$ |
| **Picrosirius Red (Fluorescent)** | | | | | | | | |
| Skin | $14.19 \pm 1.08$ *b | $27.8 \pm 1.9$ *b | $33{,}481 \pm 20{,}453$ *b | $0.07 \pm 0.01$ | $54.0 \pm 41.7$ | $2322 \pm 1382$ **b | $95.4 \pm 32.4$ ****b | $933.5 \pm 35.1$ |
| Lung | $16.71 \pm 2.61$ | $32.5 \pm 5.0$ | $18{,}818 \pm 2972$ | $0.07 \pm 0.04$ | $35.0 \pm 16.7$ | $18{,}818 \pm 2972$ *c | $530.0 \pm 129.0$ | $31.1 \pm 7.0$ ****a |
| LV | $12.66 \pm 1.11$ **b | $24.7 \pm 2.1$ **b | $30{,}346 \pm 11{,}839$ **b | $0.07 \pm 0.02$ | $62.5 \pm 26.8$ | $30{,}346 \pm 11{,}839$ ****a | $136.4 \pm 29.0$ ****b | $59.3 \pm 35.7$ ****a |

[1] Data in micrometers (μm). [2] High Density Matrix (%). All data are shown as Mean $\pm$ standard deviation (S.D.). Statistical differences between (a) skin-ECM, (b) lung-ECM and (c) LV-ECM are highlighted and their significance shown: * $p < 0.05$; ** $p < 0.01$; *** $p < 0.001$ and **** $p < 0.0001$ according to one-way ANOVA and post hoc Tukey after robust regression and outlier removal (ROUT).

### 3.6.2. Branch Points and End Points

The number of immuno-stained collagen type I branch points and end points did not differ between the organ ECM hydrogels (Figure 7, Table 2). The number of immuno-stained elastin branch points and end points were both lower in the skin-ECM and the lung-ECM, compared to the LV-ECM (Figure 7, Table 2). Histochemical picrosirius fluoro-micrographs also showed similar numbers of fiber branch points in all three ECM hydrogels (Figure 8, Table 2). Yet, the number of end points was lower in the skin-ECM than in the lung-ECM and the LV-ECM. Additionally, the lung-ECM had less end points than the LV-ECM (Figure 8, Table 2).

### 3.6.3. Lacunarity and High-Density Matrix (HDM)

The lacunarity of the collagen A1 distribution (Figure 7) did not differ between the ECM hydrogels from the skin, lung and LV, while the high-density matrix was larger in the skin-ECM than both the lung-ECM and the LV-ECM. The lacunarity of the elastin distribution (Figure 7) was higher in the skin-ECM and the lung-ECM compared to the LV-ECM. In contrast, the high-density matrix was smaller in both the skin-ECM and the lung-ECM compared to the LV-ECM. In the Picrosirius red-stained fluoro-micrographs

(Figure 8), the lacunarity was smaller in the skin-ECM and the LV-ECM than in the lung-ECM. In the Picrosirius red-stained micrographs, the high-density matrix was larger in the skin-ECM than in the lung-ECM and the LV-ECM.

### 3.7. Ultrastrucure

The ultrastructure of hydrogels, visualized with SEM, showed qualitative differences between the skin-ECM and both the lung-ECM and the LV-ECM. In both the lung-ECM and the LV-ECM, most of the surface displayed a sheet-like organization, with randomly scattered openings. Beneath these sheets, a fibrous network could be discerned. In contrast, the skin-ECM lacked these condensed sheets and was comprised of a fibrous network with irregularly shaped pores with fibrils of a similar thickness (Figure 9).

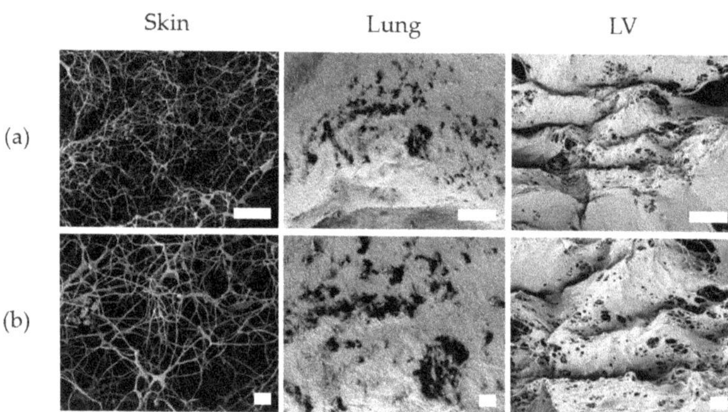


**Figure 9.** Surface microarchitecture of skin-ECM, lung-ECM and LV-ECM. (**a**) Magnification at 5000×; (**b**) Magnification, 10,000×. Scale bars: 10 (**a**) and 2 μm (**b**).

## 4. Discussion

Our study shows that the ECM hydrogels from the skin, lung and LV of the heart, at an equal protein concentration, differ distinctly in composition, gelling kinetics and viscoelasticity. The mechanical properties and ultrastructure of the lung-ECM and the LV-ECM were similar. The content of sGAG was higher in the skin-ECM than the LV-ECM, while sGAGs were below the detection limit in the lung-ECM. Turbidity assays demonstrated that the skin-ECM had a higher starting absorbance and a faster sol-gel transition than in the lung-ECM and the LV-ECM. The skin-ECM hydrogels had a lower elastic modulus and faster stress relaxation than the lung-ECM and the LV-ECM. The surface topography of the skin-ECM hydrogels was more porous than the lung-ECM and the LV-ECM and the TWOMBLI analyses illustrated differences in the collagen and elastin fiber-related parameters (length, density, number, among others).

The first indication of an organ-specific ECM composition and ultrastructure is the difference in the time needed for pepsin digestion (skin, 8 h; lung and LV, 48 h). Depending on the composition, crosslinking and, consequently, the ultrastructure of the ECM, it will require more or less "cutting" by pepsin in order to be solubilized. This difference in pepsin digestion times as well as the decellularization methods required has been reported to be organ specific [45]. The decellularization process as well as the pepsin solubilization have been shown to impact the composition and the mechanical properties of ECM hydrogels and need to be optimized to each tissue or organ [38,39].

All the organ-derived ECM hydrogels had a unique turbidity profile and kinetic parameters, which implies that the composition of these ECM hydrogels differ with respect to their assembly mechanism and kinetics. The differences in the composition of the ECM can impact the gelling process, where the addition of the proteoglycan decorin increased the lag phase of collagen 1 gelation and collagen V can regulate the fiber diameter of collagen [46,47]. The increased total gelling time in the lung and LV-ECM could be due to the loss of collagen 1 telopeptides with longer pepsin digestion times, which is observed [30,48]. Turbidity analyses provide an insight into gelling kinetics but do not give information on the molecular assembly and fibril ultrastructure, the endpoint of which we evaluated using SEM and staining.

The elastic modulus of the skin-ECM hydrogels was 50% lower than the lung-ECM and the LV-ECM, consistent with previously reported data [26,31]. The stress relaxation of the skin-ECM was an order of magnitude faster than the lung-ECM and the LV-ECM. Stress dissipation occurs via the displacement of water and also via polymer network rearrangements. The speed and kinetics of each process can be mathematically modelled using Maxwell analysis [49]. Each rearrangement process is represented by a Maxwell element active for a definite time period (tau; $\tau$) and has a definite contribution to the overall relaxation process, i.e., the relative importance ($R_i$). A previous study showed that the stiffness and viscoelastic properties of ECM hydrogels still resemble that of the organ of origin [31]. The incorporation of cells into hydrogels may help bridge the gap, the small difference between tissue and ECM hydrogel mechanics, as well as to explain their contribution to the viscoelastic relaxation process of organs and tissues. Interestingly, the ECM hydrogels from the two organs that experience continuous rhythmic mechanical stresses had similar stiffnesses and viscoelasticity, in contrast to skin.

Maxwell analyses showed that the skin-ECM had only two Maxwell elements, while the lung-ECM and the LV-ECM had three. The skin-ECM's first and second Maxwell elements (fast and intermediate) had a greater $R_i$ with a lower $\tau$ than in the lung-ECM and the LV-ECM. These elements are associated with the fastest stress dissipating components of the ECM, such as water and small molecules (e.g., growth factors). GAGs are among the ECM molecules that strongly bind, and thus resist the displacement of, water [50]. The quantification of sGAGs indicated a greater presence in the skin-ECM than in the LV-ECM but was below detection level in the lung-ECM.

Information about the role of (s)GAGs in organ and tissue viscoelasticity is limited. One study showed that stress relaxation decreases in GAG-depleted arteries [24]. Depleting sGAGs such as chondroitin sulphate, dermatan sulfate and heparan sulfate in lung tissue was shown to increase stress relaxation, whereas depleting hyaluronic acid had no effect [51]. No sGAGs were detected in the lung-ECM according to the DMMB assay, but Alcian Blue staining did detect the presence of GAGs in general. This finding might indicate that the non-sulfated GAG hyaluronic acid (HA) is an abundant GAG in lung-ECM hydrogels. In native lung tissue, HA is the most abundant GAG [52]. HA would not be detected using the DMMB assay owing to its non-sulfated nature [33]. Other authors have reported that detergent-based lung decellularization causes significant loss of both types of GAGs, as well as the depletion of specific sulfation patterns [53]. Our findings also indicate a loss in lung-ECM-derived sGAGs. Further studies should expand on the proteomic characterization of an organ-derived ECM, both from the original organ, as well as the resulting ECM after decellularization through mass-spectrometry. This information might give insights to the contribution of individual matrix components to viscoelasticity.

The differences among the hydrogels may not only depend on the composition but also on the architecture and conformation of the matrix. TWOMBLI analyses of the immunohistochemistry showed that COL1A1 had a higher density in the skin-ECM hydrogels than the lung-ECM and the LV-ECM. In Elastin, the TWOMBLI analyses showed differences in the fiber number, length, amount and density, among others, in the LV-ECM, compared to the skin-ECM and the lung-ECM. Observing the polymer network under SEM demonstrated that the surface microarchitecture of the LV-ECM and the lung-ECM hydrogels had a condensed layer that displayed a sheet-like organization. Localized regions showed the presence of pores on the surface. In contrast, the skin-ECM hydrogels had an open fiber structure with large fibers and pores. These findings correlate to the mechanical properties observed, as the lung-ECM and the LV-ECM had similar viscoelastic and surface architecture properties, while the skin-ECM was highly porous with lower stiffness and faster stress relaxation. Porosity represents a percentage of void space in a solid [54,55], where an excess of voids can compromise the mechanical stability of materials [56]. The swelling of hydrogels might have resulted in a different water content depending on the organ source of ECM. We recognize that not evaluating this is a limitation of our study. GAGs and proteoglycans contribute to the water retention of the ECM. The compositional differences in these proteins may affect the swelling and, consequently, also the mechanical properties of the ECM hydrogels.

Overall, these findings demonstrate that organ-specific ECM composition idiosyncrasies remain present after decellularization. Discrepancies exist with our colleagues, where a fourth Maxwell element in LV-ECM hydrogels is reported [26]. Compared to the protocol from our colleagues, we employed two hours more for pepsin digestion [26]. The pepsin digestion time influences organ-derived ECM hydrogel mechanics, which could explain the differences observed [29].

The analyses of hydrogel architecture through SEM can produce artifacts because of the required sample drying during the preparation process that sample fixation before freeze-drying may not prevent [57–59]. To corroborate the SEM results, we used fluorescent imaging of the picrosirius red stained sections. A possible way to image the ultrastructure of ECM hydrogels in a wet state would be 5(6)-Carboxytetramethylrhodamine N-succinimidylester (TAMRA-SE) staining in combination with confocal laser scanning microscopy [60,61]. Limitations concerning the LLCT have been addressed in the past by our research group [31].

## 5. Conclusions

In conclusion, organ-derived ECM hydrogels retain their specific composition and, with that, the accompanying mechanical properties and ultrastructure. Organ- or tissue-specific ECM hydrogels provide opportunities for simulating the organ or tissue microenvironment, opening possibilities for use in tissue engineering and as model systems for understanding disease underlying mechanisms. Organ ECM hydrogels enable the generation of novel models for mimicking and incorporating a native organ ECM in a research environment. Further characterizing the ECM composition of organs and ECM hydrogels will allow us to discern how different ECM proteins influence the mechanics, as well as what compositional elements ECM hydrogels need to more completely mimic the native environment. With organ-specific ECM hydrogels, we can explore cell–matrix interactions, which are dynamic and reciprocal. Both factors influence the biomechanical outcome of the cellular environment, leading to changes in both cell fate and ECM composition and mechanics. The crosstalk between cells and the ECM is often dysregulated in disease and, with organ-specific ECM hydrogels, we might elucidate the precise role of the ECM in the pathophysiology.

**Author Contributions:** F.D.M.-G. and R.H.J.d.H. (Conceptualization, methodology, investigation, data curation, visualization, formal analysis, writing—original draft preparation). P.K.S. (Resources, software and validation, formal analysis, writing—review and editing). T.B. (Methodology, investigation, software, data curation). M.N.H. (Supervision, writing—review and editing). J.K.B. and M.C.H. (Conceptualization, supervision, resources, writing—review and editing). F.D.M.-G. and R.H.J.d.H. contributed equally to this work. All authors have read and agreed to the published version of the manuscript.

**Funding:** F.D.M.-G. was funded by the Mexican National Council of Science and Technology (CONACyT; CVU-695528), and the Stichting de Cock-Hadders (2019-45 and 2021-16). This work was supported by ZonMW Grant project number 114021507 (M.N.H), an unrestricted research grant from Astra Zeneca (M.N.H). J.K.B is funded by a Rosalind Franklin Fellowship co-funded by The University of Groningen and European Union.

**Institutional Review Board Statement:** Ethical review and approval were waived for this study, as all materials used were slaughter house waste products.

**Informed Consent Statement:** Not applicable.

**Data Availability Statement:** Data are contained within the article and are available upon reasonable request.

**Acknowledgments:** Part of the work has been performed in the UMCG Microscopy and Imaging Center (UMIC), sponsored by ZonMW grant 91111.006. The authors would like to thank Ben Giepmans and Jeroen Kuipers for their support and technical assistance in SEM. The authors acknowledge the support of Henk Moorlag and Matthijs Blömer for their support in the sectioning of histological material.

**Conflicts of Interest:** The authors declare no conflict of interest. The funders had no role in the design of the study; in the collection, analyses, or interpretation of data; in the writing of the manuscript, or in the decision to publish the results.

# References


1. Theocharis, A.D.; Skandalis, S.S.; Gialeli, C.; Karamanos, N.K. Extracellular Matrix Structure. *Adv. Drug Deliv. Rev.* **2016**, *97*, 4–27. [CrossRef]
2. Bonnans, C.; Chou, J.; Werb, Z. Remodelling the Extracellular Matrix in Development and Disease. *Nat. Rev. Mol. Cell Biol.* **2014**, *15*, 786–801. [CrossRef]
3. Jana, S.; Hu, M.; Shen, M.; Kassiri, Z. Extracellular Matrix, Regional Heterogeneity of the Aorta, and Aortic Aneurysm. *Exp. Mol. Med.* **2019**, *51*, 1–15. [CrossRef]
4. Burgess, J.K.; Mauad, T.; Tjin, G.; Karlsson, J.C.; Westergren-Thorsson, G. The Extracellular Matrix—The Under-recognized Element in Lung Disease? *J. Pathol.* **2016**, *240*, 397–409. [CrossRef]
5. Hastings, J.F.; Skhinas, J.N.; Fey, D.; Croucher, D.R.; Cox, T.R. The Extracellular Matrix as a Key Regulator of Intracellular Signalling Networks: ECM Regulation of Intracellular Signalling Networks. *Br. J. Pharmacol.* **2019**, *176*, 82–92. [CrossRef]
6. Cox, T.R.; Erler, J.T. Remodeling and homeostasis of the extracellular matrix: Implications for fibrotic diseases and cancer. *Dis. Model. Mech.* **2011**, *4*, 165–178. [CrossRef]
7. Elosegui-Artola, A. The extracellular matrix viscoelasticity as a regulator of cell and tissue dynamics. *Curr. Opin. Cell Biol.* **2021**, *72*, 10–18. [CrossRef]
8. Wang, K. Die Swell of Complex Polymeric Systems. In *Viscoelasticity—From Theory to Biological Applications*; De Vicente, J., Ed.; InTech: Rijeka, Croatia, 2012; ISBN 978-953-51-0841-2.
9. Guimarães, C.F.; Gasperini, L.; Marques, A.P.; Reis, R.L. The stiffness of living tissues and its implications for tissue engineering. *Nat. Rev. Mater.* **2020**, *5*, 351–370. [CrossRef]
10. Chaudhuri, O.; Cooper-White, J.; Janmey, P.A.; Mooney, D.J.; Shenoy, V.B. Effects of extracellular matrix viscoelasticity on cellular behaviour. *Nature* **2020**, *584*, 535–546. [CrossRef]
11. Chaudhuri, O.; Gu, L.; Klumpers, D.; Darnell, M.; Bencherif, S.A.; Weaver, J.C.; Huebsch, N.; Lee, H.-P.; Lippens, E.; Duda, G.N.; et al. Hydrogels with tunable stress relaxation regulate stem cell fate and activity. *Nat. Mater.* **2015**, *15*, 326–334. [CrossRef] [PubMed]
12. Harjanto, D.; Zaman, M.H. Modeling Extracellular Matrix Reorganization in 3D Environments. *PLoS ONE* **2013**, *8*, e52509. [CrossRef]
13. Ahmed, E.M. Hydrogel: Preparation, characterization, and applications: A review. *J. Adv. Res.* **2013**, *6*, 105–121. [CrossRef] [PubMed]
14. Lee, J.-H.; Kim, H.-W. Emerging properties of hydrogels in tissue engineering. *J. Tissue Eng.* **2018**, *9*, 2041731418768285. [CrossRef] [PubMed]
15. Lee, K.Y.; Mooney, D. Hydrogels for Tissue Engineering. *Chem. Rev.* **2001**, *101*, 1869–1880. [CrossRef]
16. Laronda, M.M.; Rutz, A.; Xiao, S.; Whelan, K.A.; Duncan, F.E.; Roth, E.W.; Woodruff, T.K.; Shah, R.N. A bioprosthetic ovary created using 3D printed microporous scaffolds restores ovarian function in sterilized mice. *Nat. Commun.* **2017**, *8*, 15261. [CrossRef]
17. Habib, A.; Sathish, V.; Mallik, S.; Khoda, B. 3D Printability of Alginate-Carboxymethyl Cellulose Hydrogel. *Materials* **2018**, *11*, 454. [CrossRef]
18. Doyle, A.D.; Carvajal, N.; Jin, A.; Matsumoto, K.; Yamada, K.M. Local 3D matrix microenvironment regulates cell migration through spatiotemporal dynamics of contractility-dependent adhesions. *Nat. Commun.* **2015**, *6*, 8720. [CrossRef] [PubMed]
19. Sapudom, J.; Rubner, S.; Martin, S.; Kurth, T.; Riedel, S.; Mierke, C.T.; Pompe, T. The phenotype of cancer cell invasion controlled by fibril diameter and pore size of 3D collagen networks. *Biomaterials* **2015**, *52*, 367–375. [CrossRef]
20. Ivankova, E.; Dobrovolskaya, I.P.; Popryadukhin, P.; Kryukov, A.; Yudin, V.E.; Morganti, P. In-situ cryo-SEM investigation of porous structure formation of chitosan sponges. *Polym. Test.* **2016**, *52*, 41–45. [CrossRef]
21. Cantini, M.; Donnelly, H.; Dalby, M.; Salmeron-Sanchez, M. The Plot Thickens: The Emerging Role of Matrix Viscosity in Cell Mechanotransduction. *Adv. Health Mater.* **2019**, *9*, e1901259. [CrossRef] [PubMed]

22. Park, S.-N.; Park, J.-C.; Kim, H.O.; Song, M.J.; Suh, H. Characterization of porous collagen/hyaluronic acid scaffold modified by 1-ethyl-3-(3-dimethylaminopropyl)carbodiimide cross-linking. *Biomaterials* **2001**, *23*, 1205–1212. [CrossRef]
23. Annabi, N.; Mithieux, S.; Boughton, E.A.; Ruys, A.J.; Weiss, A.; Dehghani, F. Synthesis of highly porous crosslinked elastin hydrogels and their interaction with fibroblasts in vitro. *Biomaterials* **2009**, *30*, 4550–4557. [CrossRef]
24. Mattson, J.M.; Turcotte, R.; Zhang, Y. Glycosaminoglycans contribute to extracellular matrix fiber recruitment and arterial wall mechanics. *Biomech. Model. Mechanobiol.* **2016**, *16*, 213–225. [CrossRef]
25. Choudhury, D.; Tun, H.W.; Wang, T.; Naing, M.W. Organ-Derived Decellularized Extracellular Matrix: A Game Changer for Bioink Manufacturing? *Trends Biotechnol.* **2018**, *36*, 787–805. [CrossRef]
26. Liguori, G.R.; Liguori, T.T.A.; De Moraes, S.R.; Sinkunas, V.; Terlizzi, V.; Van Dongen, J.A.; Sharma, P.; Moreira, L.F.P.; Harmsen, M.C. Molecular and Biomechanical Clues from Cardiac Tissue Decellularized Extracellular Matrix Drive Stromal Cell Plasticity. *Front. Bioeng. Biotechnol.* **2020**, *8*, 520. [CrossRef]
27. Pouliot, R.A.; Link, P.; Mikhaiel, N.S.; Schneck, M.B.; Valentine, M.S.; Gninzeko, F.J.K.; Herbert, J.A.; Sakagami, M.; Heise, R.L. Development and characterization of a naturally derived lung extracellular matrix hydrogel. *J. Biomed. Mater. Res. Part A* **2016**, *104*, 1922–1935. [CrossRef] [PubMed]
28. Fernández-Pérez, J.; Ahearne, M. The impact of decellularization methods on extracellular matrix derived hydrogels. *Sci. Rep.* **2019**, *9*, 1–12. [CrossRef]
29. Pouliot, R.A.; Young, B.; Link, P.; Park, H.E.; Kahn, A.R.; Shankar, K.; Schneck, M.B.; Weiss, D.J.; Heise, R.L. Porcine Lung-Derived Extracellular Matrix Hydrogel Properties Are Dependent on Pepsin Digestion Time. *Tissue Eng. Part C: Methods* **2020**, *26*, 332–346. [CrossRef] [PubMed]
30. Kreger, S.T.; Bell, B.J.; Bailey, J.; Stites, E.; Kuske, J.; Waisner, B.; Voytik-Harbin, S.L. Polymerization and matrix physical properties as important design considerations for soluble collagen formulations. *Biopolymers* **2010**, *93*, 690–707. [CrossRef] [PubMed]
31. De Hilster, R.H.J.; Sharma, P.; Jonker, M.R.; White, E.S.; A Gercama, E.; Roobeek, M.; Timens, W.; Harmsen, M.; Hylkema, M.N.; Burgess, J.K. Human lung extracellular matrix hydrogels resemble the stiffness and viscoelasticity of native lung tissue. *Am. J. Physiol. Cell. Mol. Physiol.* **2020**, *318*, L698–L704. [CrossRef]
32. Van Dongen, J.A.; Getova, V.; Brouwer, L.A.; Liguori, G.R.; Sharma, P.; Stevens, H.P.; Van Der Lei, B.; Harmsen, M.C. Adipose tissue-derived extracellular matrix hydrogels as a release platform for secreted paracrine factors. *J. Tissue Eng. Regen. Med.* **2019**, *13*, 973–985. [CrossRef]
33. Coulson-Thomas, V.; Gesteira, T. Dimethylmethylene Blue Assay (DMMB). *Bio-Protocol* **2014**, *4*, e1236. [CrossRef]
34. Farndale, R.W.; Sayers, C.A.; Barrett, A. A Direct Spectrophotometric Microassay for Sulfated Glycosaminoglycans in Cartilage Cultures. *Connect. Tissue Res.* **1982**, *9*, 247–248. [CrossRef] [PubMed]
35. Lai, M.; Lü, B. Tissue Preparation for Microscopy and Histology. In *Comprehensive Sampling and Sample Preparation*; Elsevier: Amsterdam, The Netherlands, 2012; pp. 53–93. ISBN 978-0-12-381374-9.
36. Junqueira, L.C.U.; Bignolas, G.; Brentani, R.R. Picrosirius staining plus polarization microscopy, a specific method for collagen detection in tissue sections. *J. Mol. Histol.* **1979**, *11*, 447–455. [CrossRef]
37. Foot, N.C. The Masson Trichrome Staining Methods in Routine Laboratory Use. *Stain. Technol.* **1933**, *8*, 101–110. [CrossRef]
38. Vogel, B.; Siebert, H.; Hofmann, U.; Frantz, S. Determination of collagen content within picrosirius red stained paraffin-embedded tissue sections using fluorescence microscopy. *MethodsX* **2015**, *2*, 124–134. [CrossRef] [PubMed]
39. Schindelin, J.; Arganda-Carreras, I.; Frise, E.; Kaynig, V.; Longair, M.; Pietzsch, T.; Preibisch, S.; Rueden, C.; Saalfeld, S.; Schmid, B.; et al. Fiji: An open-source platform for biological-image analysis. *Nat. Chem. Biol.* **2012**, *9*, 676–682. [CrossRef] [PubMed]
40. Wershof, E.; Park, D.; Barry, D.J.; Jenkins, R.P.; Rullan, A.; Wilkins, A.; Schlegelmilch, K.; Roxanis, I.; I Anderson, K.; Bates, A.P.; et al. A FIJI macro for quantifying pattern in extracellular matrix. *Life Sci. Alliance* **2021**, *4*, e202000880. [CrossRef] [PubMed]
41. Landini, G. Quantitative analysis of the epithelial lining architecture in radicular cysts and odontogenic keratocysts. *Head Face Med.* **2006**, *2*, 4. [CrossRef] [PubMed]
42. Motulsky, H.J.; E Brown, R. Detecting outliers when fitting data with nonlinear regression–a new method based on robust nonlinear regression and the false discovery rate. *BMC Bioinform.* **2006**, *7*, 123–220. [CrossRef] [PubMed]
43. Ghasemi, A.; Zahediasl, S. Normality Tests for Statistical Analysis: A Guide for Non-Statisticians. *Int. J. Endocrinol. Metab.* **2012**, *10*, 486–489. [CrossRef]
44. Das, K.R. A Brief Review of Tests for Normality. *Am. J. Theor. Appl. Stat.* **2016**, *5*, 5–12. [CrossRef]
45. Saldin, L.T.; Cramer, M.C.; Velankar, S.S.; White, L.J.; Badylak, S.F. Extracellular matrix hydrogels from decellularized tissues: Structure and function. *Acta Biomater.* **2016**, *49*, 1–15. [CrossRef]
46. Birk, D.; Fitch, J.; Babiarz, J.; Doane, K.; Linsenmayer, T. Collagen fibrillogenesis in vitro: Interaction of types I and V collagen regulates fibril diameter. *J. Cell Sci.* **1990**, *95*, 649–657. [CrossRef]
47. Brightman, A.O.; Rajwa, B.P.; Sturgis, J.E.; McCallister, M.E.; Robinson, J.P.; Voytik-Harbin, S.L. Time-Lapse Confocal Reflection Microscopy of Collagen Fibrillogenesis and Extracellular Matrix Assembly in Vitro. *Biopolymers* **2000**, *54*, 222–234. [CrossRef]
48. Gelman, R.; Poppke, D.; Piez, K. Collagen fibril formation in vitro. The role of the nonhelical terminal regions. *J. Biol. Chem.* **1979**, *254*, 11741–11745. [CrossRef]
49. Peterson, B.; van der Mei, H.C.; Sjollema, J.; Busscher, H.J.; Sharma, P.K. A Distinguishable Role of eDNA in the Viscoelastic Relaxation of Biofilms. *mBio* **2013**, *4*, e00497-13. [CrossRef]

50. Colliec-Jouault, S. Skin tissue engineering using functional marine biomaterials. In *Functional Marine Biomaterials*; Elsevier: Amsterdam, The Netherlands, 2015; pp. 69–90. ISBN 978-1-78242-086-6.
51. Al Jamal, R.; Roughley, P.J.; Ludwig, M.S. Effect of glycosaminoglycan degradation on lung tissue viscoelasticity. *Am. J. Physiol. Cell. Mol. Physiol.* **2001**, *280*, L306–L315. [CrossRef]
52. Papakonstantinou, E.; Karakiulakis, G. The 'sweet' and 'bitter' involvement of glycosaminoglycans in lung diseases: Pharmacotherapeutic relevance. *Br. J. Pharmacol.* **2009**, *157*, 1111–1127. [CrossRef]
53. Uhl, F.E.; Zhang, F.; Pouliot, R.A.; Uriarte, J.J.; Enes, S.R.; Han, X.; Ouyang, Y.; Xia, K.; Westergren-Thorsson, G.; Malmström, A.; et al. Functional role of glycosaminoglycans in decellularized lung extracellular matrix. *Acta Biomater.* **2019**, *102*, 231–246. [CrossRef]
54. Karageorgiou, V.; Kaplan, D. Porosity of 3D biomaterial scaffolds and osteogenesis. *Biomaterials* **2005**, *26*, 5474–5491. [CrossRef]
55. León, C.A.L.Y. New perspectives in mercury porosimetry. *Adv. Colloid Interface Sci.* **1998**, *76–77*, 341–372. [CrossRef]
56. Roosa, S.M.M.; Kemppainen, J.M.; Moffitt, E.N.; Krebsbach, P.H.; Hollister, S.J. The pore size of polycaprolactone scaffolds has limited influence on bone regeneration in anin vivomodel. *J. Biomed. Mater. Res. Part A* **2010**, *92*, 359–368. [CrossRef] [PubMed]
57. Koch, M.; Włodarczyk-Biegun, M.K. Faithful scanning electron microscopic (SEM) visualization of 3D printed alginate-based scaffolds. *Bioprinting* **2020**, *20*, e00098. [CrossRef]
58. McKinlay, K.J.; Allison, F.J.; Scotchford, C.; Grant, D.; Oliver, J.M.; King, J.R.; Wood, J.V.; Brown, P. Comparison of environmental scanning electron microscopy with high vacuum scanning electron microscopy as applied to the assessment of cell morphology. *J. Biomed. Mater. Res.* **2004**, *69*, 359–366. [CrossRef] [PubMed]
59. Jia, W.; Gungor-Ozkerim, P.S.; Zhang, Y.S.; Yue, K.; Zhu, K.; Liu, W.; Pi, Q.; Byambaa, B.; Dokmeci, M.R.; Shin, S.R.; et al. Direct 3D bioprinting of perfusable vascular constructs using a blend bioink. *Biomaterials* **2016**, *106*, 58–68. [CrossRef] [PubMed]
60. Doyle, A.D.; Yamada, K.M. Mechanosensing via cell-matrix adhesions in 3D microenvironments. *Exp. Cell Res.* **2015**, *343*, 60–66. [CrossRef] [PubMed]
61. Fischer, T.; Hayn, A.; Mierke, C.T. Fast and reliable advanced two-step pore-size analysis of biomimetic 3D extracellular matrix scaffolds. *Sci. Rep.* **2019**, *9*, 1–10. [CrossRef]

*Review*

# Application of Computational Method in Designing a Unit Cell of Bone Tissue Engineering Scaffold: A Review

Nur Syahirah Mustafa [1], Nor Hasrul Akhmal [1,*], Sudin Izman [1], Mat Hussin Ab Talib [1], Ashrul Ishak Mohamad Shaiful [2], Mohd Nazri Bin Omar [2], Nor Zaiazmin Yahaya [2] and Suhaimi Illias [2]

[1] Faculty of Engineering, School of Mechanical Engineering, Universiti Teknologi Malaysia, Johor Bahru, Johor 81310, Malaysia; nsyahirah83@graduate.utm.my (N.S.M.); izman@utm.my (S.I.); mathussin@utm.my (M.H.A.T.)

[2] Faculty of Mechanical Engineering Technology, Universiti Malaysia Perlis, Arau 02600, Malaysia; mshaiful@unimap.edu.my (A.I.M.S.); nazriomar@unimap.edu.my (M.N.B.O.); zaiazmin@unimap.edu.my (N.Z.Y.); suhaimi@unimap.edu.my (S.I.)

* Correspondence: norhasrul@utm.my

**Abstract:** The design of a scaffold of bone tissue engineering plays an important role in ensuring cell viability and cell growth. Therefore, it is a necessity to produce an ideal scaffold by predicting and simulating the properties of the scaffold. Hence, the computational method should be adopted since it has a huge potential to be used in the implementation of the scaffold of bone tissue engineering. To explore the field of computational method in the area of bone tissue engineering, this paper provides an overview of the usage of a computational method in designing a unit cell of bone tissue engineering scaffold. In order to design a unit cell of the scaffold, we discussed two categories of unit cells that can be used to design a feasible scaffold, which are non-parametric and parametric designs. These designs were later described and being categorised into multiple types according to their characteristics, such as circular structures and Triply Periodic Minimal Surface (TPMS) structures. The advantages and disadvantages of these designs were discussed. Moreover, this paper also represents some software that was used in simulating and designing the bone tissue scaffold. The challenges and future work recommendations had also been included in this paper.

**Keywords:** numerical analysis; computational method; tissue engineering scaffold design; mechanical strength; simulation software

**Citation:** Mustafa, N.S.; Akhmal, N.H.; Izman, S.; Ab Talib, M.H.; Shaiful, A.I.M.; Omar, M.N.B.; Yahaya, N.Z.; Illias, S. Application of Computational Method in Designing a Unit Cell of Bone Tissue Engineering Scaffold: A Review. *Polymers* **2021**, *13*, 1584. https://doi.org/10.3390/polym13101584

Academic Editors: Andrada Serafim and Stefan Ioan Voicu

Received: 23 April 2021
Accepted: 12 May 2021
Published: 14 May 2021

**Publisher's Note:** MDPI stays neutral with regard to jurisdictional claims in published maps and institutional affiliations.



## 1. Introduction

An engineered tissue can be a huge aid in the future, especially in clinical application. The relationship that brings life sciences and engineering together as an application to be a great help in understanding the structure and function of a mammalian tissue can be described as tissue engineering [1]. Not only offering help in understanding the structure and function of tissue of a human being, but it is also helping the researchers to understand the necessity of developing an engineered tissue. An engineered tissue can help in restoring, maintaining, repairing and improving the damaged tissue's condition, which is caused by numerous diseases such as disabilities and injuries [2]. Zhang et al. stated in their paper that the current implementation of tissue engineering had faced many issues, including ethical and technical issues [3]. Despite many challenges faced in the field of tissue engineering, it is a fast-paced developing field since it can be a great help in providing treatments that can generate most of the tissue and organ of the human being [3].

One of the components that is crucially needed to be studied is the scaffold of the engineered tissue. This is due to the function of the scaffold that provides a suitable environment and structure in order to enable the cells to attach, proliferate, differentiate and secrete their own extra-cellular matrix (ECM) [3]. It is important to ensure the scaffold to have a proper environment and structure so that it promotes a good rate of the formation of tissue. In order to produce an appropriate scaffold, it must be ensured to possess a few characteristics so that it will not be harmful to the body. The characteristics included biocompatibility, biodegradable, bioactivity, scaffold architecture and mechanical properties. Turnbull et al. reported that the manufacturing of a scaffold should be compatible with the human body so that it will not trigger any immune response while it is implanted in the body [4]. In order to comply with these conditions, the materials used in manufacturing the scaffold should non-toxic and easy to eliminate from the body.

Another important feature in manufacturing the scaffold is that the degradation rate of the scaffold is needed to be properly controlled. This is to ensure that the scaffold does not suffer mechanical failure. In addition, the scaffold's structure and architecture need to be considered when manufacturing a scaffold because it provides viability and encourages tissue ingrowth. The human body is a very sensitive creation where all the parts are subjected to a certain value of strength and provide sufficient endurance of pressure. Therefore, it is crucial that the mechanical properties of the manufactured scaffold achieve the same or properly adjusted to the original tissue so that it can have no negative effect on the host tissue [5,6]. All these features will help in promoting tissue growth and avoiding a negative response of the immune system if the scaffold can be produced in a proper manner [4].

The 'trial-and-error' method has been adopted by most common researchers in enhancing the tissue engineering field. The method which involved the modification of the current or existing design of a scaffold can cause many unwanted factors. This conventional method is very expensive and does not have a precise control due to the repeatable modification. It is also time-consuming since the production of an improved model of the scaffold will take too long. Therefore, a computational approach needs to be used.

Besides that, although much research has been done on a scaffold of tissue engineering computationally, there is still a lack of research that focuses on fluid properties and designs of the scaffolds. In addition, the designs and fluid properties of a scaffold play a crucial role in facilitating the growth of the bone tissue.

The porosity and mechanical strength of the scaffold have an inversely proportional relationship. However, bone scaffold needs to be manufactured porously in order to enable cell proliferation and transportation of nutrients, oxygen and metabolites in the blood [7–9]. Yet, up to this date, there is very little research done that can computationally produce a scaffold with good design and possessing excellent mechanical and fluid properties, simultaneously.

Therefore, various great efforts have been done by many researchers in order to help in producing a scaffold model that possesses all of the ideal characteristics. One of the ways to develop a scaffold model that can cater to the needs to encourage a high rate of tissue formation is by using computational methods, which consist of simulating, modelling and 3D printing techniques. Current research studies show that the computational method has been a great help in order to expedite the implementation of tissue engineering in the near future.

### 1.1. Bone Tissue Engineering

The most major structural and connective tissue of the body is bone tissue [10]. There are two types of bones that can be identified, which are cortical bones and cancellous bone. The outer part of the cone that is denser and has low porosity is called cortical bone, while the inner part and spongy-like material is called cancellous bone. The porosities of the cancellous bone should be in the range of 50% to 80% [10–13]. The bone is one of the parts that is having high mechanical strength. A cortical bone possesses a high modulus of elasticity and compressive strength as compared to the cancellous bones [14]. Although bone has high mechanical strength, bone can be subjected to many traumas and diseases such as injuries. Therefore, researchers have come to a solution which to produce regenerative medicine in terms of tissue engineering. Despite many organs that can be regenerated by the method of tissue engineering, bone tissue engineering is a widely studied field.

A large bone defect is treated with the current conventional method, which is by using autografting. Autografting technique required the usage of bone from a non-load-bearing site of the patient to be transplanted into the damaged part [10]. Nowadays, various researchers have led to the implementation of bone tissue engineering in the future since it can overcome the problems faced by current clinical treatments [15].

### 1.2. Scaffold of Bone Tissue Engineering

A biomaterial porous structure that helps in providing support and a suitable extracellular matrix is called a scaffold [16,17]. In designing a scaffold, one must consider the strength and porosity of the materials so that it will regenerate the properties of the bone that is comparable to the original bone. Nowadays, there are many applications of technology to develop a scaffold, such as additive manufacturing, which includes 3D printing. Additive manufacturing has provided a platform which helps in customizing and developing a suitable design that can be used in biomedical application.

The scaffolds must be designed and developed based on a few characteristics that will provide the best condition for the bone to regenerate. The characteristics include biocompatibility, bioactivity, biodegradability, mechanical properties and scaffold architecture [18]. Biocompatibility can be defined as non-toxic and non-inflammatory so that it will not bring harm to the body [5,6,19]. A biodegradable scaffold should be able to eliminate itself from the body easily once the tissue has fully restored. Adequate mechanical properties should be possessed by the scaffold so that it will be able to withstand any forces and loads during the restoration time in the implantation site [19].

Designing a scaffold with a proper architecture is important due to it will affect the mechanical and biocompatibility properties of the scaffold [20,21]. Scaffold architecture should be able to provide a large surface area to volume ratio so that cell migration can occur. The porosity must be sufficient so that it will allow cell and nutrition migration for the restoration of the tissue. However, it must not compromise with the mechanical strength of the scaffold [22].

### 1.3. Significance of Computational Method in Bone Tissue Engineering

The successful production of bone tissue engineering scaffold can help to contribute to ensuring the tissue formation goes smoothly, especially bone tissue. This is due to the computational method to help in generating precise properties of the scaffold. Besides that, the usage of the computational method, which is the simulation aimed to provide good support in implementing the usage of tissue engineering in the future, especially in clinical application. This is because the simulation can help in reducing the intervention of humans in manufacturing the appropriate scaffold model. The risks will be minimized due to the involvement of automation that will eventually produce fewer damaged organs.

## 2. Computational Method in Designing a Scaffold

Computational modelling has been the most common approach that had been done by the researchers. This is due to its ability to simulate the behaviour of the scaffold under certain loadings. In addition, it is proved to reduce time and experiments since it is not time-consuming and cheaper [23].

Moreover, the computational modelling technique has been adopted by the researchers in order to improve the performance of scaffolds while maintaining certain important parameters [3]. It is also a great predictive tool that can help in predicting the scaffold properties before manufacturing them. Some uncommon properties, such as stress-strain distribution, can also be predicted by using the computational method.

According to Bocaccio et al., the computational method has allowed the approximation of how the mechanical environment is affecting the differentiation of tissue and bone regeneration [24]. It also helps in understanding the mechanisms that will enhance the reliability of the function of scaffolds. Zhang et al. stated that the function of computational modelling includes designing and simulation that had been a great aid in 3D printing technique [4].

Recently, the usage of the computational method in assessing the properties of a scaffold's structure has been progressively studied. With the aid of computational programs, such as Finite Elements Analysis (FEA), the properties of the scaffold can be easily predicted. Thus, it helps in reducing the time and energy to find the most feasible scaffold by eliminating the modification step of an existing scaffold.

### 2.1. Unit Cell of a Scaffold Structure

The unit cell is the basic structure of a scaffold. It can be divided into two types of designs which are non-parametric and parametric. Non-parametric designs consist of a unit cell which is designed by using structural and geometric shape. Meanwhile, the parametric designs have to be produced by using specific algorithms. There are many advantages and disadvantages regarding each design which will be furthered discussed.

### 2.2. Non-Parametric Design

A non-parametric design is a simple structure that is designed based on geometry. The most common non-parametric designs are the circular, cubic and honeycomb designs as such in Table 1. However, there are a lot of other designs, which are produced based on certain geometries, such as hexagonal and octet. There are many advantages of these designs as compared to the parametric designs. One of them is that the non-parametric designs are easy to be produced since it does not engage to any specific algorithms. In addition, there are many ways that can be used to fabricate the designs. These designs are mostly being fabricated via subtractive manufacturing such as machining. However, since most of the researchers are looking forward to utilizing additive manufacturing, the production of these designs is very much possible to pursue, especially by using Selective Laser Melting (SLM) additive manufacturing. The characteristics of non-parametric designs can be described, as in Table 1.

### 2.2.1. Circular Design

Many studies have been done that adopted circular pore shape as their scaffold model. This is due to the ability of the circular pore to avoid stress concentration point; therefore, it relatively would possess a high bearing stress capacity [18]. The study conducted by Sun et al. showed that the circular shape produced a more uniform axial deformation. Thus, it gives a smaller strain risk when subjected to a uniform stress concentration [25]. Bocaccio et al. suggested that the circular design exhibits a greater Young's Modulus when they are subjected to a certain amount of pressure [26].

In terms of the porosity of the scaffold, the pore size of the scaffold plays an important role in determining the mechanical properties of the scaffold. Boccaccio et al. suggested that the circular pore demonstrated a certain amount of mechanical properties when its porosity distribution law is varied [27]. A circular pore with a low amount of porosity also helps in ensuring a high amount of mechanical strength [28]. Although the porosity amount of a circular-shaped pore is high, its mechanical strength was lower in a study carried out by Jahir-Hussain et al. [29]. Therefore, the circular-shaped pore needs to possess a high porosity amount so that it can enhance mechanical and morphological properties [30]. Gomez et al., in their study, described that the circular-shaped scaffold needs to possess a porosity amount in the range of 70–90% in order to obtain a high mechanical strength [31].

**Table 1.** Non-parametric design and its characteristics.

| Non-Parametric Design | Description | Advantages | Disadvantages | Ref. |
|---|---|---|---|---|
| Circular [18] | A scaffold with a circular-shaped pore is a structure, which is commonly used in investigating the behaviour of the scaffold in terms of mechanical and fluidic. | • Simple design—easy to be produced<br>• Less high-stress concentration points<br>• Exhibits stable resistance for fatigue damage<br>• Easy to fabricate using both conventional method and additive manufacturing | • May cause underestimations of the behaviour of the scaffold<br>• High tendency to cause pore blockage, which affects bone growth by disrupting transportation of nutrients, oxygen and waste of the scaffold | [18,25–33] |
| Square [18] | A square-shaped pore structure, which is reliable in producing high mechanical strength and adequate amount of porosity but also high in the stress concentration area | • Simple design—easy to be produced<br>• Exhibits high proliferation rate<br>• Easy to fabricate using both conventional method and additive manufacturing | • May cause underestimations of the behaviour of the scaffold<br>• Contains a high-stress concentration point | [19,29,34–37] |
| Honeycomb [38] | A structure that imitates the shape of the beeswax that exhibits excellent properties in terms of lightweight, stiffness and porosity | • Simple design—easy to be produced<br>• Good mechanical stability<br>• Better at avoiding shrinkage of the scaffold during cell growth<br>• Promotes high cell proliferation | • May cause underestimations of the behaviour of the scaffold<br>• Limitation on the fabrication based on the adjustable pore size, spatial arrangement and reproducible architectures | [21,39–46] |

### 2.2.2. Square Design

Although the square-shaped designs have a high-stress concentration region, they are also still relevant to be studied by the researchers since they possess a high mechanical strength while maintaining an adequate amount of porosity. In order to avoid this problem, researchers came out with a solution where they modified the square scaffold by adding a few struts. The struts help to improve the stiffness of the porous structure as well as reduces the stress concentration at the joints [37]. In a study conducted by Jahir-Hussain et al., they varied the pore shape of the scaffold, resulting in high mechanical strength but a low amount of porosity [29]. When the porosity of the scaffold is increased to be more than 50%, the square-shaped scaffold can exhibit mechanical properties similar to that of the host tissue [36]. Habib et al. modified the square-shaped scaffold by increasing the porosity but could maintain the mechanical properties of the scaffold [19]. The failure mechanism of a square-shaped scaffold is that it failed in the unidirectional failure according to the direction of loading subjected to it.

### 2.2.3. Honeycomb Design

The honeycomb structure was designed based on the hexagonal prismatic wax cells, which are built by honey bees. In the engineering field, the honeycomb structure was first introduced to the aerospace discipline. However, it gets the attention of the other fields' researchers, including the biomedical field, since it can be found naturally in biomedical structure. Moreover, it is light-weighted with a high amount of stiffness and porosity.

By using Finite Element Analysis, the mechanical properties of the honeycomb structures can be simulated. The Young's Modulus of the honeycomb structure can be controlled by varying the porosity of the structures. This would cause the honeycomb design to be able to fit in between cortical and cancellous bone properties [21,45]. However, the honeycomb tends to fail in multi-directions when it is exposed to certain loadings [44]. There are a few designs that were generated by modifying the original shape of the honeycomb structure. The new design of the honeycomb structure has the ability to demonstrate the mechanical properties of a cancellous bone [46].

In conclusion, we can say that a non-parametric design can still be adopted into various research works due to its ability to demonstrate the desired properties of a bone tissue engineering scaffold. However, there is a need to modify the design in order to match the properties of the host tissues. A circular design exhibits an excellent characteristic in terms of fewer stress concentration points as compared to other non-parametric designs. Meanwhile, a square-shaped pore promotes a high rate of cell proliferation due to its ability to possess a high amount of porosity. The honeycomb structure has the ability to maintain excellent mechanical stability by reducing the risk of scaffold shrinkage during cell growth.

### 2.3. Parametric Design

Since tissue engineering is highly related to the usage of additive manufacturing (AM), the researchers tend to shift from using simple structure to using complex structure since the AM technology has the ability to produce a complex structure [47]. The simpler shape is more likely to face some issues such as strut thickness [48], interface mismatch [49] and surface smoothness [50]. According to Chen et al., there are two main methods that were used to generate a parametric structure, named Triply Periodic Minimal Surfaces (TPMS) and Voronoi Tessellation, as shown in Table 2 [8].

**Table 2.** Parametric design and its characteristics.

| Parametric Design | | Description | Advantages | Disadvantages | Ref. |
|---|---|---|---|---|---|
| Triply Periodic Minimal Surfaces (TPMS) | Schwarz P (Primitives) | Schwarz P (Primitives) is one of the earliest Triply Periodic Minimal Surfaces (TPMS) structures that was proposed by Schwarz in the 1860s. It belongs to the stretching surface structure. Its shape is strictly governed by a mathematical equation, as shown in Table 3. | • Promotes high cell attachment, migration and proliferation<br>• Has the ability to possess the natural bone's properties<br>• Helps in avoiding stress shielding<br>• High in mechanical strength as compared to the Schwarz D structure<br>• Possess a mechanical strength that complies to that of cortical bone when subjected to a low amount of porosity<br>• Simplest form of a TPMS structure | • Complex shape that can only be generated through a mathematical equation<br>• Has a high concentration region in the neck of the structure<br>• Possess a low specific surface area as compared to the bending surfaces TPMS<br>• Can easily be fabricated via additive manufacturing, but not by using conventional method due to the complex shape | [51–55] |
| | Schwarz D(Diamond) | Schwarz D (Diamond) can be categorised as a bending surface Triply Periodic Minimal Surfaces (TPMS) structure that was proposed by Schwarz in the 1860s. The Schwarz D shape can be generated via a mathematical equation, as shown in Table 3. | • Promotes high cell attachment, migration and proliferation<br>• Has the ability to possess the natural bone's propertiesHelps in avoiding stress shielding<br>• Possess a mechanical strength that complies to that of cortical bone when subjected to a low amount of porosity<br>• High in specific surface area as compared to the Schwarz P structure, thus it promotes high bone in growth rate<br>• Good structure for uniform stress distribution when subjected to ultimate pressure | • Complex shape that can only be generated through a mathematical equation<br>• Low load-bearing capacity when subjected to uniaxial loading<br>• Can be easily fabricated via additive manufacturing, but not by using conventional method due to the complex shape | [51,52,55–57] |

**Table 2.** *Cont.*

| Parametric Design | | Description | Advantages | Disadvantages | Ref. |
|---|---|---|---|---|---|
| Triply Periodic Minimal Surfaces (TPMS) | Gyroid | A gyroid structure was first introduced by Schoen, which became the most assessed Triply Periodic Minimal Surfaces (TPMS) structure in researches. Its shape is strictly governed by a mathematical equation, as shown in Table 3. | • Promotes high cell attachment, migration and proliferation<br>• Has the ability to possess the natural bone's properties<br>• Helps in avoiding stress shielding<br>• Mechanically better than a solid structure due to the consistency in the amount of porosity<br>• High in permeability, thus it promotes high bone in growth rate<br>• Good structure for uniform stress distribution when subjected to ultimate pressure | • Complex shape that can only be generated through a mathematical equation<br>• Low load-bearing capacity when subjected to uniaxial loading<br>• Can easily be fabricated via additive manufacturing, but not by using conventional method due to the complex shape | [51,55–58] |
| | I-WP | An I-WP is one of the most assessed Triply Periodic Minimal Surfaces (TPMS) structure in research. The mathematical equation that is used to produce an I-WP structure can be found in Table 3 | • Promotes high cell attachment, migration and proliferation<br>• Has the ability to possess the natural bone's properties<br>• Helps in avoiding stress shielding<br>• Possess excellent mechanical properties | • Complex shape that can only be generated through a mathematical equation<br>• Has a high concentration region in the upper part of the structure<br>• Can be easily fabricated via additive manufacturing, but not by using conventional method due to the complex shape | [48,56,59–61] |
| | Voronoi [62] | A structure that is generated through a specific algorithm that creates random discrete points that turn into a network structure. The shape can easily imitate the structure of the host tissue. Therefore, it helps in bone ingrowth. | • Has the ability to imitate the host tissue shape<br>• Matched the properties of the natural bone; thus, it helps in expediting the tissue growth rate<br>• Can be integrated with the usage of computed tomography (CT) scan; therefore, it has a good possibility to be used in real-life clinical application | • Complex shape that is governed by a specific algorithm<br>• Time-consuming<br>• Might have difficulties in repeating the experiments<br>• Intersection of strut leads to stress changes | [63–67] |

### 2.3.1. Triply Periodic Minimal Surfaces (TPMS)

Triply Periodic Minimal Surfaces (TPMS) is a smooth infinite and non-self-intersecting periodic structure in three principal directions associated with crystallographic space group symmetry [68,69]. In 1865–1883, Schwarz and Neovius had introduced some TPMS structures, which are Schwarz P (Primitives), Schwarz D (Diamond), Schwarz H (Hexagonal) and Neovius. In 1970, Schoen described the most popular TPMS structure, which is the Gyroid and also a few other TPMS structures [70]. The most common TPMS structures that were studied by the researchers are Gyroid, Diamond and Primitives [71]. This is because the structures can be easily found in nature, such as butterfly wing scales and sea urchins [72]. TPMS structures are likely to be favoured by the researchers since it promotes higher cell attachment, migration and proliferation as compared to the scaffold with sharp edges [47,73].

TPMS structure can be classified into two types which are skeletal TPMS and sheet TPMS. Most of the researchers tend to assess the properties of TPMS via the sheet typed TPMS. Therefore, there is a lack of research on the skeletal TPMS structure. In research conducted by Barba et al., they found that the Gyroid skeletal TPMS shows a feasible design of a scaffold, which is superior in terms of manufacturability, mechanical properties and bone ingrowth [74]. However, Cai et al. stated that the compressive strength of the skeletal TPMS is much lower as compared to the sheet TPMS [75].

**Table 3.** Mathematical Equations of Triply Periodic Minimal Surfaces (TPMS) Structures.

| TPMS Structure | Equation | Ref. |
|---|---|---|
| Schwarz P (Primitives) | $\cos(x) + \cos(y) + \cos(z) = t$ | [70,77] |
| Schwarz D (Diamond) | $\sin(x)\sin(y)\sin(z) + \sin(x)\cos(y)\cos(z) + \cos(x)\sin(y)\cos(z) + \cos(x)\cos(y)\sin(z) = t$ | [77,78] |
| Neovius | $3[\cos(x) + \cos(y) + \cos(z)] + 4[\cos(x)\cos(y)\cos(z)] = t$ | [70,77] |
| Gyroid | $\cos(x)\sin(y) + \cos(y)\sin(z) + \cos(z)\sin(x) = t$ | [70,77,78] |
| I-WP | $2[\cos(x)\cos(y) + \cos(y)\cos(z) + \cos(z)\cos(x)] - [\cos(2x) + \cos(2y) + \cos(2z)] = t$ | [70,78] |
| Fisher-Koch S | $\cos(2x)\sin(y)\cos(z) + \cos(x)\cos(2y)\sin(z) + \sin(x)\cos(y)\cos(2z) = t$ | [76,78,79] |
| Fisher-Koch Y | $\cos(x)\cos(y)\cos(z) + \sin(x)\sin(y)\sin(z) + \sin(2x)\sin(y) + \sin(2y)\sin(z) + \sin(x)\sin(2z) + \sin(2x)\cos(z) + \cos(x)\sin(2y) + \cos(y)\sin(2z) = t$ | [76,78] |

Meanwhile, the sheet TPMS structure is widely assessed in the literature as it shows a superior design as compared to the skeletal TPMS. There are many types of TPMS structures that can be generated through a mathematical equation that controls the TPMS structure. By using a few software such as Minisurf, the TPMS structure can easily be generated [76]. The built-in equations can be tabulated in Table 3.

However, the most assessed sheet-TPMS structure in the previous studies are Schwarz P (Primitives), I-WP, Schwarz D (Diamond) and Gyroid. This is due to their ability to match the properties of the host tissue of cancellous and cortical bone. According to Bobbert et al., these structures are able to avoid stress shielding by possessing high yield stress and low Young's Modulus [50]. These TPMS can be categorized into two main categories that are based on their deformation mechanisms, which are stretching surface and bending surface [68]. Schwarz P (Primitives) and I-WP belong to the stretching surface, while Schwarz D (Diamond) and Gyroid are the bending surface TPMS.

By using the computational method, which is adopting the Finite Element Analysis (FEA) to assess the properties of the TPMS scaffold, Shi et al. found out that the TPMS scaffold possessed excellent scaffold properties that are matched with the bone tissue properties [80]. In terms of porosity, Castro et al. had reported that the gyroid TPMS can be used in clinical practices in the bone tissue engineering field. They had carried out both numerical and experimental methods to assess the mechanical properties of two gyroids with 50% and 70% porosity, respectively [81]. In a research carried out by

Yang et al., they found out that the Young's Modulus of Schwarz P, Schwarz D, I-WP and Gyroid were matched to the Young's Modulus of the cancellous bone. However, this only happened when they were subjected to a high amount of porosity [82]. In addition, the compressive properties of Schwarz P and I-WP was higher than the cancellous bone [54]. This fact is supported by Montezarian et al., when they also reported that the compressive strength was higher in the Schwarz P and I-WP structures than the cancellous bone [48,61]. Meanwhile, at a low amount of porosity in the range of 5–10%, the Schwarz P and Schwarz D scaffolds possessed Young's Modulus properties similar to that of the cortical bone [83].

Most of the researchers tend to compare the mechanical properties of these TPMS structures in order to determine the suitable application of the structure in clinical practice in the future. For example, Afshar et al. reported that the Schwarz P structure showed better mechanical properties as compared to the Schwarz D structure [84]. In addition, Maskery et al. also had stated the same conclusion since they found that the stretching surface TPMS has twice the Young's modulus of bending surface TPMS by using Finite Element Analysis [85]. In order to find the mechanical properties, Finite Element Analysis showed the failure mechanism while simulating the behaviour of the structure. In their research, Maskery et al. had suggested that the stretching surface TPMS failed because the stress concentration region was located at the Schwarz P neck, which is situated at the top surface of the whole structure [85]. This has shown that the structure would start to fail layer by layer when it is subjected to loadings [86,87]. However, the bending surface TPMS would start to fail once the scaffold is subjected to ultimate pressure due to the uniform stress distribution in bending surface TPMS by showing a shear band [48,54,84,88,89]. From these studies, we can see that the stretching structure would fail due to the axial deformation while the bending structure would fail once the shearing linkages appear on the structure. Thus, the stretching structures possess a high capacity of load-bearing as compared to the bending structures when a uniaxial loading is subjected to them [68].

Although the porosity of the TPMS structure can be predetermined by varying certain parameters in the mathematical equation, the actual porosity amount of the structure was consistent with the TPMS design. This is due to some studies showed that the solid structure such as cube has lower mechanical properties as compared to the TPMS structure. In a study conducted by Zaharin et al., they discovered that the gyroid structure is mechanically better than the cubic structure [58]. The strut-based structure is also happened to possess lower mechanical properties as compared to the TPMS structure. This fact is supported by Al-Ketan et al. in their study since they stated that the TPMS structure exhibits excellent mechanical properties [72]. Nonetheless, Du Plessis et al. realized that there is not much significant difference when comparing the mechanical properties of TPMS structures and strut-based structures [90]. According to Guo et al., although there is not much difference in the mechanical properties of the TPMS and strut-based structures, the TPMS structure showed a more uniform and smooth transition of stress distribution [56]. From these studies, we can see that the TPMS structure kind of possess the same mechanical properties as the strut-based structures.

In the matter of bone ingrowth, the researchers would take permeability and specific surface area as a prediction tool [68]. The specific surface area helps in predicting the cell absorption area, while permeability indicated the ability of the scaffold to facilitate the transportation of oxygen, nutrients and waste. Schwarz D had the highest specific surface area, while Gyroid had the highest permeability [11,48,56,91]. Therefore, we can say that the Schwarz D and Gyroid might be the suitable TPMS for bone ingrowth. However, it needs to be furthered verified by biological experiments. Schwarz P has a high manufacturing accuracy as compared to other TPMS since it has the simplest geometry rather than the other.

In conclusion, we can say that the TPMS structure can be a suitable design for bone tissue scaffold. The stretching structure and bending structure both have different advantages. Stretching TPMS has excellent properties while the bending structure possesses high permeability properties. In general, these TPMS showed properties that are similar to that of the natural bone.

### 2.3.2. Voronoi Tessellation

The Voronoi structure is said to be similar to the host tissue in terms of morphology. This is a need in bone tissue engineering since it should be able to copy the natural bone properties [92,93]. The Voronoi structure can be produced when a mesh structure is generated based on random discrete points, which are then connected and performed a network structure [94]. In 2010, Kou and Tan had introduced the Voronoi method by creating irregular and random scaffolds, which were merged with Voronoi cells [95]. They used B-spline curves in order to indicate the irregularly shaped pores' boundaries. When this method was adopted, it can be seen that the shape of the scaffold was kind of similar to that of the shape of the bone structure [96]. After the Voronoi method has been proposed by Kou and Tan, researches related to the usage of the Voronoi method has been widened progressively, which includes the reverse engineering method that adopted computed tomography (CT) scan method to extract its data. Yang and Zhao stated that the Voronoi method could be used to recreate a bone-like-shaped scaffold by utilizing the data obtained from a computed tomography (CT) scan [67]. Although the Voronoi structure can be generated via the tessellation method, which employs some indices such as trabecular thickness and bone volume to total volume ratio, it is still time-consuming, long-cycle and mostly unrepeatable experiments [31,97,98].

In the computational method, the Finite Element Analysis was used to indicate the stress of the Voronoi structure. The study carried out by Wei et al. showed that the stress gradient of the Voronoi structure increase when the amount of porosity is low and vice versa [99]. In terms of fluid properties, Gomez et al. suggested that the Voronoi structure is depending on the amount of porosity and the bone surface area, which is very much favourable [31]. Therefore, it helps in bone ingrowth. Although the structure is having a good resemblance with the properties of the cancellous bone, Maliaris and Sarafis discovered that the intersection of struts was exposed to a stress change [100].

### 2.3.3. Other Parametric Design

Besides the most common two designs of a parametric scaffold, there are also other parametric designs, which help in finding the suitable scaffold shape. This is due to the demand of the bone tissue engineering scaffold, which needs them to be able to possess excellent mechanical and fluidic properties in terms of permeability. Naturally, many structures in our environment possess high compressive strength. For example, Achrai and Wagner discovered that the turtle shell structure might help in producing a feasible scaffold design [101].

### 2.3.4. Method of Anatomical Features (MAF)

In addition, the B-spline curve method had also been adopted by Vitkovic et al. when they produced a scaffold via reverse engineering method for mandible tissue scaffold. In their study, they identified the mathematical equation that is governing each point in the shape of the damaged bone. By doing this, the shape of the scaffold that resembled the damaged bone shape can easily be reproduced. They also found that at a certain amount of porosity, the mechanical properties of the scaffold can match with its host tissue [102].

Besides that, there is also a parametric method that can be defined as the new approach to describe the geometry of human bones, which is based on anatomical landmarks. Since the researchers found difficulties in tailoring the bone substitute with the geometry of host tissue for a specific patient, the Method of Anatomical Features (MAF) was introduced by Vidosav et al. in their paper [103,104]. For example, the anatomical landmarks for femur bone are the Centre of Femoral Head. MAF has been a huge aid in determining the 3D model of the bone by using reverse engineering. In addition, it is also reliable in producing the predictive model of the bone or simply known as a parametric model of the bone [105].

The Method of Anatomical Features (MAF) consists of a few steps that are necessary to obtain the parametric model, and the most important step is to define the Referential Geometrical Entities (RGEs). Referential Geometrical Entities (RGEs) can be defined as the basic prerequisite in order to develop a successful reverse engineering modelling of the human bone as well as the predictive model of the human bone [106]. Planes, axes, curves, surfaces and points are examples of RGEs. All of the elements of the human bone must be referred to as the defined RGEs. In a study conducted by Stojkovic et al., they carried out MAF on the femur bone of a human, and they defined some of the RGEs of the femur bone of a human [107]. Anterior–Posterior (A–P) plane and Lateral–Medial (L–M) plane are the crucial views that needed to be defined precisely in order to develop the reversed model of the human femur bone successfully.

In order to understand the procedure of generating a mathematical equation by using the Method of Anatomical Features (MAF), the following flowchart in Figure 1 can be referred.

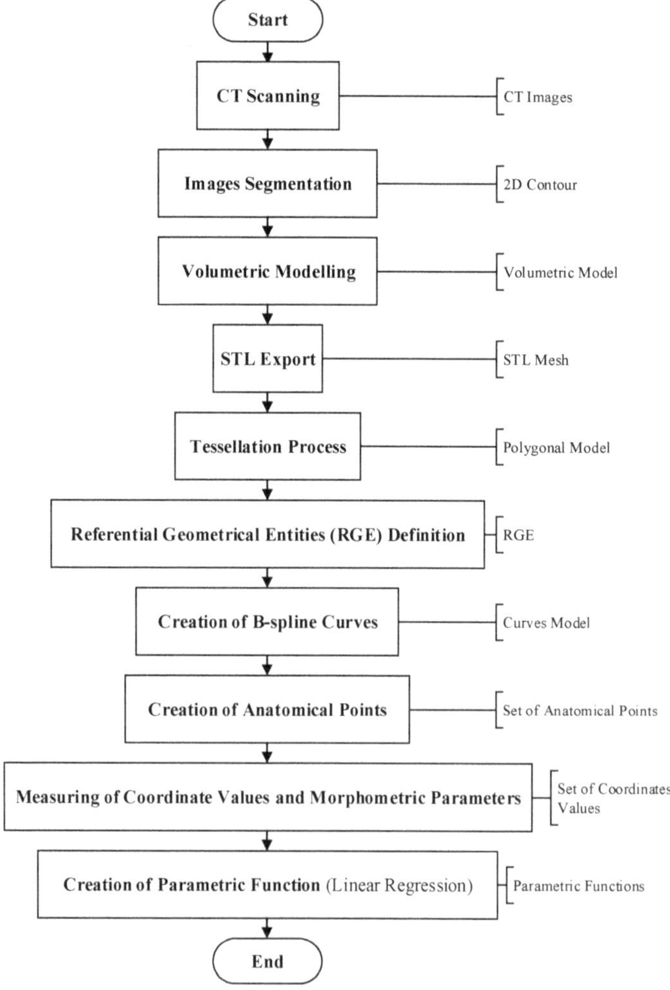


**Figure 1.** Flowchart of the Method of Anatomical Features (MAF).

### Step 1: CT Scanning

In a real-world application, the CT scanning technique is used to identify any abnormalities of an organ. This method is carried out by scanning the organ so that the defective part of the organ can be detected.

### Step 2: Volumetric Modelling

The volumetric model of the bone will be created in order to identify the initial geometry of the bone, which is required in order to locate the missing part of the bone. This model is created morphologically and anatomically in order to define the descriptive model of human bone. The model will then be saved in the STL file. The STL file will be exported into CAD software.

### Step 3: Tessellation Process

This process is crucial in order to determine the polygonal model of the scanned bone. The tessellated model helps in identifying and filling any gaps that are found in the scanned bone during the STL mesh obtained.

### Step 4: Referential Geometrical Entities (RGE) Definition

Referential Geometrical Entities or RGE includes the characteristics, points, planes, directions and views of the bone. These entities are defined in order to create a successful reverse engineering model of the bone.

### Step 5: Creation of B-Spline Curves

These curves are created by using the referential geometrical entities (RGE) created earlier. However, a few additional curves might be needed in order to create curves that can fit the shape of anatomical features precisely.

### Step 6: Creation of Anatomical Points

The anatomical points can be generated on the curves created in the previous step or can be created on anatomical landmarks. By creating the points on spline curves, they will be distributed evenly on the curves. Meanwhile, the points that are created on the anatomical landmarks will be positioned in correspondence to the landmarks such as the distal part of the femur. These points defined the boundary of the anatomical regions on the polygonal model. The process of defining RGE and the creation of B-spline and anatomical points are repeated for each part of the damaged bone.

### Step 7: Measuring of Coordinate Values and Morphometric Parameters

Values of coordinates are measured on each part of the bone model in 3D. The morphometric parameters are also measured in the same 3D model.

### Step 8: Creation of Parametric Function (Linear Regression)

The parametric functions can be generated by defining the relationship between morphometric parameters and coordinate values. The parametric model of the bone will be created, which consists of multiple parametric function. This model is used as the predictive model for the bone.

Generally, a parametric design is very much reliable in providing a feasible tissue engineering scaffold. Since a parametric design consists of a complex shape, it is very much compatible with the additive manufacturing sector. However, it is governed by a specific algorithm, therefore making it is difficult to be produced. Triply Periodic Minimal Surfaces (TPMS) is an excellent structure that provides the scaffold with an adequate amount of mechanical properties along while possessing a good amount of porosity. Meanwhile, a Voronoi structure is a top-notch structure that possesses excellent scaffold properties since it matches the properties of the host tissue and is able to imitate the actual structure of the host tissue.

*2.4. Summary of Scaffold Design*

Based on the previous discussion, there are a lot of unit cell scaffold designs that have been studied by the researchers. The designs can be categorized under various categories, which eventually bring many different benefits to each other. Table 4 summarizes the designs that have been adopted by the researchers in their studies.

Based on Table 4, the varieties of designs produced by the researchers has showed that the study that is involving the design of scaffold has been rapidly increasing especially with the aid of the emerging technologies. Non-parametric designs were chosen by the researchers previously due to their simpler design process as compared to the parametric designs. However, scaffold which possess a high amount of porosity will facilitate the tissue growth process, and it can be seen that the non-parametric designs can possess 60–80% porosity and exhibits lower elastic modulus values. Meanwhile, the parametric designs can increase their elastic modulus varies from 0.8 GPa up until 3.92 GPa with an adequate amount of porosity. This shows that the parametric designs can easily imitate the properties of the host tissue [108]. In addition, the usage of additive manufacturing and also 3D scanner has shown a great impact in contributing to the complex designs of scaffold structure. Therefore, there is a need to produce a design that is not just limited to imitating the bone structure, but also utilizing the designs that are available naturally.

## 3. Computational Software Used in Simulation of Tissue Engineering Scaffold

There is a lot of software that can be used to design and simulate the behaviour of the tissue engineering scaffold. The software can be utilized based on the function that is embedded in the software. Table 5 describes the characteristics of the software.

**Table 4.** Scaffold design and its properties.

| Type of Design | | Material | Porosity (%) | Mechanical Properties | | | Software | Ref. |
|---|---|---|---|---|---|---|---|---|
| | | | | Elastic Modulus (GPa) | Young's Modulus (MPa) | Compressive Strength (MPa) | | |
| Non-Parametric Design | Circular | Poly(L-lactic-co-glycolic acid) (PLGA), type I collagen, and nano-hydroxyapatite (nHA) | 54.3–65.2 | 4.03–5.67 | - | - | COMSOL Multiphysics | [109] |
| | | Polylactic Acid (PLA) | 80 | - | - | 0.163 | Creo Simulate | [19] |
| | | Poly-L-Lactic Acid (PLLA) | 70–97 | - | - | 0.2–0.35 | - | [110] |
| | Square | User-defined | 64.8 | - | 0.5–1.0 | | Abaqus | [111] |
| | | User-defined | 60 | 0.16 | - | | Ansys Fluent | [36] |
| | | Polylactic Acid (PLA) | 80 | - | - | 0.186 | Creo Simulate | [19] |
| | | Polyamide (PA) | Graded 0.74–0.89 | 0.01 | - | - | Abaqus | [54] |
| | Hexagonal | Poly-D-L-Lactic Acid (PDLLA) | 55–70 | - | 274–1514 | - | Ansys Fluent | [112] |
| | | Glass Ceramic | 60 | 2.4 | - | - | - | [22] |
| | | | | - | - | 139 | - | |
| | Octet | User-defined | 60 | 6 | - | - | Ansys | [36] |
| Parametric Design | Triply Periodic Minimal Surfaces (TPMS) | Schwarz P (Primitives) | | | | | | |
| | | Photopolymer Resin | 30 | - | 150 | - | Abaqus | [84,87] |
| | | Photopolymer Resin | 60 | - | 490 | - | Abaqus | [84,87] |
| | | Photopolymer Resin | Graded 30–60 | - | 350 | - | Abaqus | [84,87] |
| | | Visijet M3 Crystal | 70 | - | 103.54 | - | Abaqus | [113] |
| | | Schwarz D(Diamond) | | | | | | |
| | | Photopolymer Resin | 30 | - | 336 | - | Abaqus | [84,87] |
| | | Photopolymer Resin | 60 | - | 79.5 | - | Abaqus | [84,87] |
| | | Visijet M3 Crystal | 70 | - | 171.37 | - | Abaqus | [113] |
| | | Gyroid | | | | | | |
| | | Poly-D-L-Lactic Acid (PDLLA) | 55–70 | - | 181–1011 | - | Ansys Fluent | [112] |
| | | Visijet M3 Crystal | 70 | - | 145.05 | - | Abaqus | [113] |
| | | I-WP | | | | | | |
| | | Photopolymer Resin | Graded 40–60 | - | 170 | - | Abaqus | [84,87] |
| | Voronoi | Poly-D-Lactic Acid (PDLA) | 75–85 | 0.3–0.5 | - | - | - | [31] |
| | | Titanium Alloy | 70 | - | 3920 | - | Grasshopper | [114] |
| | Other | Titanium Alloy | 60–90 | - | - | 11.4 MPa | - | [115] |
| | | Titanium Alloy | 30–70 | 2.3–8.6 | - | - | - | [116] |

**Table 5.** Commonly Used Software for Tissue Engineering Scaffold Simulation.

| Software | Description | Advantages | Disadvantages | Ref. |
|---|---|---|---|---|
| Solidworks | Computer-aided software that acts as a platform to design a scaffold model. | • User-friendly interface<br>• Easy to be utilized<br>• Wider range of rendering options<br>• Capable of designing a parametric model<br>• Helps validate the products in terms of performance and safety | • Lack of built-in library option<br>• Inefficient design in terms of model elements | [29,117–120] |
| Catia | Computer-aided software that acts as a platform to design a scaffold model. It is also known to be a wide-ranging software that provides kinematic simulation. | • Has a wider range of built-in library option<br>• Provides a kinematic solution<br>• Efficiently design elements of a model | • Difficult to be immediately utilized by beginner<br>• Each program is dedicated to different industries | [73,102,121–123] |
| Abaqus | Abaqus helps in modelling and carries out a Finite Element Analysis. It facilitates in visualizing the behaviour of the scaffold design in terms of mechanical properties by providing the failure mechanism of the scaffold based on the boundary condition that has been subjected to the scaffold. | • Able to simulate explicit/implicit model<br>• User-friendly interface<br>• Able to mesh a design accurately<br>• Excellent software for Finite Element Analysis (FEA) | • Unable to modify orphan mesh<br>• Less reliable in simulating fluid properties as compared to Ansys Fluent | [35,113,124–126] |
| Ansys Fluent | Computer-aided engineering software that is reliable to simulate and visualize a Computational Fluid Dynamics Analysis (CFD). The scaffold can be subjected to various conditions that are including fluid flow analysis, etc. The software is capable of visualizing the behaviour of the scaffold in terms of fluidic properties. | • Produce very good mesh properties<br>• User-friendly interface<br>• Excellent software for Computational Fluid Dynamics (CFD) Analysis | • Limited mesh options<br>• Incapable of Finite Element Analysis (FEA) simulation without Computational Fluid Dynamics (CFD) Analysis | [127–130] |

From Table 5, there are four common software programs that have been utilized by the researchers in order to determine the properties of the scaffold of bone tissue engineering. From the description, we can see that Solidworks and Catia belong to the modelling and designing part of the simulation process. The software programs are reliable in producing an accurate design of the scaffold, which then will be used in the properties' simulation. Besides that, the Finite Element Analysis (FEA) is carried out mostly by using Abaqus software which is capable of visualizing the behaviour and failure mechanism of the scaffold model. Meanwhile, Ansys Fluent is used to simulate the Computational Fluid Dynamics (CFD) Analysis of the scaffold behaviour. It is able to simulate and visualize the scaffold behaviour under various conditions, especially in fluid flow analysis.

*Other Software*

There is also other software that was used in simulating the behaviour of the scaffold mechanically and fluidic. For example, COMSOL Multiphysics was used by Uth et al. in their study in order to validate and optimize the design parameters of a scaffold [109]. Sahin et al. had also used COMSOL Multiphysics to carry out Finite Element Analysis (FEA) simulation [131]. Apart from that, Creo Simulate was also adopted by researchers since it is capable of designing a scaffold model [19,132]. It is also reliable performing a numerical analysis of an anatomical model [133].

From these trends, we can see that there is a variety of software that can be used to simulate the behaviour of the scaffold. However, the software chosen to be adopted in the study should match the objectives and able to carry out the desired simulation.

## 4. Challenges and Future Work Recommendation

With the 3D printing technology nowadays, it seems like computational methods have been attracting many researchers' attention in producing many studies that can fully unleash the potential of computational modelling in the future. However, as we know, there are no technologies that are perfectly developed. In computational modelling, there are still challenges that needed to be solved by the researchers. The limitation of the usage of computational modelling in designing a feasible scaffold of tissue engineering includes.

- The accuracy of the simulation technique. A model that is designed through computational method tends to be simplified in the computer-aided design (CAD) software. The structure of the scaffolds might be not fully accurate when it comes to comparing the simulated model and fabricated parts.
- The simulation of the scaffold's behaviour can only be done by simulating uniaxial loadings in most studies. However, in real-life conditions, the scaffolds can be subjected to many loadings that are much more complicated as compared to uniaxial loadings.
- The simulation can only focus on small-scale models [3]. This is due to the constraints that are involving the technologies, such as computer power and application simulating time.
- In the future, it is advisable if the research can contribute to:
- Increase the accuracy of the simulation when it comes to the fabrication process. This process can be achieved by adopting image-based modelling such as images from a 3D scanner.
- The need to simulate the scaffolds models under various types of loading is crucial since many loads can be exerted on it, physically.
- To expand the study on using simulation method that can reduce the effect of size of the small-scale scaffold model on the large-scale scaffold model.
- Improvise the 3D printing technique is crucial since it can affect the surface of the scaffold.
- The studies can integrate artificial intelligence in the computational method.

## 5. Conclusions

This paper has reviewed the studies that comprise the application of the computational method in the area of bone tissue engineering. The computational method can be used to simulate the properties of the scaffold of bone tissue engineering. Moreover, the simulation technique can also be used to predict the design of the scaffold model. In order to produce a scaffold with good mechanical properties, many studies have been carried out to simulate the mechanical properties of the scaffold. It is desired that the scaffold possesses high compressive strength so that it can withstand the load exerted on it when it is planted into the body of a human or an animal. Since the porosity and mechanical strength have an inversely proportional relationship, most researchers came out with integrating the optimization process and simulation process, which produced the optimal scaffold model with good mechanical and fluid properties. Furthermore, the design of the scaffold was also simulated by using computational software. The types of designs that can be generated by using the computational method have varied. From the discussion, we can see that the parametric designs have attracted researchers' attention since it exhibits a good balance between mechanical and fluid properties of the scaffold. Moreover, the parametric designs had also shown huge potential in terms of imitating the properties of the host tissue. With this review, it can be concluded that the computational method has great potential to be adopted in future studies due to its ability to predict the properties of the scaffolds. Moreover, the computational method is less time-consuming and very much reliable than the conventional method.

**Author Contributions:** N.S.M. reviewed the literature on the designs of the scaffold; M.H.A.T. reviewed the literature on the software used in tissue engineering; A.I.M.S. and S.I. (Suhaimi Illias) reviewed the literature on bone tissue engineering. N.H.A. provided expertise in tissue engineering and designs. N.Z.Y. and M.N.B.O. revised the paper according to the designs. S.I. (Sudin Izman) coordinated the team, contributed to the development and planning of the work and its revision with a particular focus on the designs of the scaffold in bone tissue engineering applications. All authors have read and agreed to the published version of the manuscript.

**Funding:** This research was funded by the Universiti Teknologi Malaysia (UTM) and the Universiti Malaysia Perlis (UNIMAP) under the Collaborative Research Grant (CRG) funding number 08G22 and 4B447 (UTM) and 9023-00007 (UNIMAP), the UTM R&D Fund (4J506) and the Fundamental Research Grant Scheme (FRGS) funding number 5F188.

**Institutional Review Board Statement:** Not applicable.

**Informed Consent Statement:** Not applicable.

**Data Availability Statement:** No new data were created or analyzed in this study. Data sharing is not applicable to this article.

**Acknowledgments:** The authors would like to thank the Ministry of Higher Education (MOHE), the Universiti Teknologi Malaysia (UTM) and the Universiti Malaysia Pahang (UNIMAP) for financial support to this work through the Collaborative Research Grant (CRG) funding number 08G22 and 4B447 (UTM) and 9023-00007 (UNIMAP), the UTM R&D Fund (4J506) and the Fundamental Research Grant Scheme (FRGS) funding number 5F188.

**Conflicts of Interest:** The authors declare no conflict of interest

## References

1. Eltom, A.; Zhong, G.; Muhammad, A. Scaffold Techniques and Designs in Tissue Engineering Functions and Purposes: A Review. *Adv. Mater. Sci. Eng.* **2019**, *2019*, 3429527. [CrossRef]
2. Murizan, N.I.S.; Mustafa, N.S.; Ngadiman, N.H.A.; Mohd Yusof, N.; Idris, A. Review on Nanocrystalline Cellulose in Bone Tissue Engineering Applications. *Polymers* **2020**, *12*, 2818. [CrossRef] [PubMed]
3. Zhang, S.; Vijayavenkataraman, S.; Lu, W.F.; Fuh, J.Y.H. A review on the use of computational methods to characterize, design, and optimize tissue engineering scaffolds, with a potential in 3D printing fabrication. *J. Biomed. Mater. Res. Part B* **2019**, *107*, 1329–1351. [CrossRef] [PubMed]
4. Turnbull, G.; Riches, P.; Jia, L.; Clarke, J.; Han, F.; Li, B.; Shu, W. 3D bioactive composite scaffolds for bone tissue engineering. *Bioact. Mater.* **2018**, *3*, 278–314. [CrossRef] [PubMed]

5.  Moreno Madrid, A.P.; Vrech, S.M.; Sanchez, M.A.; Rodriguez, A.P. Advances in additive manufacturing for bone tissue engineering scaffolds. *Mater. Sci. Eng. C* **2019**, *100*, 631–644. [CrossRef]

6.  Kazimierczak, P.; Benko, A.; Nocun, M.; Przekora, A. Novel chitosan/agarose/hydroxyapatite nanocomposite scaffold for bone tissue engineering applications: Comprehensive evaluation of biocompatibility and osteoinductivity with the use of osteoblasts and mesenchymal stem cells. *Int. J. Nanomed.* **2019**, *14*, 6615–6630. [CrossRef]

7.  Carluccio, D.; Xu, C.; Venezuela, J.; Cao, Y.; Kent, D.; Bermingham, M.; Demir, A.G.; Previtali, B.; Ye, Q.; Dargusch, M. Additively manufactured iron-manganese for biodegradable porous load-bearing bone scaffold applications. *Acta Biomater.* **2020**, *103*, 346–360. [CrossRef]

8.  Chen, Z.; Yan, X.; Yin, S.; Liu, L.; Liu, X.; Zhao, G.; Ma, W.; Qi, W.; Ren, Z.; Liao, H.; et al. Influence of the pore size and porosity of selective laser melted Ti6Al4V ELI porous scaffold on cell proliferation, osteogenesis and bone ingrowth. *Mater. Sci. Eng. C* **2020**, *106*, 110289. [CrossRef]

9.  Ngadiman, N.H.A.; Noordin, M.Y.; Idris, A.; Kurniawan, D. A review of evolution of electrospun tissue engineering scaffold: From two dimensions to three dimensions. *Proc. Inst. Mech. Eng. Part H J. Eng. Med.* **2017**, *231*, 597–616. [CrossRef]

10. Zhang, X.; Fang, G.; Zhou, J. Additively Manufactured Scaffolds for Bone Tissue Engineering and the Prediction of their Mechanical Behavior: A Review. *Materials* **2017**, *10*, 50. [CrossRef]

11. Ali, D.; Ozalp, M.; Blanquer, S.B.G.; Onel, S. Permeability and fluid flow-induced wall shear stress in bone scaffolds with TPMS and lattice architectures: A CFD analysis. *Eur. J. Mech. B Fluids* **2020**, *79*, 376–385. [CrossRef]

12. Ngadiman, N.H.A.; Mohd, N.; Idris, A.; Misran, E.; Kurniawan, D. Development of highly porous biodegradable $\gamma$-Fe$_2$O$_3$/polyvinyl alcohol nano fi ber mats using electrospinning process for biomedical application. *Mater. Sci. Eng. C* **2017**, *70*, 520–534. [CrossRef]

13. Wang, S.; Liu, L.; Li, K.; Zhu, L.; Chen, J.; Hao, Y. Pore functionally graded Ti6Al4V scaffolds for bone tissue engineering application. *Mater. Des.* **2019**, *168*, 107643. [CrossRef]

14. Sheikh, Z.; Najeeb, S.; Khurshid, Z.; Verma, V.; Rashid, H.; Glogauer, M. Biodegradable Materials for Bone Repair and Tissue Engineering Applications. *Materials* **2015**, *8*, 5744–5794. [CrossRef] [PubMed]

15. Rao, S.H.; Harini, B.; Shadamarshan, R.P.K.; Balagangadharan, K.; Selvamurugan, N. Natural and Synthetic Polymers/Bioceramics/Bioactive Compounds-mediated Cell Signaling in Bone Tissue Engineering. *Int. J. Biol. Macromol.* **2018**, *110*, 88–96. [CrossRef] [PubMed]

16. Castilho, M.; Dias, M.; Vorndran, E. Fabrication of computationally designed scaffolds by low temperature 3D printing. *Biofabrication* **2013**, *5*, 035012. [CrossRef]

17. Egan, P.F.; Shea, K.A.; Ferguson, S.J.; Egan, P.F. Simulated tissue growth for 3D printed scaffolds. *Biomech. Model. Mechanobiol.* **2018**, *17*, 1481–1495. [CrossRef] [PubMed]

18. Noordin, M.A.; Saad, A.P.; Ngadiman, N.H.A.; Mustafa, N.S.; Noordin, M.Y.; Ma'aram, A. Finite Element Analysis of Porosity Effects on Mechanical Properties For Tissue Engineering Scaffold. *Biointerface Res. Appl. Chem.* **2021**, *11*, 8836–8843.

19. Habib, F.N.; Nikzad, M.; Masood, S.H.; Saifullah, A.B.M. Design and Development of Scaffolds for Tissue Engineering Using Three-Dimensional Printing for Bio-Based Applications. *3D Print. Addit. Manuf.* **2016**, *3*, 119–127. [CrossRef]

20. Ali, D.; Sen, S. Finite Element Analysis of Mechanical Behavior, Permeability and Fluid Induced Wall Shear Stress of High Porosity Scaffolds with Gyroid and Lattice-Based Architectures. *J. Mech. Behav. Biomed. Mater.* **2017**, *75*, 262–270. [CrossRef]

21. Zhao, H.; Li, L.; Ding, S.; Liu, C.; Ai, J. Effect of porous structure and pore size on mechanical strength of 3D-printed comby scaffolds. *Mater. Lett.* **2018**, *223*, 21–24. [CrossRef]

22. Roohani-Esfahani, S.-I.; Newman, P.; Zreiqat, H. Design and Fabrication of 3D printed Scaffolds with a Mechanical Strength Comparable to Cortical Bone to Repair Large Bone Defects. *Sci. Rep.* **2016**, *6*, 1–8. [CrossRef]

23. Campos Marin, A.; Lacroix, D. Computational Simulation of Cell Seeding in a Tissue Engineering Scaffold. In *Multiscale Mechanobiology in Tissue Engineering*; Lacroix, D., Brunelli, M., Perrault, C., Baldit, A., Shariatzadeh, M., Campos Marin, A., Castro, A., Barreto, S., Eds.; Springer: Singapore, 2019; pp. 81–104. [CrossRef]

24. Boccaccio, A.; Uva, A.E.; Fiorentino, M.; Bevilacqua, V.; Pappalettere, C.; Monno, G. A Computational Approach to the Design of Scaffolds for Bone Tissue Engineering. In *Lecture Notes in Bioengineering, Proceedings of the Advances in Bionanomaterials: Selected Papers from the 2nd Workshop in Bionanomaterials, BIONAM 2016, Salerno, Italy, 4–7 October 2016*; Piotto, S., Rossi, F., Concilio, S., Reverchon, E., Cattaneo, G., Eds.; Springer International Publishing: Cham, Switzerland, 2018; pp. 111–117. [CrossRef]

25. Sun, K.; Li, R.; Li, H.; Fan, M.; Li, H. Analysis and Demonstration of a Scaffold Finite Element Model for Cartilage Tissue Engineering. *ACS Omega* **2020**, *5*, 32411–32419. [CrossRef] [PubMed]

26. Boccaccio, A.; Uva, A.E.; Fiorentino, M.; Lamberti, L.; Monno, G. A Mechanobiology-based Algorithm to Optimize the Microstructure Geometry of Bone Tissue Scaffolds. *Int. J. Biol. Sci.* **2016**, *12*, 1–17. [CrossRef]

27. Boccaccio, A.; Uva, A.E.; Fiorentino, M.; Mori, G.; Monno, G. Geometry Design Optimization of Functionally Graded Scaffolds for Bone Tissue Engineering: A Mechanobiological Approach. *PLoS ONE* **2016**, *11*, e0146935. [CrossRef] [PubMed]

28. Kamboj, N.; Aghayan, M.; Rodrigo-Vazquez, C.S.; Rodríguez, M.A.; Hussainova, I. Novel silicon-wollastonite based scaffolds for bone tissue engineering produced by selective laser melting. *Ceram. Int.* **2019**, *45*, 24691–24701. [CrossRef]

29. Jahir-Hussain, M.; Maaruf, N.; Esa, N.; Jusoh, N. The effect of pore geometry on the mechanical properties of 3D-printed bone scaffold due to compressive loading. In *IOP Conference Series: Materials Science and Engineering*; IOP Publishing Ltd.: Bristol, UK, 2021; Volume 1051, p. 012016.

30. Noordin, M.A.; Rahim, R.A.A.; Roslan, A.N.H.; Ali, I.A.; Syahrom, A.; Saad, A.P.M. Controllable Macroscopic Architecture of Subtractive Manufactured Porous Iron for Cancellous Bone Analogue: Computational to Experimental Validation. *J. Bionic Eng.* **2020**, *17*, 357–369. [CrossRef]

31. Gómez, S.; Vlad, M.D.; López, J.; Fernández, E. Design and Properties of 3D Scaffolds for Bone Tissue Engineering. *Acta Biomater.* **2016**, *42*, 341–350. [CrossRef]

32. Deng, F.; Liu, L.; Li, Z.; Liu, J. 3D printed Ti6Al4V bone scaffolds with different pore structure effects on bone ingrowth. *J. Biol. Eng.* **2021**, *15*, 1–13. [CrossRef]

33. Gong, B.; Cui, S.; Zhao, Y.; Sun, Y.; Ding, Q. Strain-controlled fatigue behaviors of porous PLA-based scaffolds by 3D-printing technology. *J. Biomater. Sci. Polym. Ed.* **2017**, *28*, 2196–2204. [CrossRef]

34. Aliabouzar, M.; Lee, S.-J.; Zhou, X.; Zhang, G.L.; Sarkar, K. Effects of scaffold microstructure and low intensity pulsed ultrasound on chondrogenic differentiation of human mesenchymal stem cells. *Biotechnol. Bioeng.* **2018**, *115*, 495–506. [CrossRef]

35. Tang, M.; Kadir, A.A.; Ngadiman, N. Simulation analysis of different bone scaffold porous structures for fused deposition modelling fabrication process. In *IOP Conference Series: Materials Science and Engineering*; IOP Publishing Ltd.: Bristol, UK, 2020; Volume 788, p. 012023.

36. Egan, P.F.; Engensperger, M.; Ferguson, S.J.; Shea, K. Design and Fabrication of 3D Printed Tissue Scaffolds Informed by Mechanics and Fluids Simulations. In Proceedings of the ASME 2017 International Design Engineering Technical Conferences and Computers and Information in Engineering Conference, Cleveland, OH, USA, 6–9 August 2017; pp. 1–10.

37. Ren, X.; Xiao, L.; Hao, Z. Multi-property cellular material design approach based on the mechanical behaviour analysis of the reinforced lattice structure. *Mater. Des.* **2019**, *174*, 107785. [CrossRef]

38. Hernandez, I.; Kumar, A.; Joddar, B. A Bioactive Hydrogel and 3D Printed Polycaprolactone System for Bone Tissue Engineering. *Gels* **2017**, *3*, 26. [CrossRef]

39. Bui, V.-T.; Thuy, L.T.; Tran, Q.C.; Nguyen, V.-T.; Dao, V.-D.; Choi, J.S.; Choi, H.-S. Ordered honeycomb biocompatible polymer films via a one-step solution-immersion phase separation used as a scaffold for cell cultures. *Chem. Eng. J.* **2017**, *320*, 561–569. [CrossRef]

40. Chang, Y.-H.; Wu, K.-C.; Wang, C.-C.; Ding, D.-C. Enhanced chondrogenesis of human umbilical cord mesenchymal stem cells in a gelatin honeycomb scaffold. *J. Biomed. Mater. Res. Part A* **2020**, *108*, 2069–2079. [CrossRef]

41. Itoh, H.; Aso, Y.; Furuse, M.; Noishiki, Y.; Miyata, T. A Honeycomb Collagen Carrier for Cell Culture as a Tissue Engineering Scaffold. *Artif. Organs* **2001**, *25*, 213–217. [CrossRef]

42. Liu, K.; Li, W.; Chen, S.; Wen, W.; Lu, L.; Liu, M.; Zhou, C.; Luo, B. The design, fabrication and evaluation of 3D printed gHNTs/gMgO whiskers/PLLA composite scaffold with honeycomb microstructure for bone tissue engineering. *Compos. Part B Eng.* **2020**, *192*, 108001. [CrossRef]

43. Paun, I.A.; Popescu, R.C.; Mustaciosu, C.C.; Zamfirescu, M.; Calin, B.S.; Mihailescu, M.; Dinescu, M.; Popescu, A.; Chioibasu, D.; Soproniy, M. Laser-direct writing by two-photon polymerization of 3D honeycomb-like structures for bone regeneration. *Biofabrication* **2018**, *10*, 025009. [CrossRef]

44. Choy, S.Y.; Sun, C.-N.; Leong, K.F.; Wei, J. Compressive properties of Ti-6Al-4V lattice structures fabricated by selective laser melting: Design, orientation and density. *Addit. Manuf.* **2017**, *16*, 213–224. [CrossRef]

45. Golodnov, A.I.; Loginov, Y.N.; Stepanov, S.I. Numeric loading simulation of titanium implant manufactured using 3d printing. In *Solid State Phenomena*; Trans Tech Publications Ltd.: Bäch, Switzerland, 2018; pp. 380–385.

46. Peng, W.-M.; Liu, Y.-F.; Jiang, X.-F.; Dong, X.-T.; Jun, J.; Baur, D.A.; Xu, J.-J.; Pan, H.; Xu, X. Bionic mechanical design and 3D printing of novel porous Ti6Al4V implants for biomedical applications. *J. Zhejiang Univ. Sci. B* **2019**, *20*, 647–659. [CrossRef] [PubMed]

47. Nazir, A.; Abate, K.M.; Kumar, A.; Jeng, J.-Y. A state-of-the-art review on types, design, optimization, and additive manufacturing of cellular structures. *Int. J. Adv. Manuf. Technol.* **2019**, *104*, 3489–3510. [CrossRef]

48. Montazerian, H.; Mohamed, M.; Montazeri, M.M.; Kheiri, S.; Milani, A.; Kim, K.; Hoorfar, M. Permeability and mechanical properties of gradient porous PDMS scaffolds fabricated by 3D-printed sacrificial templates designed with minimal surfaces. *Acta Biomater.* **2019**, *96*, 149–160. [CrossRef] [PubMed]

49. Yuan, L.; Ding, S.; Wen, C. Additive manufacturing technology for porous metal implant applications and triple minimal surface structures: A review. *Bioact. Mater.* **2019**, *4*, 56–70. [CrossRef]

50. Bobbert, F.; Lietaert, K.; Eftekhari, A.A.; Pouran, B.; Ahmadi, S.; Weinans, H.; Zadpoor, A. Additively manufactured metallic porous biomaterials based on minimal surfaces: A unique combination of topological, mechanical, and mass transport properties. *Acta Biomater.* **2017**, *53*, 572–584. [CrossRef] [PubMed]

51. Kadkhodapour, J.; Montazerian, H.; Raeisi, S. Investigating internal architecture effect in plastic deformation and failure for TPMS-based scaffolds using simulation methods and experimental procedure. *Mater. Sci. Eng. C* **2014**, *43*, 587–597. [CrossRef] [PubMed]

52. Santos, J.; Pires, T.; Gouveia, B.P.; Castro, A.P.; Fernandes, P.R. On the permeability of TPMS scaffolds. *J. Mech. Behav. Biomed. Mater.* **2020**, *110*, 103932. [CrossRef] [PubMed]

53. Wang, S.; Liu, L.; Zhou, X.; Zhu, L.; Hao, Y. The design of Ti6Al4V Primitive surface structure with symmetrical gradient of pore size in biomimetic bone scaffold. *Mater. Des.* **2020**, *193*, 108830. [CrossRef]

54. Maskery, I.; Aremu, A.; Parry, L.; Wildman, R.; Tuck, C.; Ashcroft, I. Effective design and simulation of surface-based lattice structures featuring volume fraction and cell type grading. *Mater. Des.* **2018**, *155*, 220–232. [CrossRef]

55. Restrepo, S.; Ocampo, S.; Ramírez, J.A.; Paucar, C.; García, C. Mechanical properties of ceramic structures based on Triply Periodic Minimal Surface (TPMS) processed by 3D printing. *J. Phys. Conf. Ser.* **2017**, *935*, 012036. [CrossRef]

56. Guo, X.; Zheng, X.; Yang, Y.; Yang, X.; Yi, Y. Mechanical behavior of TPMS-based scaffolds: A comparison between minimal surfaces and their lattice structures. *SN Appl. Sci.* **2019**, *1*, 1–11. [CrossRef]

57. Lu, Y.; Zhao, W.; Cui, Z.; Zhu, H.; Wu, C. The anisotropic elastic behavior of the widely-used triply-periodic minimal surface based scaffolds. *J. Mech. Behav. Biomed. Mater.* **2019**, *99*, 56–65. [CrossRef] [PubMed]

58. Zaharin, H.A.; Abdul Rani, A.M.; Azam, F.I.; Ginta, T.L.; Sallih, N.; Ahmad, A.; Yunus, N.A.; Zulkifli, T.Z.A. Effect of unit cell type and pore size on porosity and mechanical behavior of additively manufactured Ti6Al4V scaffolds. *Materials* **2018**, *11*, 2402. [CrossRef] [PubMed]

59. Abu Al-Rub, R.K.; Lee, D.-W.; Khan, K.A.; Palazotto, A.N. Effective anisotropic elastic and plastic yield properties of periodic foams derived from triply periodic Schoen's I-WP minimal surface. *J. Eng. Mech.* **2020**, *146*, 04020030. [CrossRef]

60. Feng, J.; Fu, J.; Shang, C.; Lin, Z.; Li, B. Porous scaffold design by solid T-splines and triply periodic minimal surfaces. *Comput. Methods Appl. Mech. Eng.* **2018**, *336*, 333–352. [CrossRef]

61. Montazerian, H.; Davoodi, E.; Asadi-eydivand, M.; Kadkhodapour, J.; Solati-hashjin, M. Materials & Design Porous scaffold internal architecture design based on minimal surfaces: A compromise between permeability and elastic properties. *Mater. Des.* **2017**, *126*, 98–114. [CrossRef]

62. Sharma, V.; Grujovic, N.; Zivic, F.; Slavkovic, V. Influence of Porosity on the Mechanical Behavior during Uniaxial Compressive Testing on Voronoi-Based Open-Cell Aluminium Foam. *Materials* **2019**, *12*, 1041. [CrossRef]

63. Chen, H.; Liu, Y.; Wang, C.; Zhang, A.; Chen, B.; Han, Q.; Wang, J. Design and properties of biomimetic irregular scaffolds for bone tissue engineering. *Comput. Biol. Med.* **2021**, *130*, 104241. [CrossRef]

64. Suárez, A.F.; Hubert, E. Scaffolding skeletons using spherical Voronoi diagrams: Feasibility, regularity and symmetry. *Comput. Aided Des.* **2018**, *102*, 83–93. [CrossRef]

65. Deering, J.; Dowling, K.I.; DiCecco, L.-A.; McLean, G.D.; Yu, B.; Grandfield, K. Selective Voronoi tessellation as a method to design anisotropic and biomimetic implants. *J. Mech. Behav. Biomed. Mater.* **2021**, *116*, 104361. [CrossRef] [PubMed]

66. Liu, S.; Chen, J.; Chen, T.; Zeng, Y. Fabrication of trabecular-like beta-tricalcium phosphate biomimetic scaffolds for bone tissue engineering. *Ceram. Int.* **2021**, *47*, 13187–13198. [CrossRef]

67. Yang, H.; Zhao, Y. A new method for designing porous implant. In Proceedings of the DS 87-5 21st International Conference on Engineering Design (ICED 17) Vol 5: Design for X, Design to X, Vancouver, BC, Canada, 21–25 August 2017; pp. 337–344.

68. Han, L.; Che, S. An Overview of Materials with Triply Periodic Minimal Surfaces and Related Geometry: From Biological Structures to Self-Assembled Systems. *Adv. Mater.* **2018**, *30*, 1705708. [CrossRef]

69. Feng, J.; Fu, J.; Lin, Z.; Shang, C.; Niu, X. Layered infill area generation from triply periodic minimal surfaces for additive manufacturing. *Comput. Aided Des.* **2019**, *107*, 50–63. [CrossRef]

70. Schoen, A.H. *Infinite Periodic Minimal Surfaces without Self-Intersections*; National Aeronautics and Space Administration: Washington, DC, USA, 1970.

71. Andersson, S.; Hyde, S.; Larsson, K.; Lidin, S. Minimal surfaces and structures: From inorganic and metal crystals to cell membranes and biopolymers. *Chem. Rev.* **1988**, *88*, 221–242. [CrossRef]

72. Al-Ketan, O.; Abu Al-Rub, R.K. Multifunctional Mechanical Metamaterials Based on Triply Periodic Minimal Surface Lattices. *Adv. Eng. Mater.* **2019**, *21*, 1900524. [CrossRef]

73. Ambu, R.; Morabito, A.E. Porous scaffold design based on minimal surfaces: Development and assessment of variable architectures. *Symmetry* **2018**, *10*, 361. [CrossRef]

74. Barba, D.; Alabort, E.; Reed, R. Synthetic bone: Design by additive manufacturing. *Acta Biomater.* **2019**, *97*, 637–656. [CrossRef] [PubMed]

75. Cai, Z.; Liu, Z.; Hu, X.; Kuang, H.; Zhai, J. The effect of porosity on the mechanical properties of 3D-printed triply periodic minimal surface (TPMS) bioscaffold. *Bio Des. Manuf.* **2019**, *2*, 242–255. [CrossRef]

76. Hsieh, M.-T.; Valdevit, L. Minisurf—A minimal surface generator for finite element modeling and additive manufacturing. *Softw. Impacts* **2020**, *6*, 100026. [CrossRef]

77. Wohlgemuth, M.; Yufa, N.; Hoffman, J.; Thomas, E.L. Triply periodic bicontinuous cubic microdomain morphologies by symmetries. *Macromolecules* **2001**, *34*, 6083–6089. [CrossRef]

78. Michielsen, K.; Kole, J. Photonic band gaps in materials with triply periodic surfaces and related tubular structures. *Phys. Rev. B* **2003**, *68*, 115107. [CrossRef]

79. Blanquer, S.B.G.; Werner, M.; Hannula, M.; Sharifi, S.; Lajoinie, G.P.R.; Eglin, D.; Hyttinen, J.; Poot, A.A.; Grijpma, D.W. Surface curvature in triply-periodic minimal surface architectures as a distinct design parameter in preparing advanced tissue engineering scaffolds. *Biofabrication* **2017**, *9*, 25001. [CrossRef] [PubMed]

80. Shi, J.; Zhu, L.; Li, L.; Li, Z.; Yang, J.; Wang, X. A TPMS-based method for modeling porous scaffolds for bionic bone tissue engineering. *Sci. Rep.* **2018**, *8*, 1–10. [CrossRef] [PubMed]

81. Castro, A.; Ruben, R.; Gonçalves, S.; Pinheiro, J.; Guedes, J.; Fernandes, P. Numerical and experimental evaluation of TPMS Gyroid scaffolds for bone tissue engineering. *Comput. Methods Biomech. Biomed. Eng.* **2019**, *22*, 567–573. [CrossRef]
82. Yang, L.; Mertens, R.; Ferrucci, M.; Yan, C.; Shi, Y.; Yang, S. Continuous graded Gyroid cellular structures fabricated by selective laser melting: Design, manufacturing and mechanical properties. *Mater. Des.* **2019**, *162*, 394–404. [CrossRef]
83. Yan, C.; Hao, L.; Hussein, A.; Young, P. Ti–6Al–4V triply periodic minimal surface structures for bone implants fabricated via selective laser melting. *J. Mech. Behav. Biomed. Mater.* **2015**, *51*, 61–73. [CrossRef]
84. Afshar, M.; Anaraki, A.P.; Montazerian, H.; Kadkhodapour, J. Additive manufacturing and mechanical characterization of graded porosity scaffolds designed based on triply periodic minimal surface architectures. *J. Mech. Behav. Biomed. Mater.* **2016**, *62*, 481–494. [CrossRef]
85. Maskery, I.; Sturm, L.; Aremu, A.; Panesar, A.; Williams, C.; Tuck, C.; Wildman, R.D.; Ashcroft, I.; Hague, R.J. Insights into the mechanical properties of several triply periodic minimal surface lattice structures made by polymer additive manufacturing. *Polymer* **2018**, *152*, 62–71. [CrossRef]
86. Abueidda, D.W.; Elhebeary, M.; Shiang, C.-S.A.; Pang, S.; Al-Rub, R.K.A.; Jasiuk, I.M. Mechanical properties of 3D printed polymeric Gyroid cellular structures: Experimental and finite element study. *Mater. Des.* **2019**, *165*, 107597. [CrossRef]
87. Afshar, M.; Anaraki, A.P.; Montazerian, H. Compressive characteristics of radially graded porosity scaffolds architectured with minimal surfaces. *Mater. Sci. Eng. C* **2018**, *92*, 254–267. [CrossRef]
88. Yang, L.; Yan, C.; Cao, W.; Liu, Z.; Song, B.; Wen, S.; Zhang, C.; Shi, Y.; Yang, S. Compression–compression fatigue behaviour of gyroid-type triply periodic minimal surface porous structures fabricated by selective laser melting. *Acta Mater.* **2019**, *181*, 49–66. [CrossRef]
89. Gawronska, E.; Dyja, R. A Numerical Study of Geometry's Impact on the Thermal and Mechanical Properties of Periodic Surface Structures. *Materials* **2021**, *14*, 427. [CrossRef] [PubMed]
90. Du Plessis, A.; Yadroitsava, I.; Yadroitsev, I.; le Roux, S.; Blaine, D. Numerical comparison of lattice unit cell designs for medical implants by additive manufacturing. *Virtual Phys. Prototyp.* **2018**, *13*, 266–281. [CrossRef]
91. Yang, L.; Yan, C.; Han, C.; Chen, P.; Yang, S.; Shi, Y. Mechanical response of a triply periodic minimal surface cellular structures manufactured by selective laser melting. *Int. J. Mech. Sci.* **2018**, *148*, 149–157. [CrossRef]
92. Zhang, J.; Wang, Z.; Zhao, L. Dynamic response of functionally graded cellular materials based on the Voronoi model. *Compos. Part B Eng.* **2016**, *85*, 176–187. [CrossRef]
93. Fantini, M.; Curto, M. Interactive design and manufacturing of a Voronoi-based biomimetic bone scaffold for morphological characterization. *Int. J. Interact. Des. Manuf.* **2018**, *12*, 585–596. [CrossRef]
94. Liu, T.; Guessasma, S.; Zhu, J.; Zhang, W. Designing Cellular Structures for Additive Manufacturing Using Voronoi–Monte Carlo Approach. *Polymers* **2019**, *11*, 1158. [CrossRef]
95. Kou, X.; Tan, S. A simple and effective geometric representation for irregular porous structure modeling. *Comput. Aided Des.* **2010**, *42*, 930–941. [CrossRef]
96. Kou, X.; Tan, S. Microstructural modelling of functionally graded materials using stochastic Voronoi diagram and B-Spline representations. *Int. J. Comput. Integr. Manuf.* **2012**, *25*, 177–188. [CrossRef]
97. Du, Y.; Liang, H.; Xie, D.; Mao, N.; Zhao, J.; Tian, Z.; Wang, C.; Shen, L. Design and statistical analysis of irregular porous scaffolds for orthopedic reconstruction based on voronoi tessellation and fabricated via selective laser melting (SLM). *Mater. Chem. Phys.* **2020**, *239*, 121968. [CrossRef]
98. Wang, G.; Shen, L.; Zhao, J.; Liang, H.; Xie, D.; Tian, Z.; Wang, C. Design and compressive behavior of controllable irregular porous scaffolds: Based on voronoi-tessellation and for additive manufacturing. *ACS Biomater. Sci. Eng.* **2018**, *4*, 719–727. [CrossRef]
99. Chen, W.; Dai, N.; Wang, J.; Liu, H.; Li, D.; Liu, L. Personalized design of functional gradient bone tissue engineering scaffold. *J. Biomech. Eng.* **2019**, *141*, 111004. [CrossRef] [PubMed]
100. Maliaris, G.; Sarafis, E. Mechanical behavior of 3D printed stochastic lattice structures. In *Solid State Phenomena*; Trans Tech Publications Ltd.: Bäch, Switzerland, 2017; pp. 225–228.
101. Achrai, B.; Wagner, H.D. The turtle carapace as an optimized multi-scale biological composite armor—A review. *J. Mech. Behav. Biomed. Mater.* **2017**, *73*, 50–67. [CrossRef] [PubMed]
102. Vitkovic, N.; Stojkovic, M.; Majstorovic, V.; Trajanovic, M.; Milovanovic, J. Novel design approach for the creation of 3D geometrical model of personalized bone scaffold. *CIRP Ann. Manuf. Technol.* **2018**, *67*, 177–180. [CrossRef]
103. Vidosav, M.; Trajanovic, M.; Vitkovic, N.; Stojkovic, M. Reverse engineering of human bones by using method of anatomical features. *CIRP Ann. Manuf. Technol.* **2013**, *62*, 167–170. [CrossRef]
104. Skallevold, H.E.; Rokaya, D.; Khurshid, Z.; Zafar, M.S. Bioactive Glass Applications in Dentistry. *Int. J. Mol. Sci.* **2019**, *20*, 5960. [CrossRef] [PubMed]
105. Husain, K.; Stojkovic, M.; Vitkovic, N.; Milovanovic, J.; Trajanovic, M.; Rashid, M.; Milovanović, A. Procedure for Creating Personalized Geometrical Models of the Human Mandible and Corresponding Implants. *Teh. Vjesn.* **2019**, *26*, 1044–1051. [CrossRef]
106. Vitković, N.; Mitić, J.; Manić, M.; Trajanović, M.; Husain, K.; Petrović, S.; Arsić, S. The Parametric Model of the Human Mandible Coronoid Process Created by Method of Anatomical Features. *Comput. Math. Methods Med.* **2015**, *2015*, 574132. [CrossRef]

107. Stojkovic, M.; Trajanovic, M.; Vitkovic, N.; Milovanovic, J.; Arsic, S.; Mitkovic, M. Referential geometrical entities for reverse modeling of geometry of femur. In Proceedings of the VIPIMAGE2009—Second Thematic Conference on Computational Vision and Medical Image Processing, Porto, Portugal, 14–16 October 2009; pp. 189–194.

108. Gu, X.-N.; Zheng, Y.-F. A review on magnesium alloys as biodegradable materials. *Front. Mater. Sci. China* **2010**, *4*, 111–115. [CrossRef]

109. Uth, N.; Mueller, J.; Smucker, B.; Yousefi, A.-M. Validation of scaffold design optimization in bone tissue engineering: Finite element modeling versus designed experiments. *Biofabrication* **2017**, *9*, 015023. [CrossRef] [PubMed]

110. Wang, X.; Lou, T.; Zhao, W.; Song, G. The effect of fiber size and pore size on cell proliferation and infiltration in PLLA scaffolds on bone tissue engineering. *Biomater. Appl.* **2016**, *30*, 1545–1551. [CrossRef]

111. Liu, L.; Shi, Q.; Chen, Q.; Li, Z. Mathematical modeling of bone in-growth into undegradable porous periodic scaffolds under mechanical stimulus. *J. Tissue Eng.* **2019**, *10*, 1–13. [CrossRef] [PubMed]

112. Olivares, A.L.; Marsal, È.; Planell, J.A.; Lacroix, D. Finite element study of scaffold architecture design and culture conditions for tissue engineering. *Biomaterials* **2009**, *30*, 6142–6149. [CrossRef] [PubMed]

113. Castro, A.; Pires, T.; Santos, J.; Gouveia, B.; Fernandes, P. Permeability versus design in TPMS scaffolds. *Materials* **2019**, *12*, 1313. [CrossRef] [PubMed]

114. Liang, H.; Yang, Y.; Xie, D.; Li, L.; Mao, N.; Wang, C.; Tian, Z.; Jiang, Q.; Shen, L. Trabecular-like Ti-6Al-4V scaffolds for orthopedic: Fabrication by selective laser melting and in vitro biocompatibility. *J. Mater. Sci. Technol.* **2019**, *35*, 1284–1297. [CrossRef]

115. Zhao, L.; Pei, X.; Jiang, L.; Hu, C.; Sun, J.; Xing, F.; Zhou, C. Bionic design and 3D printing of porous titanium alloy scaffolds for bone tissue repair. *Compos. Part B* **2019**, *162*, 154–161. [CrossRef]

116. Torres-sanchez, C.; Mclaughlin, J.; Fotticchia, A. Porosity and pore size effect on the properties of sintered Ti35Nb4Sn alloy scaffolds and their suitability for tissue engineering applications. *J. Alloys Compd.* **2018**, *731*, 189–199. [CrossRef]

117. Jusoh, N. Analysis of 3D-Printed Hexagon Pore For Scaffold Fabrication: Nursyafiqah Amani Maulat Che Omar, Nurul Azlin Zakaria, Wan Sabrina Wan Safuan, Tee Ee Ling, Jeysheni Shree Pupathi, Nur Elia Insyirah Zaini, Musfirah Jiyanah Jahir Hussain, Norhana Jusoh. *J. Tomogr. Syst. Sens. Appl.* **2020**, *3*, 42–47.

118. Koski, C.; Onuike, B.; Bandyopadhyay, A.; Bose, S. Starch-hydroxyapatite composite bone scaffold fabrication utilizing a slurry extrusion-based solid freeform fabricator. *Addit. Manuf.* **2018**, *24*, 47–59. [CrossRef] [PubMed]

119. Syuhada, G.; Ramahdita, G.; Rahyussalim, A.; Whulanza, Y. Multi-material poly (lactic acid) scaffold fabricated via fused deposition modeling and direct hydroxyapatite injection as spacers in laminoplasty. *AIP Conf. Proc.* **2018**, *1933*, 020008.

120. Wang, J.; Nor Hidayah, Z.; Razak, S.I.A.; Kadir, M.R.A.; Nayan, N.H.M.; Li, Y.; Amin, K.A.M. Surface entrapment of chitosan on 3D printed polylactic acid scaffold and its biomimetic growth of hydroxyapatite. *Compos. Interfaces* **2019**, *26*, 465–478. [CrossRef]

121. Begum, S.R.; Kumar, M.S.; Pruncu, C.; Vasumathi, M.; Harikrishnan, P. Optimization and fabrication of customized scaffold using additive manufacturing to match the property of human bone. *J. Mater. Eng. Perform.* **2021**, 1–12. [CrossRef]

122. Martin, V.; Ribeiro, I.A.; Alves, M.M.; Gonçalves, L.; Claudio, R.A.; Grenho, L.; Fernandes, M.H.; Gomes, P.; Santos, C.F.; Bettencourt, A.F. Engineering a multifunctional 3D-printed PLA-collagen-minocycline-nanoHydroxyapatite scaffold with combined antimicrobial and osteogenic effects for bone regeneration. *Mater. Sci. Eng. C* **2019**, *101*, 15–26. [CrossRef]

123. Vishnurajan, P.; Karuppudaiyan, S.; Singh, D.K. Design and Analysis of Feature Primitive Scaffold Manufactured Using 3D-Printer—Fused Deposition Modelling (FDM). In *Trends in Mechanical and Biomedical Design*; Springer: Berlin/Heidelberg, Germany, 2021; pp. 577–588.

124. Bhardwaj, T.; Singh, S.P.; Shukla, M. Finite element modeling and analysis of implant scaffolds. In Proceedings of the 2017 International Conference on Advances in Mechanical, Industrial, Automation and Management Systems (AMIAMS), Allahabad, India, 3–5 February 2017; pp. 358–362.

125. Monshi, M.; Esmaeili, S.; Kolooshani, A.; Moghadas, B.K.; Saber-Samandari, S.; Khandan, A. A novel three-dimensional printing of electroconductive scaffolds for bone cancer therapy application. *Nanomed. J.* **2020**, *7*, 138–148.

126. Wahid, Z.; Ariffin, M.; Baharudin, B.; Ismail, M.; Mustapha, F. Abaqus simulation of different critical porosities cubical scaffold model. In *IOP Conference Series: Materials Science and Engineering*; IOP Publishing Ltd.: Bristol, UK, 2019; Volume 530, p. 012018.

127. Bogu, V.P.; Madhu, M.; Kumar, Y.R.; Asit, K. Design and Analysis of Various Homogeneous Interconnected Scaffold Structures for Trabecular Bone. In *Mechanical Engineering for Sustainable Development*; Apple Academic Press: Cambridge, MA, USA, 2019; p. 91.

128. Sahai, N.; Saxena, K.K.; Gogoi, M. Modelling and simulation for fabrication of 3D printed polymeric porous tissue scaffolds. *Adv. Mater. Process. Technol.* **2020**, *6*, 530–539. [CrossRef]

129. Shi, C.; Lu, N.; Qin, Y.; Liu, M.; Li, H.; Li, H. Study on mechanical properties and permeability of elliptical porous scaffold based on the SLM manufactured medical Ti6Al4V. *PLoS ONE* **2021**, *16*, e0247764. [CrossRef] [PubMed]

130. Pires, T.; Santos, J.; Ruben, R.B.; Gouveia, B.P.; Castro, A.P.; Fernandes, P.R. Numerical-experimental analysis of the permeability-porosity relationship in triply periodic minimal surfaces scaffolds. *J. Biomech.* **2021**, *117*, 110263. [CrossRef]

131. Sahin, M.; Tabak, A.F.; Kiziltas Sendur, G. Initial Study Towards the Integrated Design of Bone Scaffolds Based on Cell Diffusion, Growth Factor Release and Tissue Regeneration. In Proceedings of the ASME International Mechanical Engineering Congress and Exposition, Online, 16–19 November 2020; p. V005T005A003.

132. Ma, S.; Tang, Q.; Feng, Q.; Song, J.; Han, X.; Guo, F. Mechanical behaviours and mass transport properties of bone-mimicking scaffolds consisted of gyroid structures manufactured using selective laser melting. *J. Mech. Behav. Biomed. Mater.* **2019**, *93*, 158–169. [CrossRef] [PubMed]
133. Hernández-Gómez, L.H.; Beltrán-Fernández, J.A.; Ramírez-Jarquín, M.; Bantle-Chávez, I.; Alvarado-Moreno, C.; González-Rebattú y González, A.; González-Rebattú y González, M.; Flores-Campos, J.A.; Moreno-Garibaldi, P.; Pava-Chipol, N.; et al. Characterization of Scaffold Structures for the Development of Prostheses and Biocompatible Materials. In *Engineering Design Applications*; Öchsner, A., Altenbach, H., Eds.; Springer International Publishing: Cham, Switzerland, 2019; pp. 471–494. [CrossRef]

*Article*

# Antibacterial Activity and Protection Efficiency of Polyvinyl Butyral Nanofibrous Membrane Containing Thymol Prepared through Vertical Electrospinning

Wen-Chi Lu [1,2,†], Ching-Yi Chen [1,†] , Chia-Jung Cho [1,*,†], Manikandan Venkatesan [1,†], Wei-Hung Chiang [3], Yang-Yen Yu [4] , Chen-Hung Lee [5,*], Rong-Ho Lee [6] , Syang-Peng Rwei [1] and Chi-Ching Kuo [1,*]

[1] Research and Development Center of Smart Textile Technology, Institute of Organic and Polymeric Materials, National Taipei University of Technology, Taipei 10608, Taiwan; wcl2320@mail.lit.edu.tw (W.-C.L.); duckfat8@gmail.com (C.-Y.C.); manikandanchemist1093@gmail.com (M.V.); f10714@ntut.edu.tw (S.-P.R.)
[2] Department of Applied Cosmetology, Lee-Ming Institute of Technology, New Taipei City 243083, Taiwan
[3] Department of Chemical Engineering, National Taiwan University of Science and Technology, Taipei 10607, Taiwan; whchiang0102@gmail.com
[4] Department of Materials Engineering, Ming Chi University of Technology, New Taipei City 24301, Taiwan; yyyu@mail.mcut.edu.tw
[5] Division of Cardiology, Department of Internal Medicine, Chang Gung Memorial Hospital-Linkou, Chang Gung University College of Medicine, Tao-Yuan 333, Taiwan
[6] Department of Chemical Engineering, National Chung Hsing University, Taichung 402, Taiwan; rhl@dragon.nchu.edu.tw
* Correspondence: ppaul28865@mail.ntut.edu.tw (C.-J.C.); chl5265@gmail.com (C.-H.L.); kuocc@mail.ntut.edu.tw (C.-C.K.); Tel.: +886-2-2771-2171 (ext. 2407) (C.-C.K.); Fax: +886-2-27317174 (C.-C.K.)
† These authors contributed equally to this work.

**Citation:** Lu, W.-C.; Chen, C.-Y.; Cho, C.-J.; Venkatesan, M.; Chiang, W.-H.; Yu, Y.-Y.; Lee, C.-H.; Lee, R.-H.; Rwei, S.-P.; Kuo, C.-C. Antibacterial Activity and Protection Efficiency of Polyvinyl Butyral Nanofibrous Membrane Containing Thymol Prepared through Vertical Electrospinning. *Polymers* **2021**, *13*, 1122. https://doi.org/10.3390/polym13071122

Academic Editors: Andrada Serafim and Stefan Ioan Voicu

Received: 28 February 2021
Accepted: 25 March 2021
Published: 1 April 2021

**Publisher's Note:** MDPI stays neutral with regard to jurisdictional claims in published maps and institutional affiliations.

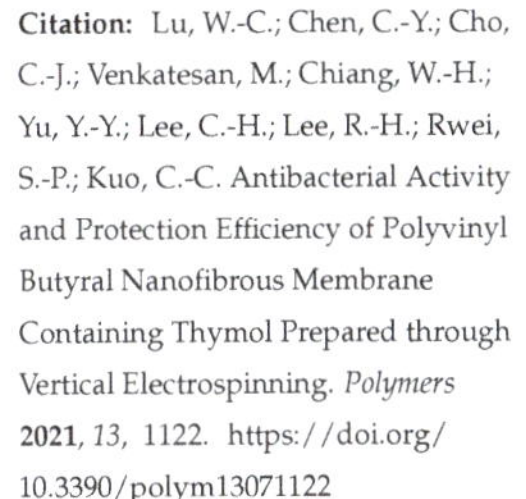



**Abstract:** Human safety, health management, and disease transmission prevention have become crucial tasks in the present COVID-19 pandemic situation. Masks are widely available and create a safer and disease transmission–free environment. This study presents a facile method of fabricating masks through electrospinning nontoxic polyvinyl butyral (PVB) polymeric matrix with the antibacterial component Thymol, a natural phenol monoterpene. Based on the results of Japanese Industrial Standards and American Association of Textile Chemists and Colorists methods, the maximum antibacterial value of the mask against Gram-positive and Gram-negative bacteria was 5.6 and 6.4, respectively. Moreover, vertical electrospinning was performed to prepare Thymol/PVB nanofiber masks, and the effects of parameters on the submicron particulate filtration efficiency (PFE), differential pressure, and bacterial filtration efficiency (BFE) were determined. Thorough optimization of the small-diameter nanofiber–based antibacterial mask led to denser accumulation and improved PFE and pressure difference; the mask was thus noted to meet the present pandemic requirements. The as-developed nanofibrous masks have the antibacterial activity suggested by the National Standard of the Republic of China (CNS 14774) for general medical masks. Their BFE reaches 99.4%, with a pressure difference of <5 mmH$_2$O/cm$^2$. The mask can safeguard human health and promote a healthy environment.

**Keywords:** electrospinning; nanofibrous membrane; antibacterial activity; protection efficiency; mask

## 1. Introduction

Particulate matter (PM) pollution, due to industrial development and living environment change, worsens air quality and causes severe health problems, such as respiratory diseases, cardiovascular diseases, and allergies [1]. In the recent years, various influenza and coronavirus diseases have become prevalent, causing harm to human health and large economic losses [2–4]. Masks can prevent droplets and PM from invading the human body and prevent respiratory infections. The types of masks include cotton masks, general medical masks, surgical masks, activated carbon masks, and N95 masks [5,6]. N95 masks have the highest ability to filter out PM with a $\leq$2.5-µm diameter (PM2.5), but because of their favorable adhesion, they can cause breathing problems and are thus unsuitable for long-term wear and general protection. To keep out epidemic virus droplets, PM with a $\leq$10-µm diameter (PM10), and some PM2.5, and thus protect the human body, surgical and medical masks must be worn correctly and be suitable for long-term wear.

General medical masks are made of polypropylene (PP), and their manufacturing method mainly involves spun bonding or melt blowing. The outer and inner layers of the mask are composed of PP spun-bond nonwoven fabric. The outer layer must repel water to prevent the penetration of blood, body fluids, and other potentially infectious substances. The inner layer is a face affinity layer, which can absorb moisture generated during breathing and thus keep the face dry and comfortable. The middle layer of the mask is a melt-blown nonwoven fabric layer, which is used to filter PM and bacteria. However, when the fibers of the middle layer are thick, the material's efficiency in filtering very small particles is low and thus high efficiency and low impedance cannot be achieved. Therefore, polytetrafluoroethylene membranes [7] or polyvinyl butyral (PVB) can be used as the matrix for preparing nanofiber membranes through electrospinning. Nanofibers have small diameter, high porosity, and internal pores [8–13], so they have favorable connectivity and high air permeability [14–18], which are conducive to capturing ultrafine particles. Nanofibers thus have high efficiency in filtering PM [19,20]. Many electrospun fiber membrane types have been manufactured as media for air filtration; the most common are those made from uniform and monostructured nanofibers, including ultrafine nylon 6 fibers [21–23], polyethylene oxide nanofibers [24], alumina nanofibers [25], and polyester nanofibers [26–29]; however, these materials cannot filter bacteria or viruses in the air.

Thymol (2-isopropyl-5-methyl phenol) is a monoterpene phenol found in the essential oils of herbal plants [30]. Many studies have shown that Thymol has antibacterial and antifungal properties [31]. Marino et al. [32] reported that *Thymus vulgaris* L. essential oil has antibacterial activity against both Gram-negative bacteria (e.g., *Escherichia coli*) and Gram-positive bacteria (e.g., *Staphylococcus aureus*). By employing the agar dilution method, Nostro et al. [33] found that Thymol has antibacterial effects and sensitivity for various types of *S. aureus*. Kavoosi et al. [34] mixed gelatin films with different concentrations of Thymol to test its antibacterial activity against *E. coli*, *Pseudomonas aeruginosa*, *Bacillus subtilis*, and *S. aureus* for potential use as an antibacterial nano wound dressing.

Some scholars have blended Thymol with polymer materials to make nanofiber membranes and tried to develop dressings for wound healing, such as by adding Thymol to poly(ε-caprolactone) and polylactic acid through electrospinning technology to prepare nanofibrous mats [11,35]. Thymol has also been added in situ to chitosan/polyethylene glycol fumarate to prepare hydrogel wound dressings. A study demonstrated that hydrogels containing Thymol have favorable mechanical properties and excellent antibacterial activity against both Gram-negative and -positive bacteria, and they can be applied in a dressing for infected wounds with a moderate amount of exudate [36].

The main mechanisms through which fibrous materials filter dust and particles are the interception effect, inertial deposition, Brownian diffusion, the electrostatic effect, and the gravity effect. Effective filtration of dust and particles by the filter materials in a mask is achieved through a combination of these five effects [37]. In the present study, Thymol was mixed with PVB through electrospinning to prepare antibacterial nanofiber membranes. Antibacterial property and protection efficiency tests were conducted to ensure that the developed medical masks meet performance specifications and have antibacterial effects.

## 2. Experimental

### 2.1. Materials and Microorganisms

Polyvinyl butyral (PVB, B18-HX, $M_W \sim$100,000, Chang Chun Co., Taipei, Taiwan) was used as the matrix of the nanofibrous membranes. Ethanol (95%) was used as a solvent purchased from I-Chang Chemical Co. ROC. Thymol (5-Methyl-2-(propan-2-yl)phenol) was supplied by Scharlab TI0080 (99% purity). The antibacterial activities of Thymol-containing PVB nanofibrous mats were evaluated using *Staphylococcus aureus* (*S. aureus*) (ATCC 6538), *Klebsiella pneumoniae* (*K. pneumoniae*) (ATCC 4352), *Escherichia coli* (*E. coli*) (ATCC 8739), which were obtained from the Bioresource Collection and Research Center (BCRC), Taipei, Taiwan.

## 2.2. Preparation of PVB Antibacterial Nanofibrous Membrane Containing Thymol

Thymol/PVB nanofibrous membranes were fabricated by vertical electrospinning technique. Primarily, 5 wt% of PVB polymer solution was prepared by dissolving 25 g of PVB powder in 475 g of ethanol solvent with magnetic stirring for 2 h at room temperature and obtained transparent solution. Subsequently, different amounts of Thymol powder were added to the above mixture with vigorous stirring to obtain the homogenous solution, various compositions of Thymol/PVB electrospinning solutions as shown in Table 1. The prepared Thymol/PVB electrospun solution was loaded into the plastic syringe containing a stainless needle (22-gauge stainless). A syringe pump (KDS-200, KD Scientific., Holliston, MA, USA) was used to feed the electrospinning solution with a fixed feeding rate of 0.72 mL/h. A high-voltage power supply (SM3030-24P1R, YOU-SHANG TECHNICAL CORP) was employed, and the applied voltage was 27 kV. The distance from tip to the collector was fixed at 10 cm, and the collecting lasted for 7 h.

**Table 1.** Various ratios of the Thymol/polyvinyl butyral (PVB) electrospinning solutions.

| Thymol:PVB (*w:w*) | Thymol (g) | PVB (g) | 5 wt% PVB (g) |
| --- | --- | --- | --- |
| 0:1 | 0 | 2.5 | 50 |
| 0.2:1 | 0.5 | 2.5 | 50 |
| 0.4:1 | 1 | 2.5 | 50 |
| 0.6:1 | 1.5 | 2.5 | 50 |
| 0.8:1 | 2 | 2.5 | 50 |
| 1:1 | 2.5 | 2.5 | 50 |

## 2.3. Preparation of Thymol/PVB Antibacterial Nanofibrous Masks

A result from the antibacterial preliminary test 0.6:1 ratio of Thymol/PVB nanofibrous membrane shows strong antibacterial effect against *S. aureus*, *K. pneumoniae*, and *E. coli*. Therefore, this optimized ratio was used in a vertical electrospinning device (SC-PME50, COSMI) to prepare the Thymol/PVB composite antibacterial nanofibrous mask. Since the vertical electrospinning device has many advantages, such as the needle can be moved left and right, collection area can be rotated, speed can be controlled, and a large-area nanofibrous membrane with better uniformity can be obtained. To this interest, the composite electrospun solution was taken into a 15 mL syringe, a stainless 22-gauge needle installed on the needle auxiliary moving mechanism. Relevant parameters were adjusted according to the experimental design, as shown in Table S1. Figure 1 represents the basic work function of three-layer facemask.

## 2.4. Surface Observation of Nanofiber Membranes

A scanning electron microscope (SEM) (JSM-6510, JEOL, Tokyo, Japan) was used to analyze the surface morphology of inner, outer, and middle layers of commercial mask fabrics and different Thymol/PVB electrospun nanofibrous membranes with varying electrospinning parameters. Diameters of uniformly distributed nanofiber were calculated from the SEM image.

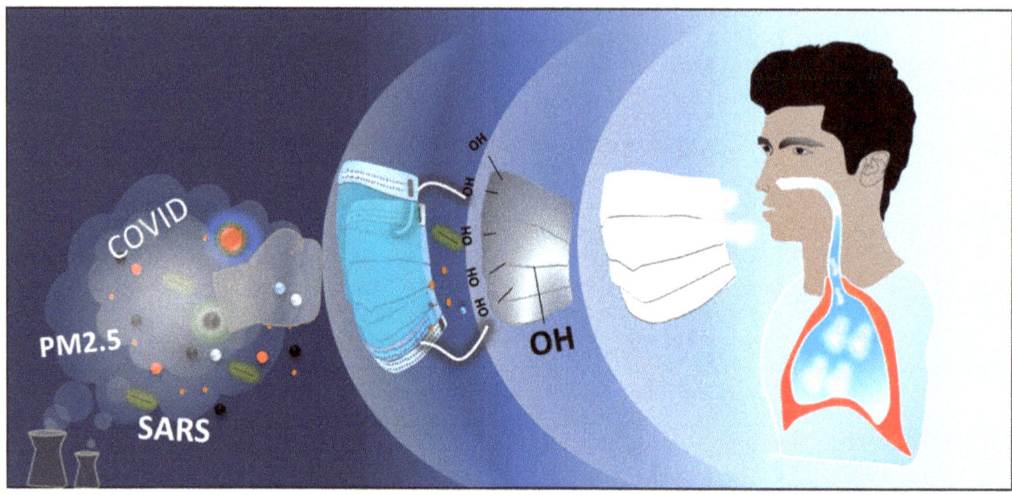


**Figure 1.** Schematic diagram of the electrospun Thymol/PVB nanofibrous facemask.

*2.5. Antibacterial Activity*

2.5.1. Antibacterial Qualitative Tests

Two qualitative methods were used to evaluate the antibacterial activities of PVB nanofibrous membranes containing Thymol, described below:

Japanese Industrial Standards (JIS) L1902 (Halo method): antibacterial qualitative tests were determined with adherence to the procedure of JIS L1902 method [38,39] (Halo method), performed by observing the zone of inhibition to evaluate the antibacterial properties of textiles. The method is suitable for textiles process by leaching antibacterial agents. The test methods were explained as follows:

A known weight of test samples was added to observe whether the Petri dish had a zone of inhibition; we measured the width of the sample (D) and the sum of the width of the sample and the zone of inhibition (T); then calculated the width of the zone of inhibition (W) with the following formula (1), and the result is expressed as an average 0.1 mm.

$$W = (T - D)/2 \tag{1}$$

American Association of Textile Chemists and Colorists (AATCC) 147 method (Parallel Streak method): the antibacterial qualitative tests were carried out in accordance with the procedures of AATCC147 method (Parallel Streak method). This method also helps to evaluate the antibacterial properties of textiles and it is suitable for textiles process by leaching antibacterial agents. The testing process was explained as follows:

(1) CZ: Clear zone of inhibition

(2) I: Inhibition of growth under the sample only

(3) NI: No inhibition of growth

2.5.2. Antibacterial Quantitative Test

JIS L1902 (Absorption method) testing method serves the quantitative information of antibacterial properties for the antibacterial textile products. The testing procedure was expressed as follows:

(1) The growth value on the control (F) should be $\geq 1.0$.

$$F = \log C_t - \log C_o \tag{2}$$

$\log C_t$: The average logarithm of the number of bacteria after 18–24 h inoculation of the control sample

$\log C_o$: The average logarithm of the number of bacteria recovered from the control at the beginning of contact time

(2) Antibacterial activity value (A) should be $\geq 2.0$, which means the sample has antibacterial effect, according to the descriptions of JIS L1902.

$$A = (\log C_t - \log C_o) - (\log T_t - \log T_o) = F - G \tag{3}$$

G: The growth value on the sample

$\log T_t$: The average logarithm of the number of bacteria after 18–24 h inoculation of the sample

$\log T_o$: The average logarithm of the number of bacteria recovered from the sample at the beginning of contact time

### 2.5.3. Particulate Filtration Efficiency (PFE)

Following the procedure of National Standards of the Republic of China (CNS) 14,755 [40], the Automated Filter Tester (Model 8130, TSI, Shoreview, Minnesota, MN, USA) was used to measure the submicron particulate filtration efficiency or penetration efficiency of the masks to understand the protection performance of the masks. The definition of submicron particulate filtration efficiency as follows:

PFE: The ratio of the aerosol concentration captured by the mask to the original upstream aerogel concentration.

$$\text{Penetration efficiency (\%)} = (\text{aerosol concentration through the mask})/(\text{upstream air aerosol concentration}) \times 100\%$$
$$\text{Protection efficiency (\%)} = 100 - \text{penetration efficiency (\%)} \tag{4}$$

Inhalation resistance: the ventilation resistance generated by a certain flow of air in the inhalation direction of the mask.

### 2.5.4. Differential Pressure of Air Exchange

According to CNS 14,777 (Test Method for Air Exchange Pressure of Medical Masks) Section 3, the tester of air exchange pressure (TTRI, TW) was used to measure the differential pressure of air exchange of the mask, to understand the breathability of the mask, and to evaluate whether it will cause breathing difficulties when wearing.

### 2.5.5. Bacterial Filtration Efficiency (BFE)

Following the procedure of CNS 14775, the Bacterial Filtration Efficiency (TTRI, TW) of the synthesized mask was calculated. A formed bacteria (*S. aureus* biological) aerogel size is approximately $(3.0 \pm 0.3)$ μm, which is controlled by the nebulizer system. The percentage of bacteria absorption before and after filtering through the mask and rate of applied pressure was estimated. The results were subjected to the following equation to calculate the percentage of BFE. The same procedure was followed for the 30 commercial masks to compare the obtained results.

$$\text{BFE (\%)} = 100 \times (C - T)/C \tag{5}$$

C: Average number of total plate count on the control.

T: Total plate count on the sample.

Calculate the average particle size of the aerosol by formula (6), which should be $(3.0 \pm 0.3)$ μm.

$$\text{Average particle size (MPS)} = \Sigma(An \times Sn)/\Sigma An = (A1 \times 7.0 + A2 \times 4.7 + A3 \times 3.3 + A4 \times 2.1 + A5 \times 1.1 + A6 \times 0.65)/(A1 + A2 + A3 + A4 + A5 + A6) \tag{6}$$

An: The number of bacteria in the petri dish of this stage

Sn: The particle size of the aerosol collected by the sampler at this stage

## 3. Results and Discussion

### 3.1. Antibacterial Qualitative Analysis for Nanofibrous Membrane of Thymol/PVB Blenders

Antibacterial qualitative analysis of the nanofibrous membranes was performed using the JIS L1902 and AATCC 147 standardized methods. Although no inhibition zones were found in samples of different compositions, to improve the inhibition zones, various methods were employed, such as adding a surfactant [41] and increasing the contact opportunities for bacterial liquid with nanofiber membrane (by using nets, pressurized glass sheets, and alumina foil). The addition of a surfactant can affect nanofiber surface morphology [42]. *S. aureus* were used as a testing agent in the JIS L1902 and AATCC 147 qualitative methods to evaluate the antibacterial effects of Thymol/PVB nanofibrous membranes. Although, as shown in Table 2 and Figure S1, spraying the tested bacteria liquid on the nanofibrous membrane or net-like architecture did not form any contact with the substrate. Even applying gentle pressure using a glass plate, this lack of interface formation results to fail the formation of the inhibition zone. In contrast, Group A and Group C showed the formation of zone inhibition on alumina foil with a high concentration ratio of Thymol in PVB (1:1). However, even though the width of the inhibition zone increased slightly, the difference was not large.

**Table 2.** Qualitative analysis of antibacterial properties of Thymol/PVB nanofibrous membranes with different proportions.

| Method | JIS L1902 (Halo Method) | | AATCC147 (Parallel Streak Method) | |
|---|---|---|---|---|
| **Carrier** | **Aluminum Foil** | **Net** | **Aluminum Foil** | **Net** |
| Pressing with glass | none | yes | yes | yes |
| group<br>Thymol:PVB (*w:w*) | A | B | C | D |
| 0:1 | [c] NI | [c] NI | [c] NI | [c] NI |
| 0.2:1 | [c] NI | [c] NI | [c] NI | [c] NI |
| 0.4:1 | [c] NI | [c] NI | [c] NI | [c] NI |
| 0.6:1 | [c] NI | [c] NI | [c] NI | [c] NI |
| 0.8:1 | [a] CZ 2.0 mm | [c] NI | [a] CZ 2.1 mm | [c] NI |
| 1:1 | [a] CZ 5.1 mm | [c] NI | [a] CZ 6.4 mm | [c] NI |

[a] CZ: Clear zone of inhibition, [c] NI: No inhibition of growth.

The inhibition zone may not have appeared because Thymol and PVB are insoluble in water, meaning that the probability of Thymol coming into contact with the inoculum was low. The weight of the sample in the antibacterial film was 4.76 times that in the nanofibrous membrane; that is, the content of the nanofiber membrane per unit area was low (Table S2). The nanofibrous membrane had a network structure, and actual contact with the bacterial liquid only occurred on the fiber mesh surface; however, the inhibition zone could only be produced when the antibacterial substance was in uniform contact with the bacterial liquid. For these reasons, the zone of inhibition was difficult to observe when the nanofibrous membranes were tested using the qualitative method.

### 3.2. Antibacterial Quantitative Analysis for Thymol/PVB Nanofibrous Membranes

The results in 3.1 showed that the qualitative method was not suitable for evaluating the antibacterial activity of Thymol/PVB nanofibrous membranes. Therefore, a quantitative determination was performed based on Absorption method from JIS L1902 to analyze the antibacterial activity of Thymol/PVB nanofibrous membranes on Gram-positive (*S. aureus*), and Gram-negative (*K. pneumoniae* and *E. coli*) bacteria.

#### 3.2.1. Staphylococcus aureus

The antibacterial activity of the Thymol/PVB nanofibrous membranes on *S. aureus* is displayed in Table S3 and Figure 2a. According to the description of JIS L1902, when the antibacterial activity value exceeds 2.0, the samples are considered to have antibacterial properties [43–46]. With the addition of Thymol to PVB, the bacterial growth was considerably reduced. For example, considering Thymol/PVB = 0.2:1 nanofiber with an antibacterial value of 2.8, this increment is attributed to the phenolic hydroxy functional group of Thymol. Consequently, a maximum amount of antibacterial value 6.4 was obtained from the optimized ratio of 0.6:1. However, further addition of Thymol does not cause any changes in it and clearly shows the saturated ratio to PVB. Therefore, enhancing the proportion of Thymol in PVB leads to increasing antibacterial activity toward *S. aureus* bacteria.

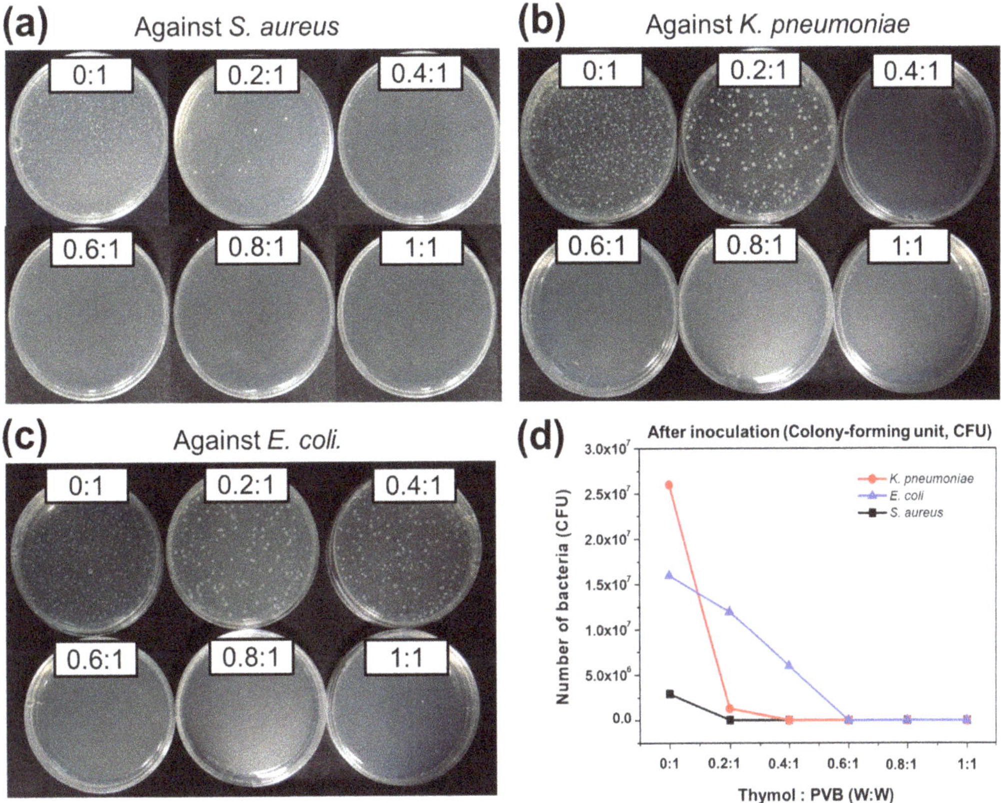


**Figure 2.** Quantitative results of antibacterial activity of Thymol/PVB nanofibrous membranes against (**a**) *Staphylococcus aureus*, (**b**) *Klebsiella pneumonia*, (**c**) *Escherichia coli*. (**d**) Comparison of the bacterial counts after culturing the three bacteria with Thymol/PVB fibrous membranes.

### 3.2.2. Klebsiella pneumoniae

The study of antibacterial effects of Thymol/PVB nanofibrous membranes against *K. pneumoniae* were monitored for 18–24 h. The control group growth value (F) should have been >1.0 and the test conditions were established. Table S4 and Figure 2b display that bacteria growth was controlled as the proportion of Thymol increased. Antibacterial activity value also reached a maximum of 6.4 when the ratio of Thymol: PVB was 0.4:1. Other proportions also provide the same result.

### 3.2.3. Escherichia coli

As shown in Table S5 and Figure 2c, the quantitative antibacterial effect of the various nanofibrous membranes against *E. coli* was in favorable agreement with the testing results for other bacteria. However, the Thymol/PVB (0.6:1) nanofibrous membrane exhibited considerable enhancement, with a value of 6.4. Therefore, enhancing the proportion of Thymol in PVB resulted in higher antibacterial activity.

A comparison of the antibacterial activity values of PVB and various-ratio Thymol/PVB nanofibrous membranes for three bacterial types is presented in Table S6. The PVB nanofiber membrane without Thymol exhibited no response to the bacterial colonization. A significant reduction in bacterial growth was discovered for the Thymol/PVB nanofiber membranes, and the reduction depended on the concentration of Thymol. Even at a low Thymol:PVB ratio of 0.2:1, the obtained antibacterial activity value against the Gram-positive *S. aureus* bacteria was 2.8, which indicated that this membrane had relevant antibacterial activity. By contrast, the antibacterial activity values of this membrane against the Gram-negative *K. pneumoniae* and *E. coli* bacteria were only 1.6 and 0.7, respectively. According to the JIS L1902 description, the membrane thus had no relevant antibacterial activity. Therefore, the growth of inhibition was observed for Gram-positive bacteria even at a low concentration of Thymol as compared with the Gram-negative bacteria. This variation was also reported in previous research [34]. The Gram-negative bacteria were protected by cell-wall lipopolysaccharides and outer membrane proteins, which restrict the diffusion of hydrophobic compounds through the lipopolysaccharide layers. However, at a higher concentration of antimicrobial agents, the polysaccharide layer was destroyed by essential oils.

As illustrated in Figure 2d, the sample with ratio 0.6:1 had the highest activity, indicating that the bacterial counts of the three test bacteria after culture were all <20 colony-forming units (CFUs), meaning that all the bacteria died. Furthermore, when the Thymol:PVB ratio was increased to 0.8 and 1, the antibacterial activity value did not change. Therefore, the Thymol:PVB ratio 0.6:1 (*w:w*) was the most suitable ratio for preparing the antibacterial nanofiber mask.

### 3.2.4. Diameter of Nanofibrous Membrane for Thymol/PVB Blenders

As shown in Figures 3 and 4, Thymol/PVB nanofiber membranes prepared with different proportions had an average fiber diameter between 500 and 700 nm, which belonged to the broad range of nanofibers.

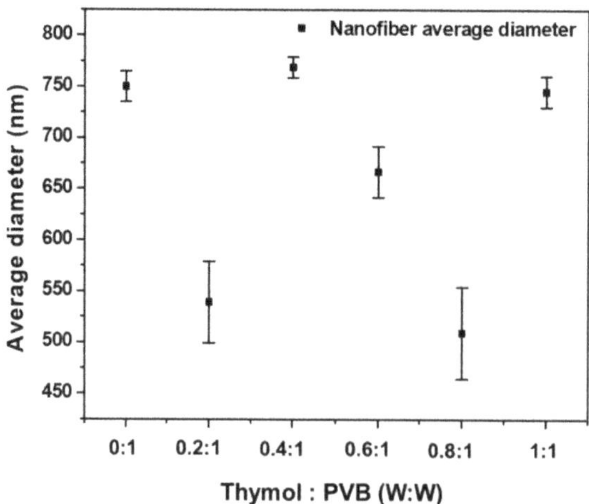


**Figure 3.** Diameter of nanofibrous membranes for Thymol/PVB blenders.

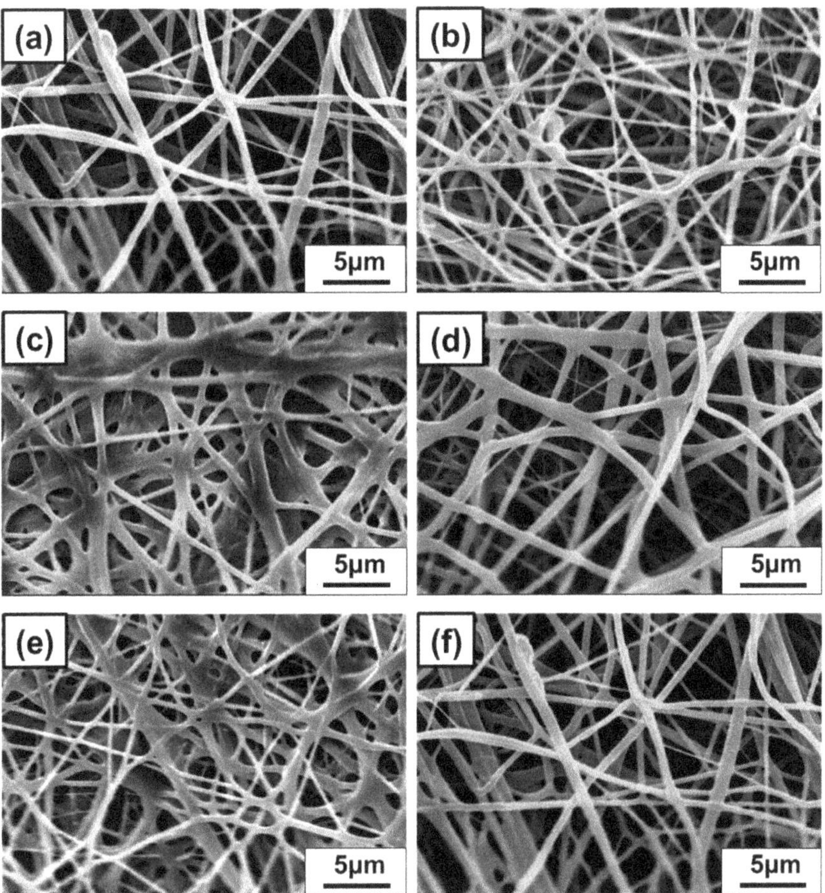

**Figure 4.** Nanofibrous membrane of Thymol/PVB blenders SEM pattern. (**a**) Thymol:PVB = 0:1, (**b**) Thymol:PVB = 0.2:1, (**c**) Thymol:PVB = 0.4:1, (**d**) Thymol:PVB = 0.6:1, (**e**) Thymol:PVB = 0.8:1, (**f**) Thymol:PVB = 1:1.

### *3.3. Analysis of the Protection Efficiency of Antibacterial Nanofiber Masks*

#### 3.3.1. Submicron Particulate Filtration Efficiency (PFE) and Pressure Difference of Air Exchange

The optimized ratio (0.6:1) Thymol/PVB antibacterial nanofiber masks were prepared using a vertical electrospinning device, and protection efficiency and pressure difference tests were conducted in accordance with the CNS 14,755 and CNS 14,777 standards. The electrospinning parameters employed are presented in Table S1, and the results are shown in Table 3. Groups C and D collected nanofiber webs at 25 kV and 5 mL/h for 45 min, and although the collection distance was different for these groups, the PFE was >80% for both. However, the pressure difference exceeded the 5 mmH$_2$O/cm$^2$ threshold specified by CNS 14,774 for surgical masks, indicating that the mask would cause breathing problems during wear. According to the parameter settings in Table S1, the influences of voltage and flow rate on the PFE, impedance, and pressure difference were explored. In Group E, electrospinning was performed at the voltage of 15 kV and flow rate of 2 mL/h. The test results revealed that the PFE was 82.7% and pressure difference was 4.5 mmH$_2$O/cm$^2$, which meets CNS 14,774 standard requirements for surgical masks.

**Table 3.** Results of protection efficiency in different electrospinning parameters.

| Group | Spinning Time (min) | CNS 14755 | | CNS 14777 |
| | | PFE (%) | Inspiratory Impedance (mmH$_2$O) | Pressure Difference (mmH$_2$O/cm$^2$) |
|---|---|---|---|---|
| C | 0 | 32.8 ± 2.5 | 3.9 ± 0.3 | 1.4 ± 0.2 |
| C | 15 | 65.0 ± 2.5 | 9.1 ± 0.9 | 3.8 ± 0.1 |
| C | 30 | 67.1 ± 2.0 | 11.4 ± 0.6 | 4.4 ± 0.1 |
| C | 45 | 89.5 ± 1.0 | 38.7 ± 0.8 | 13.7 ± 1.8 |
| C | 60 | 94.2 ± 0.3 | 54.2 ± 8.6 | 14.1 ± 4.5 |
| C | 120 | 89.5 ± 1.6 | 42.7 ± 4.9 | 12.6 ± 2.0 |
| D | 0 | 7.5 ± 0.1 | 1.8 ± 0.1 | 0.8 ± 0.1 |
| D | 45 | 85.1 ± 0.7 | 29.9 ± 0.4 | 9.7 ± 0.1 |
| E | 60 | 62.9 ± 1.2 | 7.3 ± 0.2 | 2.6 ± 0.2 |
| E | 120 | 82.7 ± 1.0 | 12.3 ± 0.4 | 4.5 ± 0.3 |

### 3.3.2. The Fiber Diameter and PFE

The fiber diameters were measured by SEM. Figure S2 and Table 4 exhibit that when the flow rate was reduced from 5 mL/h to 2 mL/h, the average fiber diameter was reduced from 836 ± 329 nm to 375 ± 69 nm. This result showed that as the flow rate decreased, the fiber diameter greatly decreased and became more uniform, and could effectively improve the impedance value and the high-pressure difference. For the nanofiber mask, the thinner and denser the fiber diameters are, the more PFE and pressure difference can meet the requirements of CNS 14,774 specifications for general medical masks [47].

**Table 4.** Comparison of fiber diameters in the middle layer of nanofiber masks.

| Group/Spinning Time | D/45 min | E/2 h |
|---|---|---|
| Voltage | 25 kV | 15 kV |
| Flow rate | 5 mL/h | 2 mL/h |
| PFE(%) | 85.1 | 82.7 |
| Inspiratory impedance (mmH$_2$O) | 29.9 | 12.3 |
| Pressure difference (mmH$_2$O/cm$^2$) | 9.7 | 4.5 |
| Average diameter (nm) | 836 | 375 |
| Standard deviation | 329 | 69 |

In the construction of general commercial masks, PP spun-bond is used as the outer and inner layers with a diameter of 20~30 μm. The middle layer is a melt-blown non-woven fabric layer with 2–10 μm diameters. The fiber diameter of each nonwoven fabric was observed by SEM (Figure S2 and Table S7). However, most of the airborne viruses (Corona) spread through the aerosol are in nanometers [48–51]. The fiber diameter of the nanofiber prepared in this study was 375 + 69 nm, and the filter efficiency was improved by reducing the fiber diameter. Hence, the antibacterial nanofibers are well replaceable for the nonwoven fabric layers.

### 3.3.3. Spinning Parameters and PFE

Although the electrospinning parameters of Group E can meet the specifications of CNS 14,774 mask, at 15 kV, the spinning solution was not able to form a stable tailor cone due to the droplets block in the needle nozzle. To overcome this, the applied voltage was increased to 18 kV (Group F) and we found that the formation of beadles continuous nanofiber. The Group F parameters are given in Table S1 (18 kV, distance 160 mm, 2 mL/h), under different spinning times. Correspondingly, the influence of PFE was further explained in Table S8, Figure 5, that as the spinning time increased, the accumulated nanofiber membranes increased. When the collection time reached 6 h, the PFE increased from 38.6% to 83.2%, and the pressure difference was fixed at 4.7 mmH$_2$O/cm$^2$, which meets the requirements of CNS 14,774 [47].

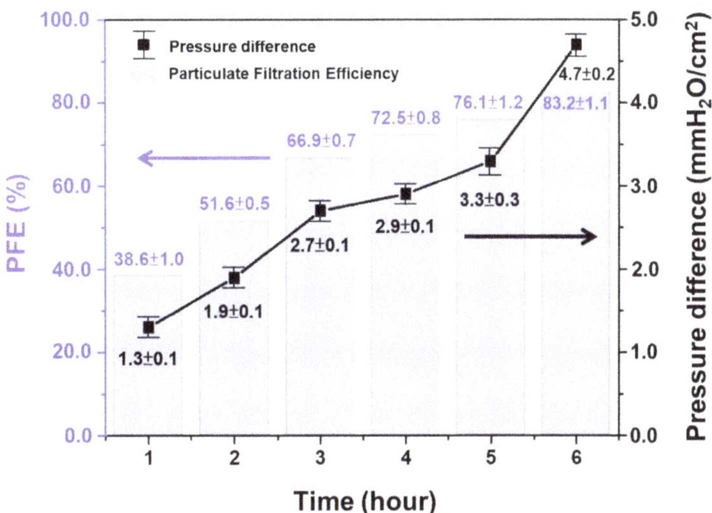



**Figure 5.** Testing results of particulate filtration efficiency (PFE) and pressure difference.

### 3.3.4. Bacteria Filtration Efficiency (BFE)

In Taiwan, general medical masks must meet the specifications of CNS 14774, which state that the BFE must be >95%. From the viewpoint of air permeation, the BFE and PFE of a filter material should be correlated. Therefore, in this study, 30 commercial masks were tested and analyzed to determine the correlation between their BFE and PFE. In addition, a Thymol/PVB nanofibrous membrane was prepared through electrospinning with the Group F parameters, with fiber collection lasting 1–6 h, and it was observed that increasing the nanofiber collecting time leads to the higher areal density nanofiber membrane, which helps to trap the bacteria aerosols. Furthermore, these BFE results were compared with the PFE data. As shown in Table S9, the correlation between BFE and PFE for the different masks were nonlinear, although when the PFE exceeded 73%, the BFE reached its maximum. Therefore, in this study, we prepared the mask with the aim of achieving the PFE of ≥80% and then conducted the BFE test. In theory, if the BFE value exceeded 95%, it could meet the CNS 14,774 specifications.

To demonstrate this, a BFE test was performed to the Group F nanofiber membrane by varying its electrospinning parameters. The results are shown in Figure 6 and Table S10. The PFE of the antibacterial nanofiber mask was 83.2%, and its BFE was 99.4%, which was consistent with the correlation in the data analysis, as shown in Figure 6h. Moreover, to support this, many other researchers also have investigated the filtration efficiency using the analytical methods on the porous fiber membranes [52,53].

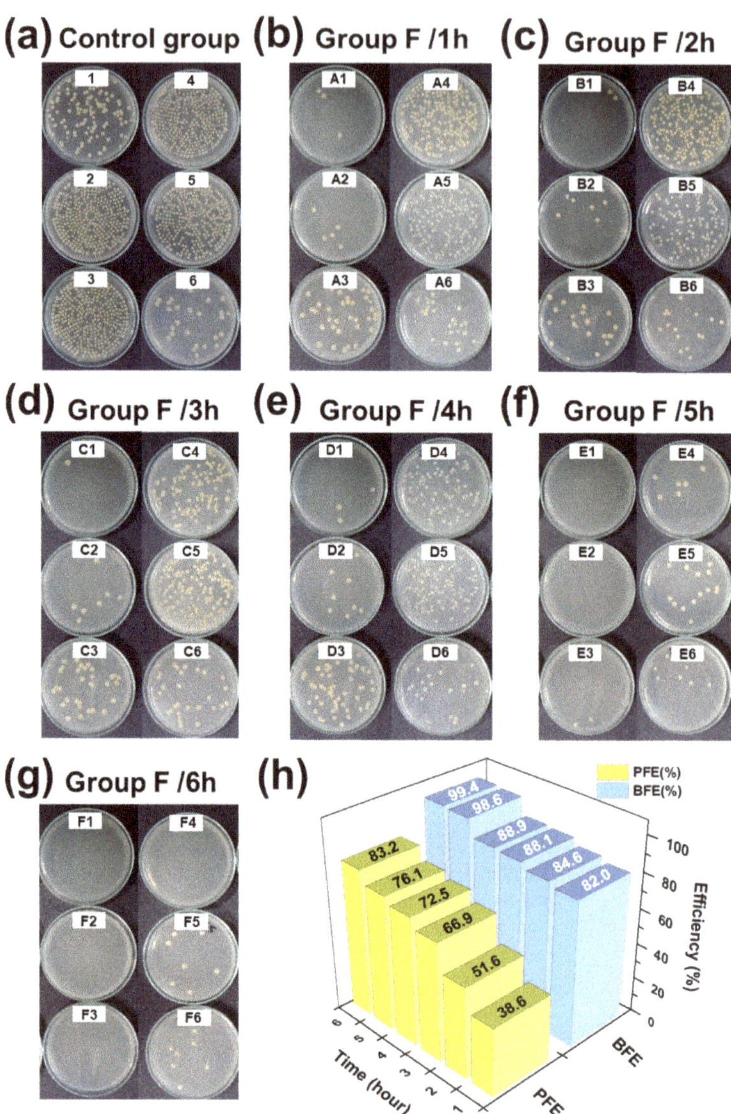



**Figure 6.** Bacterial filtration efficiency (BFE) test results of Group F spinning at different collection times. (**a**) Control group, (**b**) Group F/1h, (**c**) Group F/2h, (**d**) Group F/3h, (**e**) Group F/4h, (**f**) Group F/5h, (**g**) Group F/6h. (**h**) Correlation results of PFE and BFE tests in Group F.

## 4. Conclusions

An extension of industries may address their countries' growth, although it brings an inevitable side effect to their people. Especially, food and airborne diseases cause some serious issues. The face masks are an affordable and effective safety measure. Cost-effective bio-degradable eco-friendly masks are an inviting field of research due to increasing demands on healthcare and rising concerns of personal hygiene and safety. Through this research, we opened up a low-cost biodegradable face mask fabrication process, which can protect and sustain the human life.

In this research, a Thymol/PVB composite nanofibrous membrane was produced with various optimization conditions for the antibacterial mask. As prepared, the Thymol: PVB = 0.6:1 composite nanofibrous mask shows better performance, which meets the specifications of CNS 14774. Especially, bacterial filtration efficiency and pressure difference exhibit better performance than the commercial mask. Quantitative antibacterial activity test with JIS L1902 absorption method shows that as the content of Thymol increased, the antibacterial activity of Thymol/PVB nanofibrous membranes increased. The antibacterial activity values against Gram-positive bacteria of *S. aureus* and Gram-negative bacteria of *K. pneumoniae* and *E. coli* were 5.6 and 6.4, respectively. PFE values mainly depend on the electrospinning parameters. At 18 kV, 2 mL/h, the PFE met 83.2%, the inspiratory impedance was 10.8 mmH$_2$O/cm$^2$, and the pressure difference was 4.7 mmH$_2$O/cm$^2$, which fixes the respiration problem of the medical mask. BFE was 99.4%, which also verified the correlation between BFE and PFE data analysis inferred in this study. That is, when PFE was >73%, BFE reached ≥99% or more. Based on the results, Thymol/PVB nanofibers can be an alternate membrane layer for this pandemic environment, which is expected to be used as a multifunctional antibacterial mask against bacteria and viruses.

**Supplementary Materials:** The following are available online at https://www.mdpi.com/2073-436 0/13/7/1122/s1, Figure S1: Antibacterial qualitative results of Thymol/PVB nanofibrous membranes on Staphylococcus aureus, Figure S2: SEM image of the middle layer of the nanofibrous masks, Table S1: Parameters of the vertical electrospinning device, Table S2: Sample weight per unit area in various preparation methods, Table S3: Quantitative antibacterial properties of Thymol/PVB nanofibrous membrane against S. aureus, Table S4: Quantitative antibacterial activity of Thymol/PVB nanofibrous membrane against K. pneumonia, Table S5: Quantitative antibacterial activity of Thymol/PVB nanofibrous membrane against E. coli, Table S6: Comparison of antibacterial activity values of Thymol/PVB blenders against three bacteria strains, Table S7: Comparison of PP Spun-Bond and Melt-blown nonwoven fabrics, Table S8: Testing results of PFE and Pressure difference, Table S9: PFE and BFE test results of 30 commercial masks, Table S10: Protection efficiency of Thymol/PVB antibacterial nanofibrous masks.

**Author Contributions:** Conceptualization, C.-J.C., C.-C.K. and C.-H.L. are conceived and supervised the project. W.-C.L., C.-Y.C. and M.V. conceived designed the experiments. W.-C.L. and M.V. performed and analyzed the experiments. M.V. and C.-Y.C. assisted with sample preparation, analyzing the SEM morphology and the other measurement. W.-C.L. wrote the paper. W.-C.L., C.-C.K., C.-J.C., C.-H.L., S.-P.R., R.-H.L., W.-H.C. and Y.-Y.Y., discussed the results and revised or commented on the manuscript. All authors have read and agreed to the published version of the manuscript.

**Funding:** This research was funded by the Ministry of Science and Technology, Taiwan, (MOST 109-2622-E-027-004-CC3, MOST 109-2221-E-027-114 -MY3, and MOST 110-2222-E-027-001-MY2) and National Taipei University of Technology and Chang Gung Memorial Hospital Joint Research Program, (NTUT-CGMH-110-06) NTUT-CGMH Joint Research Program, for financial support.

**Institutional Review Board Statement:** Not applicable.

**Informed Consent Statement:** Not applicable.

**Data Availability Statement:** The data presented in this study are available on request from the corresponding author.

**Conflicts of Interest:** The authors declare no conflict of interest.

# References

1. Wang, Q.; Bai, Y.; Xie, J.; Jiang, Q.; Qiu, Y. Synthesis and filtration properties of polyimide nanofiber membrane/carbon woven fabric sandwiched hot gas filters for removal of PM 2.5 particles. *Powder Technol.* **2016**, *292*, 54–63. [CrossRef]
2. Fikenzer, S.; Uhe, T.; Lavall, D.; Rudolph, U.; Falz, R.; Busse, M.; Hepp, P.; Laufs, U. Effects of surgical and FFP2/N95 face masks on cardiopulmonary exercise capacity. *Clin. Res. Cardiol.* **2020**, *109*, 1522–1530. [CrossRef]
3. Kim, M.C.; Bae, S.; Kim, J.Y.; Park, S.Y.; Lim, J.S.; Sung, M.; Kim, S.H. Effectiveness of surgical, KF94, and N95 respirator masks in blocking SARS-CoV-2: A controlled comparison in 7 patients. *Infect. Dis.* **2020**, *52*, 908–912. [CrossRef] [PubMed]

4. Rengasamy, S.; Eimer, B.C.; Shaffer, R.E. Comparison of nanoparticle filtration performance of NIOSH-approved and CE-marked particulate filtering facepiece respirators. *Ann. Occup. Hyg.* **2009**, *53*, 117–128. [CrossRef] [PubMed]
5. Chalikonda, S.; Waltenbaugh, H.; Angelilli, S.; Dumont, T.; Kvasager, C.; Sauber, T.; Servello, N.; Singh, A.; Diaz-Garcia, R. Implementation of an Elastomeric Mask Program as a Strategy to Eliminate Disposable N95 Mask Use and Resterilization: Results from a Large Academic Medical Center. *J. Am. Coll. Surg.* **2020**, *231*, 333–338. [CrossRef]
6. Czubryt, M.P.; Stecy, T.; Popke, E.; Aitken, R.; Jabusch, K.; Pound, R.; Lawes, P.; Ramjiawan, B.; Pierce, G.N. N95 mask reuse in a major urban hospital: COVID-19 response process and procedure. *J. Hosp. Infect.* **2020**, *106*, 277–282. [CrossRef]
7. Bai, Y.; Han, C.B.; He, C.; Gu, G.Q.; Nie, J.H.; Shao, J.J.; Xiao, T.X.; Deng, C.R.; Wang, Z.L. Washable Multilayer Triboelectric Air Filter for Efficient Particulate Matter PM2.5Removal. *Adv. Funct. Mater.* **2018**, *28*, 1706680. [CrossRef]
8. Cho, C.-J.; Chen, S.-Y.; Kuo, C.-C.; Veeramuthu, L.; Au-Duong, A.-N.; Chiu, Y.-C.; Chang, S.-H. Morphology and optoelectronic characteristics of organic field-effect transistors based on blends of polylactic acid and poly(3-hexylthiophene). *Polym. J.* **2018**, *50*, 975–987. [CrossRef]
9. Jiang, D.H.; Chiu, P.C.; Cho, C.J.; Veeramuthu, L.; Tung, S.H.; Satoh, T.; Chiang, W.H.; Cai, X.; Kuo, C.C. Facile 3D Boron Nitride Integrated Electrospun Nanofibrous Membranes for Purging Organic Pollutants. *Nanomaterials* **2019**, *9*, 1383. [CrossRef]
10. Liang, F.-C.; Ku, H.-J.; Cho, C.-J.; Chen, W.-C.; Lee, W.-Y.; Chen, W.-C.; Rwei, S.-P.; Borsali, R.; Kuo, C.-C. An intrinsically stretchable and ultrasensitive nanofiber-based resistive pressure sensor for wearable electronics. *J. Mater. Chem. C* **2020**, *8*, 5361–5369. [CrossRef]
11. Lu, W.C.; Chuang, F.S.; Venkatesan, M.; Cho, C.J.; Chen, P.Y.; Tzeng, Y.R.; Yu, Y.Y.; Rwei, S.P.; Kuo, C.C. Synthesis of Water Resistance and Moisture-Permeable Nanofiber Using Sodium Alginate-Functionalized Waterborne Polyurethane. *Polymers* **2020**, *12*, 2882. [CrossRef] [PubMed]
12. Veeramuthu, L.; Li, W.-L.; Liang, F.-C.; Cho, C.-J.; Kuo, C.-C.; Chen, W.-C.; Lin, J.-H.; Lee, W.-Y.; Wang, C.-T.; Lin, W.-Y.; et al. Smart garment energy generators fabricated using stretchable electrospun nanofibers. *React. Funct. Polym.* **2019**, *142*, 96–103. [CrossRef]
13. Veeramuthu, L.; Venkatesan, M.; Liang, F.C.; Benas, J.S.; Cho, C.J.; Chen, C.W.; Zhou, Y.; Lee, R.H.; Kuo, C.C. Conjugated Copolymers through Electrospinning Synthetic Strategies and Their Versatile Applications in Sensing Environmental Toxicants, pH, Temperature, and Humidity. *Polymers* **2020**, *12*, 587. [CrossRef] [PubMed]
14. Chen, B.-Y.; Kuo, C.-C.; Cho, C.-J.; Liang, F.-C.; Jeng, R.-J. Novel fluorescent chemosensory filter membranes composed of electrospun nanofibers with ultra-selective and reversible pH and Hg 2+ sensing characteristics. *Dyes Pigment.* **2017**, *143*, 129–142. [CrossRef]
15. Cho, C.-J.; Lu, S.-T.; Kuo, C.-C.; Liang, F.-C.; Chen, B.-Y.; Chu, C.-C. Pyrene or rhodamine derivative–modified surfaces of electrospun nanofibrous chemosensors for colorimetric and fluorescent determination of Cu 2+, Hg 2+, and pH. *React. Funct. Polym.* **2016**, *108*, 137–147. [CrossRef]
16. Liang, F.C.; Kuo, C.C.; Chen, B.Y.; Cho, C.J.; Hung, C.C.; Chen, W.C.; Borsali, R. RGB-Switchable Porous Electrospun Nanofiber Chemoprobe-Filter Prepared from Multifunctional Copolymers for Versatile Sensing of pH and Heavy Metals. *ACS Appl. Mater. Interfaces* **2017**, *9*, 16381–16396. [CrossRef]
17. Liang, F.C.; Luo, Y.L.; Kuo, C.C.; Chen, B.Y.; Cho, C.J.; Lin, F.J.; Yu, Y.Y.; Borsali, R. Novel Magnet and Thermoresponsive Chemosensory Electrospinning Fluorescent Nanofibers and Their Sensing Capability for Metal Ions. *Polymers* **2017**, *9*, 136. [CrossRef]
18. Lin, C.C.; Jiang, D.H.; Kuo, C.C.; Cho, C.J.; Tsai, Y.H.; Satoh, T.; Su, C. Water-Resistant Efficient Stretchable Perovskite-Embedded Fiber Membranes for Light-Emitting Diodes. *ACS Appl. Mater. Interfaces* **2018**, *10*, 2210–2215. [CrossRef] [PubMed]
19. Chen, S.; Liu, G.-S.; He, H.-W.; Zhou, C.-F.; Yan, X.; Zhang, J.-C. Physical Structure Induced Hydrophobicity Analyzed from Electrospinning and Coating Polyvinyl Butyral Films. *Adv. Condens. Matter Phys.* **2019**, *2019*, 1–5. [CrossRef]
20. Peer, P.; Polaskova, M.; Musilova, L. Superhydrophobic poly(vinyl butyral) nanofibrous membrane containing various silica nanoparticles. *J. Text. Inst.* **2019**, *110*, 1508–1514. [CrossRef]
21. Fauzi, A.; Hapidin, D.A.; Munir, M.M.; Iskandar, F.; Khairurrijal, K. A superhydrophilic bilayer structure of a nylon 6 nanofiber/cellulose membrane and its characterization as potential water filtration media. *RSC Adv.* **2020**, *10*, 17205–17216. [CrossRef]
22. Keirouz, A.; Radacsi, N.; Ren, Q.; Dommann, A.; Beldi, G.; Maniura-Weber, K.; Rossi, R.M.; Fortunato, G. Nylon-6/chitosan core/shell antimicrobial nanofibers for the prevention of mesh-associated surgical site infection. *J. Nanobiotechnol.* **2020**, *18*, 51. [CrossRef] [PubMed]
23. Shi, L.; Zhuang, X.; Tao, X.; Cheng, B.; Kang, W. Solution blowing nylon 6 nanofiber mats for air filtration. *Fibers Polym.* **2013**, *14*, 1485–1490. [CrossRef]
24. Surgutskaia, N.S.; Martino, A.D.; Zednik, J.; Ozaltin, K.; Lovecká, L.; Bergerová, E.D.; Kimmer, D.; Svoboda, J.; Sedlarik, V. Efficient $Cu^{2+}$, $Pb^{2+}$ and $Ni^{2+}$ ion removal from wastewater using electrospun DTPA-modified chitosan/polyethylene oxide nanofibers. *Sep. Purif. Technol.* **2020**, *247*, 116914. [CrossRef]
25. Song, X.; Song, Y.; Wang, J.; Liu, Q.; Duan, Z. Insights into the pore-forming effect of polyvinyl butyral (PVB) as the polymer template to synthesize mesoporous alumina nanofibers via electrospinning. *Ceram. Int.* **2020**, *46*, 9952–9956. [CrossRef]

26. Cho, C.-J.; Chang, Y.-S.; Lin, Y.-Z.; Jiang, D.-H.; Chen, W.-H.; Lin, W.-Y.; Chen, C.-W.; Rwei, S.-P.; Kuo, C.-C. Green electrospun nanofiber membranes filter prepared from novel biomass thermoplastic copolyester: Morphologies and filtration properties. *J. Taiwan Inst. Chem. Eng.* **2020**, *106*, 206–214. [CrossRef]

27. Venkatesan, M.; Veeramuthu, L.; Liang, F.-C.; Chen, W.-C.; Cho, C.-J.; Chen, C.-W.; Chen, J.-Y.; Yan, Y.; Chang, S.-H.; Kuo, C.-C. Evolution of electrospun nanofibers fluorescent and colorimetric sensors for environmental toxicants, pH, temperature, and cancer cells—A review with insights on applications. *Chem. Eng. J.* **2020**, *397*, 125431. [CrossRef]

28. Chan, H.W.; Cho, C.J.; Hsu, K.H.; He, C.L.; Kuo, C.C.; Chu, C.C.; Chen, Y.H.; Chen, C.W.; Rwei, S.P. Smart Wearable Textiles with Breathable Properties and Repeatable Shaping in In Vitro Orthopedic Support from a Novel Biomass Thermoplastic Copolyester. *Macromol. Mater. Eng.* **2019**, *304*, 1900103. [CrossRef]

29. Hsu, K.H.; Chen, C.W.; Wang, L.Y.; Chan, H.W.; He, C.L.; Cho, C.J.; Rwei, S.P.; Kuo, C.C. Bio-based thermoplastic poly(butylene succinate-co-propylene succinate) copolyesters: Effect of glycerol on thermal and mechanical properties. *Soft Matter* **2019**, *15*, 9710–9720. [CrossRef]

30. Marchese, A.; Orhan, I.E.; Daglia, M.; Barbieri, R.; Di Lorenzo, A.; Nabavi, S.F.; Gortzi, O.; Izadi, M.; Nabavi, S.M. Antibacterial and antifungal activities of thymol: A brief review of the literature. *Food Chem.* **2016**, *210*, 402–414. [CrossRef]

31. Michalska-Sionkowska, M.; Walczak, M.; Sionkowska, A. Antimicrobial activity of collagen material with thymol addition for potential application as wound dressing. *Polym. Test.* **2017**, *63*, 360–366. [CrossRef]

32. Marino, M.; Bersani, C.; Comi, G. Antimicrobial activity of the essential oils of Thymus vulgaris L. measured using a bioimpedometric method. *J. Food Prot.* **1999**, *62*, 1017–1023. [CrossRef] [PubMed]

33. Nostro, A.; Blanco, A.R.; Cannatelli, M.A.; Enea, V.; Flamini, G.; Morelli, I.; Sudano Roccaro, A.; Alonzo, V. Susceptibility of methicillin-resistant staphylococci to oregano essential oil, carvacrol and thymol. *FEMS Microbiol. Lett.* **2004**, *230*, 191–195. [CrossRef]

34. Kavoosi, G.; Dadfar, S.M.; Purfard, A.M. Mechanical, physical, antioxidant, and antimicrobial properties of gelatin films incorporated with thymol for potential use as nano wound dressing. *J. Food Sci.* **2013**, *78*, E244–E250. [CrossRef]

35. Karami, Z.; Rezaeian, I.; Zahedi, P.; Abdollahi, M. Preparation and performance evaluations of electrospun poly(ε-caprolactone), poly(lactic acid), and their hybrid (50/50) nanofibrous mats containing thymol as an herbal drug for effective wound healing. *J. Appl. Polym. Sci.* **2013**, *129*, 756–766. [CrossRef]

36. Koosehgol, S.; Ebrahimian-Hosseinabadi, M.; Alizadeh, M.; Zamanian, A. Preparation and characterization of in situ chitosan/polyethylene glycol fumarate/thymol hydrogel as an effective wound dressing. *Mater. Sci. Eng. C Mater. Biol. Appl.* **2017**, *79*, 66–75. [CrossRef]

37. Zhu, M.; Han, J.; Wang, F.; Shao, W.; Xiong, R.; Zhang, Q.; Pan, H.; Yang, Y.; Samal, S.K.; Zhang, F.; et al. Electrospun Nanofibers Membranes for Effective Air Filtration. *Macromol. Mater. Eng.* **2017**, *302*, 1600353. [CrossRef]

38. Kaźmierczak, D.; Guzińska, K.; Dymel, M. Antibacterial Activity of PLA Fibres Estimated by Quantitative Methods. *Fibres Text. East. Eur.* **2016**, *24*, 126–130. [CrossRef]

39. Pinho, E.; Magalhães, L.; Henriques, M.; Oliveira, R. Antimicrobial activity assessment of textiles: Standard methods comparison. *Ann. Microbiol.* **2010**, *61*, 493–498. [CrossRef]

40. Lee, S.-A.; Chen, Y.-L.; Hwang, D.-C.; Wu, C.-C.; Chen, J.-K. Performance Evaluation of Full Facepiece Respirators with Cartridges. *Aerosol. Air Qual. Res.* **2017**, *17*, 1316–1328. [CrossRef]

41. Bundjaja, V.; Santoso, S.P.; Angkawijaya, A.E.; Yuliana, M.; Soetaredjo, F.E.; Ismadji, S.; Ayucitra, A.; Gunarto, C.; Ju, Y.H.; Ho, M.H. Fabrication of cellulose carbamate hydrogel-dressing with rarasaponin surfactant for enhancing adsorption of silver nanoparticles and antibacterial activity. *Mater. Sci. Eng. C Mater. Biol. Appl.* **2021**, *118*, 111542. [CrossRef]

42. Deng, L.; Kang, X.; Liu, Y.; Feng, F.; Zhang, H. Effects of surfactants on the formation of gelatin nanofibres for controlled release of curcumin. *Food Chem.* **2017**, *231*, 70–77. [CrossRef]

43. Demirdogen, R.E.; Kilic, D.; Emen, F.M.; Aşkar, Ş.; Karaçolak, A.İ.; Yesilkaynak, T.; Ihsan, A. Novel antibacterial cellulose acetate fibers modified with 2-fluoropyridine complexes. *J. Mol. Struct.* **2020**, *1204*, 127537. [CrossRef]

44. Guzinska, K.; Kazmierczak, D.; Dymel, M.; Pabjanczyk-Wlazlo, E.; Bogun, M. Anti-bacterial materials based on hyaluronic acid: Selection of research methodology and analysis of their anti-bacterial properties. *Mater. Sci. Eng. C Mater. Biol. Appl.* **2018**, *93*, 800–808. [CrossRef] [PubMed]

45. NO, H.K.; Park, N.Y.; Lee, S.H.; Meyers, S.P. Antibacterial activity of chitosans and chitosan oligomers with different molecular weights. *Int. J. Food Microbiol.* **2002**, *74*, 65–72. [CrossRef]

46. Park, S.J.; Park, Y.M. Eco-dyeing and antimicrobial properties of chlorophyllin copper complex extracted from Sasa veitchii. *Fibers Polym.* **2010**, *11*, 357–362. [CrossRef]

47. Yim, W.; Cheng, D.; Patel, S.H.; Kou, R.; Meng, Y.S.; Jokerst, J.V. KN95 and N95 Respirators Retain Filtration Efficiency despite a Loss of Dipole Charge during Decontamination. *ACS Appl. Mater. Interfaces* **2020**, *12*, 54473–54480. [CrossRef] [PubMed]

48. Chowdhury, M.A.; Shuvho, M.B.A.; Shahid, M.A.; Haque, A.; Kashem, M.A.; Lam, S.S.; Ong, H.C.; Uddin, M.A.; Mofijur, M. Prospect of biobased antiviral face mask to limit the coronavirus outbreak. *Environ. Res.* **2021**, *192*, 110294. [CrossRef] [PubMed]

49. Das, O.; Neisiany, R.E.; Capezza, A.J.; Hedenqvist, M.S.; Forsth, M.; Xu, Q.; Jiang, L.; Ji, D.; Ramakrishna, S. The need for fully bio-based facemasks to counter coronavirus outbreaks: A perspective. *Sci. Total Environ.* **2020**, *736*, 139611. [CrossRef] [PubMed]

50. El-Atab, N.; Qaiser, N.; Badghaish, H.; Shaikh, S.F.; Hussain, M.M. Flexible Nanoporous Template for the Design and Development of Reusable Anti-COVID-19 Hydrophobic Face Masks. *ACS Nano* **2020**, *14*, 7659–7665. [CrossRef]

51. Fadare, O.O.; Okoffo, E.D. Covid-19 face masks: A potential source of microplastic fibers in the environment. *Sci. Total Environ.* **2020**, *737*, 140279. [CrossRef]
52. Xiao, B.; Huang, Q.; Chen, H.; Chen, X.; Long, G. A Fractal Model for Capillary Flow through a Single Tortuous Capillary with Roughened Surfaces in Fibrous Porous Media. *Fractals* **2021**, *29*, 2150017. [CrossRef]
53. Xiao, B.; Wang, W.; Zhang, X.; Long, G.; Fan, J.; Chen, H.; Deng, L. A novel fractal solution for permeability and Kozeny-Carman constant of fibrous porous media made up of solid particles and porous fibers. *Powder Technol.* **2019**, *349*, 92–98. [CrossRef]

MDPI

St. Alban-Anlage 66

4052 Basel

Switzerland

Tel. +41 61 683 77 34

Fax +41 61 302 89 18

www.mdpi.com

*Polymers* Editorial Office

E-mail: polymers@mdpi.com

www.mdpi.com/journal/polymers